Handbook of the Circular Economy

Handbook of the Circular Economy

Transitions and Transformation

Edited by
Allen Alexander, Stefano Pascucci and Fiona Charnley

DE GRUYTER

ISBN 978-3-11-072322-9
e-ISBN (PDF) 978-3-11-072337-3
e-ISBN (EPUB) 978-3-11-072341-0

Library of Congress Control Number: 2022947282

Bibliographic information published by the Deutsche Nationalbibliothek
The Deutsche Nationalbibliothek lists this publication in the Deutsche Nationalbibliografie;
detailed bibliographic data are available on the internet at http://dnb.dnb.de.

Foreword

In bringing the Foundation's first report to the World Economic Forum in 2012, we attempted to paint the circular economy as an agent of transformative change. As a system solutions framework for an economy that is restorative and regenerative by design: one that recognises planetary boundaries and helps to tackle the global emergencies of climate change, biodiversity loss and pollution.

This book not only presents a sound academic view of the circular economy but balances this with input from thought leaders from across the globe and case studies from a range of different business types. Bringing together academics, thought leaders, business and government has been at the core of the Foundation's ability to move the thinking on the circular economy forward over the last twelve years.

This thinking was based on, and continues to be shaped by, ideas presented in the first two parts of this volume: biomimicry, products-service systems and regenerative economics, to name a few prominent examples. As circular economy has taken root in academia, business and policymaking, a deeper understanding of the core concept – with an insistence that it represents much more than recycling bolted on to a linear economy – has gone hand in hand with taking action to prove its potential.

In academia, a decade ago circular economy thinking was largely concentrated in the faculties of business, engineering and design. As well as now being found in disciplines including law, life sciences and environmental economics, it has expanded beyond teaching and research into the management of university estates and businesses. In policymaking, a necessary but limited focus on restricting end-of-pipe options, like landfilling and incineration, has shifted to placing circular economy at the heart of strategies to increase competitiveness, create jobs and tackle climate change, and to funding new circular economy enterprises on the ground.

As amply demonstrated in the third part of this book, business is also on a journey of transition. All the companies featured have displayed extraordinary levels of innovation to put themselves on a path to a regenerative and circular economic future. But that path is not always smooth or without sharp bends. That they are succeeding nonetheless is testament to their foresight and perseverance. To scale their efforts, and others like them, collaboration is crucial. Much as a multidisciplinary approach in academia yields valuable insights – the whole being greater than the sum of its parts – so collaboration between businesses along value chains, and with policymakers at all levels, brings about system-level change.

To write a handbook on circular economy might be to invite comparisons to a manual for a car or a new electrical gadget. But this book is not a set of instructions, rather it is a guide and inspiration to fellow travellers on the road to a circular economy – a road to a world in which, by design, we regenerate nature as a by-product

https://doi.org/10.1515/9783110723373-202

of economic activity and restore finite materials to the economy, fully utilising and using them instead of wasting them. A world in which we thrive.

Andrew Morlet
CEO, Ellen MacArthur Foundation

Contents

Part II: **The state of transition**

About the editors

Allen T. Alexander is Associate Professor of Innovation and Circular Economy at the University of Exeter Business School, Penryn Campus, where he is a senior researcher currently exploring various aspects of Circular Economy. His research originally focused on strategic knowledge management and the role that knowledge can play in developing enhanced commercial capability and as a source of innovation. His recent studies have explored Open Innovation, Innovation Ecosystems and Regional Systems of Innovation, and currently, he is working on exploring Circular Innovations, and the role that they can play in transitioning toward Circular Economy. He has also previously investigated the role that universities and academics play in shaping corporate innovation and entrepreneurship practices.

Fiona Charnley is Professor of Circular Innovation and Co-Director of the Exeter Centre for the Circular Economy at the University of Exeter Business School. Fiona's research interests lie within the field of design, innovation and business for a Circular Economy. She is Co-Director of the recently funded UKRI National Interdisciplinary Circular Economy Hub harnessing the UK's leading research capabilities to accelerate the transition toward a Circular Economy. Fiona collaborates with organisations from across sectors to identify new approaches to design, innovation, manufacture and business modelling to transform resource use and value creation. She has also led multiple education and executive training programmes to support future leaders in developing the skills and capabilities necessary to transform our industrial system.

Stefano Pascucci is Professor of Sustainability and Circular Economy at the University of Exeter Business School, UK. He is a research fellow at the Environment and Sustainability Institute (ESI) and a research affiliate at Exeter Centre for the Circular Economy (ECCE), University of Exeter. He is also a part-time professor at the University of Auckland Business School (NZ) and a visiting research fellow at the Department of Business Management and Organisation (BMO) at Wageningen University (NL). He holds a PhD in agricultural economics and then specialised in topics related to institutional analysis and sustainability connected to entrepreneurship, organisation studies, innovation and value chain management. His research focuses particularly on food and agribusiness, sustainability and Circular Economy. He has published, among others, in the *Organization Studies*, *Academy of Management Perspectives*, *Journal of Business Venturing*, *Journal of Business Ethics*, *Journal of Cleaner Production*, and *Agriculture and Human Values*.

https://doi.org/10.1515/9783110723373-204

List of contributors

Allen Alexander
University of Exeter

Conny Bakker
TU Delft

Fenna Blomsma
University of Hamburg

Nancy M. P. Bocken
Maastricht Sustainability Institute

Steffen Böhm
University of Exeter

Geraldine Brennan
Irish Manufacturing Research / Imperial
College London

Esther Goodwin Brown
Circle Economy

Lucy Chamberlin
University of Exeter

Fiona Charnley
University of Exeter

Ruth Cherrington
University of Exeter

Carl Dalhammar
Lund University

Domenico Dentoni
Montpellier Business School

Marta Ferri
Lancaster University

Aglaia Fischer
Wageningen University / Circle Economy

Jessica Fishburn
LUT University

Sharon Gil
United Nations Environment Programme

Daniel Guzzo
Insper Institute of Education and Research

Merryn Haines-Gadd
University of Exeter

Chia-Hao Ho
University of Exeter

Helen Holmes
University of Manchester

Georgie Hopkins
University of Exeter

Isabelle Housni
University of Exeter

Marleen Janssen Groesbeek
Avans University of Applied Sciences

Jan Konietzko
Maastricht Sustainability Institute

Ladeja Godina Kosir
Circular Change

Eléonore Maitre-Ekern
University of Oslo

Rosalind Malcolm
University of Surrey

Constantine Manolchev
University of Exeter

Janaina Mascarenhas
University of Sao Paulo

David Monciardini
University of Exeter

Saskia van den Muijsenberg
BiomimicryNL

https://doi.org/10.1515/9783110723373-205

Esteban Munoz
United Nations Environment Programme

Marijana Novak
Circle Economy

Oke Okorie
University of Exeter

Kim Poldner
The Hague University of Applied Sciences

Kate Raworth
Oxford University

Andy Rees, OBE
Welsh Government

Paavo Ritala
LUT University

Malte Rödl
Swedish University of Agricultural Sciences
(SLU)

Wouter Spekkink
Erasmus University Rotterdam

Hugo Spowers
Riversimple   .

Walter Stahel
Product-Life Institute

Alison Stowell
Lancaster University

Frances Wall
University of Exeter

Ken Webster
Cranfield University

Jamie Wheaton
University of Exeter

Gail Whiteman
University of Exeter

Diane Zandee
Trunkrs

Stefano Pascucci, Allen Alexander, Fiona Charnley,
Jessica Fishburn

The circular economy: landscape, dimensions and definitions

At the dawn of the first industrial revolution, in the eighteenth century, humanity had triggered the development of a new economic system, strongly embedded in and conditioned by both social and ecological relations. We might suggest that this economy was anchored in a wider network of socio-ecological relations. At the dawn of a socio-ecological crisis, in the early twenty-first century, humanity has fine-tuned a globalised market economy that is not only totally embedding social and ecological systems, but our environment and our societies are now consumed by a wider, world-wide network of economically driven transactions. Moreover, economic growth is fundamentally coupled with resource consumption, resulting in overwhelmingly negative societal and environmental impacts. This 'great transformation,' as Karl Polanyi (1944) would have defined it, is what the Circular Economy (CE) agenda appears to challenge and encourages a revision of. Extant scholarship often defines CE as an "industrial economy that is restorative by intention and design" (EMF, 2012, p. 14) that *"utilizes ecosystem cycles in economic cycles by respecting their natural reproduction rates"* (Korhonen et al., 2018, p. 39). Practitioners, instead, look at CE as a strategic business and political response to issues of social and environmental unsustainability (D'Amato et al., 2019; Geissdoerfer et al., 2017; Kirchherr et al., 2017). In both approaches, CE emerges as a conceptual framework, a 'worldview' and collective narrative essential to tackle both societal and environmental challenges, by transforming the twenty-first century market and consumption-driven economies (Korhonen et al., 2018; Skene, 2018).

Redesigning a now-globalised market economy entails profound social and ecological changes as well as transforming political systems and institutional regimes, and disrupting the status-quo by evoking an agenda for socio-ecological transitions, beyond the incremental changes of business strategies and practices (Fischer & Pascucci, 2017; Schulz et al., 2019) or modest, policy incentives (Morseletto, 2020; Webster, 2021). Business activities in a globalised market economy are possible after all, only when a number of forces, in the form of social norms, political processes and institutions, are in place to define the rules of the game (North, 1991). Together, these forces and tensions shape how an economy functions at any point in time. Fundamentally, any future-proof globalised market economy needs to maintain these economic forces and tensions 'within' the boundaries of socio-ecological systems, the planetary cycles that support life on planet Earth, and the social conditions to ensure a just and safe space for humanity (Leach et al., 2013; Raworth, 2017; Rockstrom et al., 2009). Accordingly, new distributive,

https://doi.org/10.1515/9783110723373-001

regenerative and restorative processes and rules need to emerge as the core of any new future-proof economy.

Over the years, particular ways in which a global market economy operates within these boundaries have timidly emerged, within and between countries and economies (Raworth, 2017). They have been informed by initiatives like the Sustainable Development Goals, for example, or the UN Global Compact, or taken the form of 'sustainability-driven,' 'one planet' strategies or 'just' socio-economic activity, referencing constructs such as 'Triple Bottom Line' evaluations or socio-economic and eco-environmental lifecycle analyses. While these initiatives and strategies indicate a step in the right direction, they have often been limited to incremental change, with still limited global impact. Instead, a socio-ecological transformation into a global CE would require novel and more disruptive frameworks to emerge (Schulz et al., 2019; Termeer & Metze, 2019), operating as a transition mechanism to enable both practitioners and scholars to mobilise ideas and practices in this arena. According to this view, CE can be seen as an emerging interdisciplinary and multifaceted field of practice and inquiry (Borrello et al., 2020b), creating the potential to change frames and perspectives on how we organise production, consumption and exchange of resources, goods and services and how we can create a more participative and distributive economy at all scales (Raworth, 2017; Webster, 2021). This is only possible through the adoption of a holistic system approach, acknowledging the role of complexity, adaptability and resilience. Despite the ambitious radical agenda of some of the CE pioneers and founding fellows, and after almost a decade of sustained effort, whilst CE is structuring as a field of inquiry and practice, a number of critical tensions and ambiguities appear to be emerging in this arena. Recently, Borrello and colleagues (2020) have unearthed and discussed these tensions, particularly looking at how scholars and practitioners position themselves in their understanding of circularity and CE. They propose to unify a CE agenda through three key insights (Borrello et al., 2020): first, to understand CE as gathering principles of other schools of thought and elaborate them in a *narrative able to inspire policy actions*. Second, interpreting CE as field of practice evoking a *socio-technical transition into multiple regimes* in which societal and material needs are fulfilled by innovative industrial systems. Finally, looking at CE as a contribution to the environmental and economic dimensions of sustainability by means of an *eco-effective approach to industrial systems*, above and beyond eco-efficiency. From a similar perspective, other scholars have pointed at the ambivalence of the CE agenda: on one hand, it suggests an acceleration towards a more disruptive and radical change process, indeed a transformation of the current economy towards a socio-ecologically embedded reality. On the other hand, the key successes of the CE agenda are in the field of scalable and implementable solutions, triggering and stimulating incremental changes and innovation, mostly led by businesses. The latter is supporting the idea of looking at CE as a toolbox rather than a worldview, or conceptual framework, therefore far away from the idea of designing an economy inspired by living systems and to create a distributive, eco-effective 'nutrient economy' through circularity (Webster, 2021).

In the following sections of this introduction, we discuss these perspectives in further detail, using them to explain how this book and its contents contribute to a unified articulation of a contemporary CE. We start by looking into the CE as a source of inspiring narratives and worldviews, stimulating thought-leaders to re-think 'the way we make things,' to quote Cradle-to-Cradle founding fathers William McDonough and Michael Braungart. We then move into discussing CE as an expanding field of practice, involving particularly businesses and corporate leaders. In the third section, we look into how CE has been seen as an opportunity to creatively combine and couple means and ends in companies looking for change, innovation and a more sustainable future – economically, socially and environmentally. Our final considerations refer to the unresolved tensions and ambiguities in this emerging field as well as future developments and agendas.

Circular economy as an inspiring narrative stimulating reconceptualisation

Our first perspective comprehends CE as an inspiring and influential narrative. This perspective builds upon rapidly elevating concerns around the compatibility between an ever-growing globalised market economy and ensuring fair and just socio-ecological conditions – a situation that will sooner or later lead to a chain of fundamental crises and eventually systemic collapse. The debate around the climate emergency is only the latest in this growing chain of crises.

Now the building blocks of CE, as a narrative capable of mobilising key economic and political actors to tackle the twenty-first century grand challenges, can be traced back to influential views. These include that presented in 'The Limits to Growth' (Meadows et al., 1972); the role of natural capital for supporting sustainable development (Costanza & Daly, 1992) and the emergence of the debate on the relationship between natural capital and other forms of capital (e.g. manufactured and technological) (Neumayer, 2003). It is within these debates that the framing of planetary eco-systems endangered by anthropic activities, the very notion of planetary boundaries, has gained prominence (Rockstrom et al., 2009), and where the more critical concept of the Anthropocene has also emerged (Gowdy & Krall, 2013; Lewis & Maslin, 2015). All these approaches converge on one simple aspect: through unprecedented socio-technological development and waves of agrarian and industrial revolutions, humanity has reached the capacity to operate at a geological scale and therefore influence the Earth's natural processes, such as the climate and atmospheric cycles, as well as other large planetary cycles, such as the water and nutrient cycles (Rockstrom et al., 2009).

The very notion of CE originated in this debate, coined to describe systems in which economic activities and the environment interact in closed loops (Pearce & Turner, 1990), but for more than two decades, this term did not make a difference. In fact, it was confined to discussions between scholars and remained a textbook notion of interest for undergraduate students dealing with subjects such as ecological economics, ecology and earth systems. It is until recently that the concept of CE has made a breakthrough, becoming the source of a rather inspiring and influential narrative. There is now a consensus in the community of CE experts and scholars that identifies a turning point of this process in the early 2010s. Particularly, recent studies have pointed out at the relevance of CE as a concept stemming out of a debate focused on 'system-level' change and large societal transformations, arguably led by international 'Think Tanks' and consultancies in the same period. Borrello and colleagues (2020), for instance, identified this turning point as the publication of the first CE report by the Ellen MacArthur Foundation (EMF) in 2012 (EMF, 2012). According to the evidence provided by these authors, this report, and the series of engagement on further publications by the EMF, has significantly reshaped the CE debate and discourse, particularly in Europe (Borrello et al., 2020). Its influence is also due to extended networking activities that advocate the concept to businesses, policy makers and academics (Geissdoerfer et al., 2017).

Even though literature rarely highlights the legacy of CE to other fields, the CE narrative framed by the EMF is actually an interesting melting pot of different schools of thoughts (Borrello et al., 2020). Influenced by such thought-leaders as Walter Stahel, Ken Webster, Kate Raworth, Michael Braungart and William McDonough, amongst others, this renewed CE narrative has quickly generated a wealth of ideas, knowledge and practices, crystallised in influential global reports and international publications (e.g. EMF, 2012–2014, 2017, 2019). As an inclusive and porous narrative, the framing of CE provided by the EMF has also worked as a catalyst for other existing frames and approaches, ranging from studies in cleaner production (Stevenson & Evans, 2004) – generally referred to as more environmentally friendly production methods – to product and organisational design innovations, including product-service systems (PSS; Borrello et al., 2020).

The CE landscape that emerged from this process quickly included influences and ideas from the Blue Economy of Gunter Pauli (2010), industrial ecology, The Natural Step framework (Bradbury & Clair, 1999), biomimicry (Benyus, 2002) and the Laws of Ecology (Commoner, 1971). Core to the emerging CE narrative has been the principles of Cradle-to-Cradle design thinking and particularly the "waste-equals-food" principle, borrowed by schools of thoughts recognising the role of nature as a model to shape production processes (Borrello et al., 2020). Similarly, the idea of taking a design perspective on restorative systems and circular products is a key notion of Cradle-to-Cradle (McDonough & Braungart, 2002) and regenerative design (Lyle, 1996). The CE narrative has also included frames from business strategy, supply chain management and engineering such as the already mentioned PSS (Tukker,

2004, 2015), reverse logistics and closed-loop management. Particularly, the CE narrative has considered servitisation as a key aspect to mobilise businesses around the idea of extended product life-cycles, and highlighting the role of cascading processes as a means to produce high value bio-materials (Berbel & Posadillo, 2018). Consistently, schools of thoughts such as the Performance economy (Stahel, 2016) and Natural capitalism (Lovins et al., 1999) have contributed to the CE narrative to increase resource productivity and usage and to combine economic gains with environmental benefits. Looking at the intellectual vitality in the emergence and formation of CE narrative, will we observe new and competing frames in the years to come? Will scholars engage in differentiating the notion of CE into multiple and diverse circular economies? Or, as for the debate around sustainability and corporate social responsibility (CSR), will CE be substituted by a new and more evocative buzz word? We have asked key experts to expose their views on these questions, and their answers, as reported in Part I of the book, trigger even more questions and suggest to engage in challenging debates.

Circular economy as a new organisational field

In parallel with its rise as a globally influential narrative, CE has quickly mobilised a wider set of political and economic actors, at the intersection of business practices and public policy-making (Fischer & Pascucci, 2017; Lieder & Rashid, 2016). CE has been key to developing an agenda, both in terms of business practices and policy-making, for stimulating more sustainable product and process designs, reducing material use, incentivising recycling and optimising waste management as well as fostering responsible innovation (RI) and the overall corporate social and environmental strategies (Cardoso, 2018). The early 2010s turning point (described in the previous section) has inspired not only academic conversations and abstract principles, but it has also helped to identify a process in which worldviews and principles could be 'translated' into business and policy-making practices (Lieder & Rashid, 2016). Until the early 2010s, the notion of CE had inspired limited intervention, and most notably it was used as a conceptual foundation for the new environmental industrial strategies in China (Geissdoerfer et al., 2017). The corporate activism of the EMF, for example, through the launch of CE100 partnership, and the subsequent mobilisation of influential actors such as the World Economic Forum and the European Commission, had quickly accelerated the translation of CE principles into practices for both a business and policy-oriented agenda (Fischer et al., 2021). What we have observed in less than a decade is the emergence of a wide landscape of think tanks, consultancies, non-governmental organisations (NGOs), multi-national corporations, small and medium-sized enterprise (SME) business networks and national and regional initiatives that, operating in different contexts, have all contributed to develop and expand a business and policy-oriented CE agenda (Cardoso, 2018; Geissdoerfer et al., 2017). Fischer and colleagues (2021) have depicted this process as the emergence and formation of a new organisational field (Schulz et al., 2019). In their view CE has initially emerged as an organisational field that is concerned with key practices, for example designing out waste; retaining resources at the highest possible value; adopting renewable energy systems; closing lifecycles of biological nutrients and in general terms cascading value, regenerating and restoring, sharing and dematerialising (Konietzko et al., 2020). Around these practices, both business strategists and policy-makers have identified a set of common practices re-enforcing a 'shared view' of the meaning of CE. This is often described as the formation of organisational logics as a "set of material practices and symbolic constructions" that prescribe and predict organisational behaviours within the nascent field domain (Thornton & Ocasio, 2008). After a period of emergence and experimentation, these organisational logics have been codified in normative acts, legal and regulatory frameworks, particularly by the European Commission (Domenech & Bahn-Walkowiak, 2019). The convergence of business-led and policy-oriented initiatives has made evident the presence of a distinct and recognisable organisational field, where organisations are co-designing and experimenting with shared principles and practices by interacting together in a "recognised area of institutional life" (DiMaggio & Powell, 1983, p. 148). Evidence of the formation of CE as an organisational field is the widespread adoption of circular business models or strategies, co-existing with other business models inspired by other principles and worldviews.

In a relatively short period, CE has influenced many corporations and inspired strategic changes. While initially CE practices have been experimented with niches of innovation, by pioneer business leaders or entrepreneurs and new ventures, they have increasingly formed and interacted in collaborative networks, progressively setting boundaries between what could be considered circular and what could not, for example, in terms of product design, value propositions and business models (Fischer et al., 2021). We are currently in a stage that, after the formation of these pioneering CE networks, we are witnessing more diffused and widespread interactions, moving from predominantly collaborative to more competitive practices and indicating the presence of competing coalitions and hierarchical relations between actors in the CE field (Fischer et al., 2021). This has also implied the emergence of a diverse set of shared principles and practices between actors and networks of actors, with the increased competition between actors in the field influencing the emergence of new principles, for instance, the idea of a regenerative and nutrient economy, different from a 'functionalistic' toolbox like CE, and a more diversified set of values and worldviews around the CE agenda and fields of practice (Webster, 2021). While these logics have been mostly defined through the interaction of private and public actors, the definition of a 'common set of rules' is still lacking (Schulz et al., 2019). For instance, clear definitions of what is considered a 'circular practice,' or the creation of industry standards, the definition of inter-organisational bodies to represent the interest of 'circular businesses,' and more in general an agreed set of regulatory instances defining, supporting and limiting the organisational boundaries of actors operating in the CE domain has not emerged fully (Fischer et al., 2021; Borrello et al., 2020a). Also as an emergent organisational field, CE seems to be influenced by different dynamics adopted by actors engaged to define new regulations and norms at business and supply chain levels (Fischer & Pascucci, 2017; Lieder & Rashid, 2016).

Hence a few questions emerge at this stage of CE organisational field formation. For example, will we observe a convergence towards a unified field of practices, with shared regulations and industry standards? Or is CE still at its infancy as an organisational field, thus still stimulating the emergence of novel practices? And is this happening quickly enough to generate solutions to socio-ecological dilemmas and struggles?

We have asked 11 teams of renowned scholars in the CE field to tackle these questions and to offer their views on this subject. They came back with a wealth of stimulating perspectives, diverging and contentious views, but they all seem to converge on the idea that rather than transformational, most of the current CE practices and strategies are still emerging with transitional or incremental features, thus offering a context for questioning their effectiveness and fairness when confronted with the solutions needed to address the socio-ecological emergencies we are facing as humanity.

Circular economy as a field of creative and innovative strategies

Despite being a source of novel narratives or worldviews to change the current globalised market economy, or a powerful toolbox to lead practitioners to change practices and strategies, CE is also perceived as a field where creativity, ingenuity, technological development and entrepreneurship can all come together to power a new wave of innovations. Considering the need to identify solutions for the systemic socio-ecological and economic crises noted previously, CE is recognised by many as a leading agenda for change (Raworth, 2017; Stahel, 2016; Webster, 2013). As indicated above, there is ample evidence of this change process, occurring at both a policy-level to stimulate top-down-driven actions and compliance (e.g. regulations, standards and legislation) as well as at a grass-root level to stimulate novel and disruptive entrepreneurial endeavours. However, regardless of the level where CE is inspiring change, there is an underlining set of tensions that could hamper a rapid transition (or transformation) to a global, regenerative and distributive CE, and namely a tendency to resist systemic and collective change, particularly when dealing with unknown scenarios, severe uncertainties and complexities. These trends have been recognised in our response to so-called grand challenges or wicked problems, such as climate change, supported by a fierce debate in the academic community around social responses in the forms of risk-aversion and cognitive biases, heuristics and simple rules (see, e.g., the debate between Gigerenzer (1991) and Tversky and Kahneman (1974)). Scholars looking into biases and heuristics as deviation from a (mostly economic) rational response to a given challenge have indicated forms of social or individual inertia, or resistance, to change in what they call optimism bias (Sharot, 2011), suggesting we might consider it 'not likely to happen to us' (as we typically overrate potential positive outcomes over potential negative outcomes) and when combined with the partisan effect (a distrust for authority and governments), a top-down approach is often challenged. Similarly, considering bottom-up 'apathy of action' and the 'status quo' bias (Kahneman et al., 1991) suggests that we don't generally like change and we hope that "someone else will do it." To coin the poster phrase of Sir Robert Scott, OBE "The greatest threat to our planet is the belief that someone else will save it" (https://www.activesustainability.com/). To compound this, if we feel we are changing and others are not, we succumb to the 'sucker effect' and assume that we are the only ones changing and thus reduce our efforts. Scholarship embracing a different view on the nature of biases and heuristics point out other mechanisms blocking our capacity to adapt to change, particularly when it is not easy to cognitively process all its aspects, it is abstract or distant, or involves collective actions. Regardless of how we fame and theorise about individual and collective response to challenges and changes, considering our current progress on the climate crisis and achieving sustainability within planetary boundaries, for example, there is an overall agreement at a scholarly and a policy level that we are not acting fast enough, potentially due to a combination of the factors noted above.

Reflecting on the nucleus of these issues, the question of 'Who leads our CE transformation?' is critical. One perspective appearing from the field of innovation management is the idea that industries can create a lead, by focusing on developing new innovations that respond to key requirements from their users, being customers but also societies (von Hippel, 2009). This would represent a middle, up/down perspective as opposed to top-down or bottom-up and there are various contributory elements to the idea of circular innovations. Indeed, if we consider the key operation of our industries, to extract and transform raw materials into products and services, from which they gain economic rents, this anchors unarguably into the key resource drivers of CE, in terms of the transformation away from take–make–dispose towards more circular resource stewardship and production practices. Exploring extant studies to understand the variety of new innovations coming forward under a CE banner, Hobson and Lynch (2016) explore the link between resources, innovation and growth (and degrowth). Grounded in environmental economics, the eco-innovation field has been growing steadily for the past 15 years, building on the concepts of change in terms of firm and societal level developments, and how these can build critical mass by blending positive action with carefully rehearsed discrete messages to users and consumers, nudging them to change their behaviour (Colombo et al., 2019, de Jesus et al., 2018). Literature on Eco-Innovations from the early 2000s onwards focuses on the environmental aspect of the innovations across the range of innovations (product, process, paradigm and positional) and also considers radical versus incremental product and service (de Jesus et al., 2018). A small-scale, pilot study carried out in 2018 (Alexander et al., 2018) exploring the EMF back catalogue of 90 circular products, services and practices, presents a simplistic classification of 'circular innovations,' in terms of the size of companies offering these 90 new products and services and assesses the relative radicalness of these offerings. Results confirm industrial suspicions that multi-national enterprises and large organisations are offering incremental and low-risk innovations, by modifying their existing extractive business models to present a new 'slightly less bad option.' Entrepreneurs and SMEs, however, were acting to create more radical offerings that provided meaning to their customers and users (Verganti, 2011). Confirmatory research results obtained from a large-scale survey of organisations, led by the Stockholm School of Economics, confirmed that companies considering the challenges during the pandemic identified that meaning-oriented business models led to loyalty and customer retention. In further exploratory work, looking to confirm and begin to assess the critical dimensions of Circular Innovations, Alexander et al. (2020) position CE ambitions towards moving from purely 100% less bad, towards a 'regenerative' focus for new products and services, called 'restorative' by Raworth (2017). This reflects how impacts can be considered in terms of environment, society and the economy – the classical Triple Bottom Line view.

The sustainable innovation agenda has also grown considerably in the past 25 years from its roots in CSR (Crane et al., 2019) with one distinct area within the sustainability literature focusing on RI (Lubberink et al., 2019). RI links Responsible Innovation to the upstream and downstream effects of the impact of the new product or service, ensuring a responsible outcome is planned. Although not expressly focused on the CE, the contribution that these fields could make to CE is extensive as there is a wealth of literature and case study data available to inform us. Theoretically, therefore, innovation in the form of new goods and services has the potential to create significant CE-oriented change. This shift in the field of CSR (noted earlier) and exercised in the innovation landscape with RI practices, we can see begin to recognise pressure to create greater rewards for stakeholders (as opposed to the traditional Schumpeterian view of shareholder reward bias) or as what Kate Raworth called "distributive rewards" (Raworth, 2017).

If we consider, however, that shifting from the traditional view of exploitative resource extraction to a restorative and environmental regenerative basis (across the economy, environment and society, as indicated by the early notions of circular innovation) there is perhaps a paradoxical relationship with the traditional Schumpeterian view of dominance of shareholder reward, gained from Ricardian rents, the sale of products and services and the revenue streams created accordingly. This paradox could shed some light on how our respective industries have reacted to competing pressures for change brought about by the realisation that our twentieth-century model of endless consumption is flawed. Figure 1 aims to illustrate this paradox, whilst offering some relative positioning to add clarity to the respective movements or extended CSR & RI, Eco-Innovations and rapid growth companies, trading on Eco-innovations (so-called Gazelle Companies).

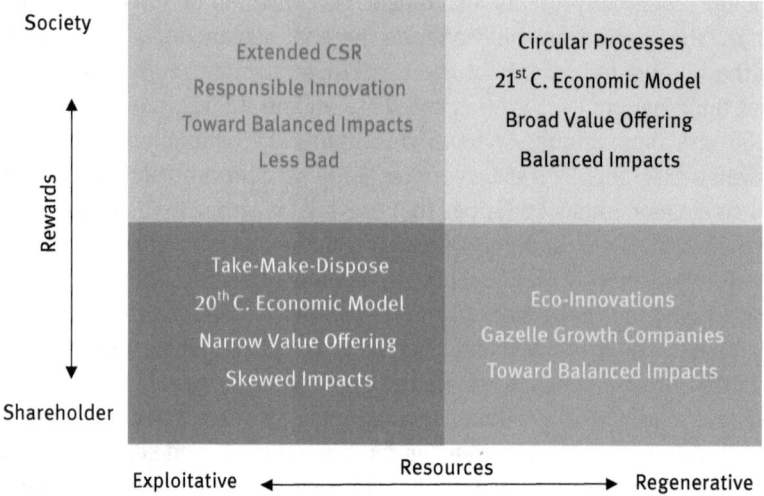

Figure 1: Mapping Circular Innovations in the twenty-first century market economy.

To make sense of this paradox, the inclusion of the concept of meaning-oriented business models and broad value offerings augments the early outline definitions of Circular Innovations: products and services created with the aim of creating impactful, regenerative change towards environmental, social and economic equity (Alexander et al., 2020).

An overview of this book: our journey in the circular economy debate

Against the articulated landscape presented so far, in this book, we investigate and discuss the critical tensions that are characterising the CE debate, and particularly its oscillations and contradictions between a transitional and practical agenda that gets support from business leaders and policy makers, and a transformational agenda inspired by eco-effectiveness, biomimicry and industrial ecology principles. We have asked a team of colleagues and experts to reflect on whether the former is open enough to allow the latter to develop and become the seeds for what recently Ken Webster (2021) has defined as a 'new nutrient and participative economy.' In the first part of the book, therefore, we have asked CE experts and scholars to reflect on key aspects and tensions characterising CE as transformational agenda for humanity and society. These are views of the big picture, a journey in the wider landscape, including its emergent traits, unknowns, and debated or contentious issues. This part of the book also offers a novel approach to articulate our understanding of CE by using an 'in conversation' approach. It is like we offer the reader the opportunity to 'grab a cup of tea,' sit with the expert and hear what he or she has to say. What is emerging from these conversations is a rather dynamic landscape of ideas, practices and visions for the future – a promising picture of where the discussion on CE can go, and how far we can push its transformative boundaries.

The second part of the book presents, instead, a more structured although still dynamic and heterogeneous view on 'what's going on' in the CE arena. Colleagues from different countries and backgrounds have connected and questioned CE from different perspectives: what is the role of design thinking in CE transitions? How do we engage with nature to design new products, processes and organisations? How will CE engage to inspire innovation and novel business models? What are the roles of key enablers and barriers such as regulation, finance and accounting structures? If we take seriously the need to put people and social change at the core of CE transitions, how do we do that? And what is the impact in terms of employment, jobs and the workplace? Finally, CE clearly indicates that transitions start from rethinking how we 'make' things, moving from a linear to a cyclical and regenerative approach. When re-thinking the way we engage with materials, resources and their meaning, what are the consequences for individuals, organisations and ultimately for the economy and society?

The final part of the book is organised to offer an inventory and library of CE practices and strategies, presented through a 'vignette' format. The idea is to give the reader the opportunity to explore and be stimulated by cases that are experimenting with the rules, principles and views discussed in the first two parts of the book. It is a snapshot of the vibrant reality of many entrepreneurs, creative minds and business activists, seeking to change the way we live and thrive on this planet. As with any inventory or library, the reader can choose what the best approach is: systematically explore all items that are offered and showcased or cherry-pick what stimulates interest. The choice, hopefully, is yours!

A final remark when looking into the future: ambiguities and tensions ahead!

Thus far we have presented the key principles, views and perspectives that arguably form the contemporary vision of CE that we aim to showcase in this book. We would like to conclude with some words of caution and a final remark when looking at how the CE agendas, movements or arenas can further develop and be articulated in the near future. In this book, we have proposed conceptual categories of CE built on an examination of existing debates and cases, where CE principles, rules and practices are being applied to different aspects of economic, ecologic and social life. These categories are presented to help the reader make sense of the emerging and diverse landscape of CE. While we are inspired by the narratives presented, the cases implemented and the perspectives discussed, we are also cautious of the need to reflect on the implications of these initiatives and the development of CE more broadly. In what follows, we identify a number of key points of attention that we believe need to be addressed if we want to fully realise a diverse, participatory and inclusive vision of CE for humanity. In short, we have identified a list of relevant ambiguities and tensions that have not been fully addressed in this book, but are worth discussing because they could shape the CE debate, and agenda, in the near future.

A first set of tensions and ambiguities is observed in relation to how CE principles can lead to the design of 'just and fair' socio-economic systems. Recently, the debate has shifted decisively on how to design an economy that is based on closed loops of resources and is also regenerative and distributive (Raworth, 2017; Johnson and Webster, 2021). However, so far, our view on the current debate seems to suggest that the application of CE views and principles tends to ignore the problematic issue of unbalanced and unfair access to resources between groups in society, gender, cultures and geographies. Also, looking at the contributions reported in this book, the ubiquitous and simultaneous use of CE as concept, worldview, and set of practices creates the risk to make it an empty and meaningless framework. In other words, it is still hard to define what is 'circular economy' and to identify a set of criteria to understand what indeed qualifies a system as being circular or not. Intuitively, such principles as closing loops of resources, using regenerative approaches or making use of current energy flows, seem core to the concept and framing of it since it is hard to envision a circular system without addressing the reuse up-cycling of materials and the wider application of renewable resources and energy. Further, the relationship between the different principles remains unclear. Namely, can we consider circular an economic system which is based on closing loops and regenerative practices but not on sharing or celebrating diversity?

Another point of ambiguity lays in the costs of the transitions, and how re-designing our economic systems, and the benefits or gains related to the re-design process. Many of the CE projects presented and discussed in literature are not yet cost-effective or are operating at such a limited scale that it is practically impossible to predict whether they can challenge the more dominant linear systems. Moreover, the costs of transitioning and re-designing using the CE principles seem to limit their application outside the corporate space, thus exposing CE to corporate exploitation issues or to projects revolving around the activities of charities, non-profit organisations and volunteers, which may be restricted to an elite of actors (Pascucci & Duncan, 2017). Similarly, the CE debate so far lacks reference to how conflicts on exclusive resources are mitigated, and how different participants in the loops and (redistributive) networks can regulate and negotiate their contributions and the gains attached to them (Webster, 2021). Although diversity and redistributive gains are starting to be mentioned in this debate, it is not clear how vulnerable actors in the economy will be protected from risk of exploitation in a globalised CE. This is particularly relevant since engagement between different types of actors with unequal relations of powers – as we might see, for example, in corporate-led initiatives – may result in so-called green-washing through CE, delegitimising the transformational potentials of this framework (Pascucci & Duncan, 2017).

These points are emerging in the current debate and all raise relevant questions pressing CE scholars and practitioners for further reflection on the actual capacity of CE principles to inspire solutions that can tackle structural issues facing our economic systems and ultimately humanity as a whole.

References

Alexander, A., Hopkinson, P., Miller, J., & Miller, M. (2018). Twenty-first century innovation: What's the name of the new game? *ISPIM innovation conference – Innovation: The name of the game*, Stockholm, Sweden, June 2020.

Alexander, A., Boehm, S., Pascucci, S., & Cherrington, R. (2020). Circular innovations: Sustainable innovations, eco-innovations and circular regeneration? *ISPIM innovation conference – Innovating in times of crisis*, Virtual, June 2020.

Benyus, J. M. (2002). *Biomimicry: Innovation inspired by nature*. New York, NY, USA: Harper Collins.

Berbel, J., & Posadillo, A. (2018). Review and analysis of alternatives for the valorisation of agro-industrial olive oil by-products. *Sustainability, 10*(1), 237.

Borrello, M., Pascucci, S., & Cembalo, L. (2020b). Three propositions to unify circular economy research: A review. *Sustainability, 12*(10), 4069.

Borrello, M., Pascucci, S., Caracciolo, F., Lombardi, A., & Cembalo, L. (2020a). Consumers are willing to participate in circular business models: A practice theory perspective to food provisioning. *Journal of Cleaner Production, 259*, 121013.

Bradbury, H., & Clair, J. A. (1999). Promoting sustainable organizations with Sweden's natural step. *Academy of Management Perspectives, 13*(4), 63–74.

Cardoso, J. L. 2018. The circular economy: historical grounds. In Changing Societies: Legacies and Challenges. Vol. iii. The Diverse Worlds of Sustainability, eds. A. Delicado, N. Domingos and L. de Sousa. Lisbon: Imprensa de Ciências Sociais, 115–127

Colombo, L. A., Pansera, M., & Owen, R. (2019). The discourse of eco innovation in the European union: an analysis of the eco-innovation action plan and horizon 2020. *Journal of Cleaner Production, 214*, 653–665.

Commoner, B. (1971). *The closing circle: Nature, man, and technology*. New York, NY, USA: Random House Inc.

Crane, A., Matten, D., Glozer, S., & Spence, L. (2019). *Business ethics: Managing corporate citizenship and sustainability in the age of globalization*. USA: Oxford University Press.

D'Amato, D., Korhonen, J., & Toppinen, A. (2019). Circular, green, and bio economy: how do companies in land-use intensive sectors align with sustainability concepts? *Ecological Economics, 158*, 116–133.

de Jesus, A., Antunes, P., Santos, R., & Mendonça, S. (2018). Eco-innovation in the transition to a circular economy: An analytical literature review. *Journal of Cleaner Production, 172*, 2999–3018.

DiMaggio, P. J., & Powell, W. W. (1983). The iron cage revisited: Institutional isomorphism and collective rationality in organizational fields. *American Sociological Review, 48*(2), 147–160.

Domenech, T., & Bahn-Walkowiak, B. (2019). Transition towards a resource efficient circular economy in Europe: Policy lessons from the EU and the member states. *Ecological Economics, 155*, 7–19.

EMF. (2012). *Towards the circular economy vol. 1: An economic and business rationale for an accelerated transition*. Isle of Wight, UK: Author.

EMF. (2017). *Achieving growth within*. Isle of Wight, UK: Ellen MacArthur Foundation.

EMF (Ellen MacArthur Foundation). (2019). *Cities and circular economy for food*. Isle of Wight, UK: EMF (Ellen MacArthur Foundation).

EMF (Ellen MacArthur Foundation). (2013). *Towards the circular economy vol. 2: Opportunities for the consumer goods sector*. Isle of Wight, UK: EMF (Ellen MacArthur Foundation).

EMF (Ellen MacArthur Foundation). (2014). *Towards the circular economy vol. 3: Accelerating the scale-up across global supply chains*. Isle of Wight, UK: EMF (Ellen MacArthur Foundation).

Fischer, A., & Pascucci, S. (2017). Institutional incentives in circular economy transition: The case of material use in the Dutch textile industry. *Journal of Cleaner Production, 155*, 17–32.

Fischer, A., Pascucci, S., & Dolfsma, W. (2021). Understanding the role of institutional intermediaries in the emergence of the circular economy. In H. Kopnina, & K. Poldner (Eds.), *Circular economy* (pp. 108–126). Routledge.

Geissdoerfer, M., Savaget, P., Bocken, N. M., & Hultink, E. J. (2017). The circular economy – A new sustainability paradigm? *Journal of Cleaner Production, 143*, 757–768.

Gigerenzer, G. (1991). How to make cognitive illusions disappear: Beyond "Heuristics and Biases". *European Review of Social Psychology, 2*(1), 83–115.

Gowdy, J., & Krall, L. (2013). The ultrasocial origin of the anthropocene. *Ecological Economics, 95*, 137–147.

Hobson, K., & Lynch, N. (2016). Diversifying and de-growing the circular economy: Radical social transformation in a resource-scarce World. *Futures, 82*, 15–25.

Kahneman, D., Knetsch, J. L., & Thaler, R. H. (1991). Anomalies: The endowment effect, loss aversion, and status Quo Bias. *Journal of Economic Perspectives, 5*(1), 193–206.

Kirchherr, J., Reike, D., & Hekkert, M. (2017). Conceptualizing the circular economy: An analysis of 114 definitions. *Resources, Conservation and Recycling, 127*, 221–232.

Konietzko, J., Bocken, N., & Hultink, E. J. (2020). Circular ecosystem innovation: An initial set of principles. *Journal of Cleaner Production, 253*, 119942.

Korhonen, J., Nuur, C., Feldmann, A., & Birkie, S. E. (2018). Circular economy as an essentially contested concept. *Journal of Cleaner Production, 175*, 544–552.

Leach, M., Raworth, K., & Rockström, J. (2013). Between social and planetary boundaries: Navigating pathways in the safe and just space for humanity.

Lewis, S. L., & Maslin, M. A. (2015). Defining the anthropocene. *Nature, 519*(7542), 171–180.

Lieder, M., & Rashid, A. (2016). Towards circular economy implementation: a comprehensive review in context of manufacturing industry. *Journal of Cleaner Production, 115*, 36–51.

Lovins, A. B., Lovins, L. H., & Hawken, P. (1999). A road map for natural capitalism. In *Understanding business environments*. Michael Lucas (ed.), Routledge: London

Lubberink, R., Blok, V., van Ophem, J., & Omta, O. (2019). Responsible innovation by social entrepreneurs: an exploratory study of values integration in innovations. *Journal of Responsible Innovation, 6*(2), 179–210.

Lyle, J. T. (1996). *Regenerative design for sustainable development*. Hoboken, NJ, USA: John Wiley & Sons.

McDonough, W., & Braungart, M. (2002). Design for the triple top line: New tools for sustainable commerce. *Corporate Environmental Strategy, 9*(3), 251–258.

Meadows, D. H., Meadows, D. L., Randers, J., & Behrens, W. W. (1972). *The limits to growth: A report for the Club of Rome's Project on the predicament of mankind*. New York, NY, USA: Universe Book.

Morseletto, P. (2020). Targets for a circular economy. *Resources, Conservation and Recycling, 153*, 104553.

Neumayer, E. (2003). *Weak versus strong sustainability: Exploring the limits of two opposing paradigms*. Cheltenham, UK: Edward Elgar Publishing.

North, D. C. (1991). Institutions. *Journal of Economic Perspectives, 5*(1), 97–112.

Pascucci, S., & Duncan, J. (2017). *From Pirate Islands to communities of hope* (pp. 186–200). New York: Routledge.

Pauli, G. A. (2010). *The blue economy: 10 years, 100 innovations, 100 million jobs*. Paradigm Publications. Taos, New Mexico

Polanyi, K. (1944). *The great transformation: The political and economic origins of our time* (pp. 360). Beacon Press. Farrar & Rinehart: New York

Raworth, K. (2017). A doughnut for the anthropocene: Humanity's compass in the twenty-first Century. *The Lancet Planetary Health, 1*(2), e48–e49.

Rockström, J., Steffen, W., Noone, K., Persson, Å., Chapin III, F. S., Lambin, E., & Foley, J. (2009). Planetary boundaries: Exploring the safe operating space for humanity. *Ecology and Society, 14*(2), 1–33.

Schulz, C., Hjaltadóttir, R. E., & Hild, P. (2019). Practising circles: Studying institutional change and circular economy practices. *Journal of Cleaner Production, 237*, 117749.

Sharot, T. (2011). The optimism bias. *Current Biology, 21*(23), R941–R945.

Skene, K. R. (2018). Circles, spirals, pyramids and cubes: Why the circular economy cannot work. *Sustainability Science, 13*(2), 479–492.

Stahel, W. R. (2016). The circular economy. *Nature News, 531*(7595), 435.

Stevenson, R. S., & Evans, J. (2004). Editorial To: Cutting across interests: Cleaner production, the unified force of sustainable development. *Journal of Cleaner Production, 12*, 185–187.

Termeer, C. J. A. M., & Metze, T. A. P. (2019). More than peanuts: Transformation towards a circular economy through a small-wins governance framework. *Journal of Cleaner Production, 240*, 118272.

Thornton, P. H., & Ocasio, W. (2008). Institutional logics. *The Sage Handbook of Organizational Institutionalism, 840*(2008), 99–128.

Tukker, A. (2004). Eight types of product–service system: Eight ways to sustainability? Experiences from SusProNet. *Business Strategy and the Environment, 13*(4), 246–260.

Tukker, A. (2015). Product services for a resource-efficient and circular economy – A review. *Journal of Cleaner Production, 97*, 76–91. London

Turner, R. K., & Pearce, D. W. (1990). *The ethical foundations of sustainable economic development.* International Institute for Environment and Development.

Tversky, A., & Kahneman, D. (1974). Judgment under uncertainty: Heuristics and biases. *Science, 185*(4157), 1124–1131.

Verganti, R. (2011). Designing breakthrough products. *Harvard Business Review.* (617)783–7500

von Hippel, E. (2009), "Democratizing Innovation: The Evolving Phenomenon of User Innovation", International Journal of Innovation Science, Vol. 1 No. 1, pp. 29–40

Webster, K. (2013). What might we say about a circular economy? Some temptations to avoid if possible. *World Futures, 69*(7–8), 542–554.

Webster, K. A Circular Economy Is About the Economy. Circular Economy and Sustainability 1, 115–126 (2021).

Part I: **Introducing transformation**

Allen Alexander
Introduction

Researchers, policy authors and consultants are working across many disciplines to link together and promote the key concepts and components of the Circular Economy (CE), but as we have established previously, the field is still close to infancy. Immature fields, by their very nature, suffer from an incomplete problematisation (Locke & Golden-Biddle, 1997) and as a result can offer confusing rhetoric or present a lack of coherence that can hamper their practical adoption.

The progression of the Circular narrative has not been without criticism, in both research and policy articles (Borrello et al., 2020), particularly where sustainable development goals juxtapose against circular principles and where authors fail to identify areas of direct alignment or suggest only tangential contributions. This lack of cohesion in terms of core concepts is, however, unsurprising. The CE has originated from many spheres of philosophical, commercial or research narratives and as such is likely underpinned by many slightly different epistemologies. In the same way as innovation studies have originated from engineering studies, science and technology studies and evolutionary economics and received contributions from sociology, psychology, management, strategy and organisation studies (Ritala et al., 2020), CE is receiving an equally diverse set of contributions.

From the original philosophical narrative of Boulding (1966), the 50+ years of contributions have presented slightly different incarnations of CE across different contexts (Geissdoerfer et al., 2017). According to Webster (2017), CE and how we currently understand it, has evolved in part from the fields of environmental sustainability, industrial ecology and ecological economics and is anchored into grand theories of the institutional, stakeholder, resource and systems theories.

One potential explanation for these part-facile misunderstandings is that as an emergent discipline, the core definitions of CE are not formalised and so can be misinterpreted. As noted by Alexander et al. (2018, pp. 3–4), there were 11 readily available, but slightly different, definitions of CE presented (see Table I.1). Some definitions take a holistic view and bring forward a systems-level perspective, which are widely accepted elements of a CE perspective. Others focus on waste and material flows; promote economic cycles of extended value-in-service; position resource scarcity against micro, meso and macro levels, and so forth.

By probing beyond the superficial, however, many of these potential conflicts or misalignment issues begin to fall away. One uniting frame in the narrative presents these as components within an umbrella perspective (Blomsma & Brennan, 2017); others urge a systems-level perspective and consider equitable use, based on boundaries of the system as our planetary boundaries (Desing et al., 2020). Another recurrent suggestion that can be applied to much of the work presented on CE is that we may be framing actions in terms of transitional, without a clear understanding of a

https://doi.org/10.1515/9783110723373-002

Table I.1: Definitions of the Circular Economy.

Authors	The Circular Economy is . . .
De Jesus and Mendonça (2018)	"A multidimensional, dynamic, integrative approach, promoting a reformed socio-technical template for carrying out economic development, in an environmentally sustainable way, by re-matching, re-balancing and re-wiring industrial processes and consumption habits into a new usage-production closed-loop system."
Korhonen et al. (2018)	"An economy constructed from societal production-consumption systems that maximizes the service produced from the linear nature-society-nature material and energy throughput flow . . . using cyclical materials flows, renewable energy sources and cascading 1-type energy flows" *and* "CE promotes high value material cycles alongside more traditional recycling and develops systems approaches to the cooperation of producers, consumers and other societal actors in sustainable development work."
Franco (2017)	"A purposefully designed, interconnected system where materials flow in a closed-loop manner in order to advance sustainability."
Geissdoerfer et al. (2017)	"A regenerative system in which resource input and waste, emission, and energy leakage are minimised by slowing, closing, and narrowing material and energy loops. This can be achieved through long-lasting design, maintenance, repair, reuse, remanufacturing, refurbishing, and recycling."
Murray et al. (20176)	"An economic model wherein planning, resourcing, procurement, production and reprocessing are designed and managed, as both process and output, to maximize ecosystem functioning and human well-being"
den Hollander et al. (2017)	where "the economic and environmental value of materials is preserved for as long as possible by keeping them in the economic system, either by lengthening the life of the products formed from them or by looping them back in the system to be reused. The notion of waste no longer exists in a CE, because products and materials are, in principle, reused and cycled indefinitely."
Sacchi Homrich et al. (2017)	"A strategy that emerges to oppose the traditional open-ended system, aiming to face the challenge of resource scarcity and waste disposal in a win-win approach with economic and value perspective."

Table I.1 (continued)

Authors	The Circular Economy is . . .
Prieto-Sandoval et al. (2017)	"An economic system that represents a change of paradigm in the way that human society is interrelated with nature and aims to prevent the depletion of resources, close energy and materials loops, and facilitate sustainable development through its implementation at the micro (enterprises and consumers), meso (economic agents integrated in symbiosis) and macro (city, regions and governments) levels. Attaining this circular model requires cyclical and regenerative environmental innovations in the way society legislates, produces and consumes."
Kirchherr et al. (2017)	"An economic system that is based on business models which replace the 'end-of-life' concept with reducing, alternatively reusing, recycling and recovering materials in production/distribution and consumption processes, thus operating at the micro level (products, companies, consumers), meso level (eco-industrial parks) and macro level (city, region, nation and beyond), with the aim to accomplish sustainable development, which implies creating environmental quality, economic prosperity and social equity, to the benefit of current and future generations."
Haas et al. (2015)	"A simple, but convincing, strategy, which aims at reducing both input of virgin materials and output of wastes by closing economic and ecological loops of resource flows."
Webster (2013)	"Increasingly built on renewables, and the endless flow of energy from the sun (energy in surplus), a Circular Economy is one which transforms materials into useful goods and services."

Source: Alexander et al. (2018).

transformed CE (perhaps akin to the Practice-Based View outlined by Bromiley & Rau, 2014).

To this end, many theories of transition exist, ranging from organisations that reinvent their products and services as a result of crisis (Bessant et al., 2005), to the more strategic view of setting targets to bring about the outcomes within the theory of change (Quinn & Cameron, 1988), or the more contemporary practice-based view of resource application dictating change (Bromiley & Rau, 2014).

Indeed, if we borrow a concept from sustainability transitions and wider transition science and technology-grounded socio-economic transition theory, we can identify a multi-level perspective, which might be useful to help ground our question of framing the CE.

Fundamental theory building work undertaken by Geels and Schot (2007) set out to understand socio-technological pathways and transitions by further developing multi-level perspectives aimed at "understanding the bi-directional interaction between industry and their environment" (Geels, 2014, p. 261). In 2021, the multi-level perspective (MLP) (Geels & Schot, 2007) dominates management sustainability transition literature.

MLP theory sets out three levels in an economy that is facing or is in transition. The macro or landscape level is set as the controlling force of policy, governance and societal control, which overarches and sets the norms for activity at the mid-level. The mid-level consists of a variety of technical and societal regimes, each dictating how trajectories for change play out. The expectation is that only incremental innovations can occur that reinforce, not undermine, the dominant trajectories for each regime, which in turn are bounded and governed by standardisation and widespread societal adoption. At the lowest or micro-level exist a complex cluster of new radical innovations, scrabbling to gain entry to markets and negotiate/compete for equal or Ricardian economic rents. To achieve this, MLP theory suggests they must enter and secure new market share, thereby displacing or disrupting an existing socio-technical regime within the mid-level. Potential for entry and thus change occurs only when seismic events at the landscape level (either single and catastrophic or stochastic and less intense) create a window of opportunity for the micro-level radical innovations to break into the socio-techno regime level – to access larger markets and to battle for greater market share. Over time, sustained innovations from the micro-level become mature and gain critical mass in the regime. This is when disruption of the regime occurs, where new trajectories establish, or entirely new socio-techno regimes become the norm.

Whilst not without criticism (Svensson & Nikoleris, 2018), the multi-level perspective has linked various concepts of economic and socio-relational theories, in an attempt to make sense of the system-state that a sustainable economy would need to establish, such that transition to a more sustainable economic worldview is possible.

This MLP perspective and the way in which it has been set out, however, raise one hugely important question in the context of CE. What is the sustainable economic world view and how can we envision it for the MLP's evolutionary view to be pertinent? The MLP theory suggests that seismic or continued, more stochastic drivers will create windows of opportunity at the landscape level – if we reflect on events surrounding climate change, food poverty, energy crises and the plethora of grand challenges, many of a magnitude not seen since the industrial revolution, if not the early part of the Anthropocene, we must be close to, if not already experiencing, an 'open window' for change.

The CE does, as we have established, offer some elements of an alternative economic worldview as a potential strategic horizon (as opposed to an evolutionary MLP-derived destination), but research also suggests CE struggles, due to being in its infancy, to present consistent definitions and key dimensions. Can we therefore envision this world economic view, as an evolutionary MLP view? What, therefore, is a transformed CE?

In the next section of the CE handbook, we have diverted from tradition. As the CE is still in its infancy in terms of research and adoption, and as the magnitude of the challenges being faced requires speed, clarity and focus, we have not requested thought leaders to submit their previous constructs and content to create defendable, robust, theoretical or empirical contributions.

We feel that predictably these will be carefully constructed, tightly worded and potentially lack slightly risky, off-the-cuff opinions. Instead, we have conducted informal interviews and thus transcribed a set of 'conversations' with leading thought-leaders from around the world, who are informing different perspectives and setting horizon-level (or landscape-level) agenda daily, from research, from industry and from policy.

The following section therefore commences by presenting thoughts on CE from one of the CE's most formative and enduring authors, who positioned a high-level systems view alongside far more practical 'product-service systems' and, at the end of the last century, led research, industrial engagement and product-development from the perspective of an architect – designing systems and components to better reflect the formative roots of the CE. This is followed by the ex-Head of Innovation for the Ellen MacArthur Foundation, who at the start of this century began to take lead of the field, pioneering CE concepts and authoring some of this century's most influential and practical guides to CE adoption. The key concept behind his and the EMF view is to create a much-needed call to action, at a policy, industry and community-level.

One troubling aspect of CE is the alignment, or perhaps misalignment, between our current economic models and alternatives required to position our production-consumption systems into one-planet bounds and to respect lower permitted levels for our social systems to prevent depravation. To inform a change in our viewpoint of CE, from an economics perspective, we have a narrative from a *Sunday Times* award-winning author.

Considering the flow of resources, from extraction to production and then waste, we then move to three narratives, hearing from a professor of applied minerology, with a focus on rare earth minerals and how these contribute to contemporary resource theories and practice. Then contemplating how to shift an established and resource-intensive technical trajectory, we hear from an entrepreneur and engineer who is re-envisioning next-generation automobile transport by adopting a systems-level redesign of the motor car, creating a novel, radically innovative car, fuelled by hydrogen. Completing the focus on materials, we then hear from a highly respected waste-management professional, recognised with an award from Her Majesty the Queen for his lifelong commitment to a circular and sustainable resource management strategy for Wales – a country ranked in global terms in the top three for waste management. Our final transcript adopts a far more people-centric perspective, bringing forward thoughts on living well and how we can transform our lives by being part of a network, both local and international, that drives change in circular practices at a national, regional and personal level.

Blending sometimes complimentary and sometimes contradictory themes, our thought-leaders bring forward perspectives covering regeneration – "do we know what nature wants?" or "how do we avoid having to do with less, but also waste nothing?" They contemplate normative business guides that urge prudence by "changing only one element of a system at a time – whilst driving for systemic change" and barriers presented by "established accounting practices and endless ambitions for economic growth" while respecting our earth resources, and in one or two cases, an interest in the longevity of the refrigerators of our past generations.

Each transcript, in one form or another, tries to contemplate what a transformed CE looks like from their perspective and provide personal insights, to inform our CE world view.

References

Alexander, A., Hopkinson, P., Miller, J., & Miller, M. (2018). Twenty-first century Innovation: What's the name of the new game? *ISPIM innovation conference*, Stockholm, Sweden.

Bessant, J., Phelps, R., & Adams, R. (2005). *External knowledge: A review of the literature addressing the role of external knowledge and expertise at key stages of business growth and development*. London: Advanced Institute for Management Research.

Blomsma, F., & Brennan, G. (2017). The emergence of circular economy: A new framing around prolonging resource productivity. *Journal of Industrial Ecology, 21*(3), 603–614.

Borrello, M., Pascucci, S., & Cembalo, L. (2020). Three propositions to unify circular economy research: A review. *Sustainability, 12*(10), 4069.

Boulding, Kenneth E. "The economics of the coming spaceship earth." New York (1966): 1–17.

Bromiley, P., & Rau, D. (2014). Towards a practice-based view of strategy. *Strategic Management Journal, 35*(8), 1249–1256.

Desing, H., Brunner, D., Takacs, F., Nahrath, S., Frankenberger, K., & Hischier, R. (2020). A circular economy within the planetary boundaries: towards a resource-based, systemic approach. *Resources, Conservation and Recycling, 155*, 104673.

Geels, F. W. (2014). Reconceptualising the co-evolution of firms-in-industries and their environments: Developing an inter-disciplinary triple embeddedness framework. *Research Policy, 43*(2), 261–277.

Geels, F. W., & Schot, J. (2007). Typology of sociotechnical transition pathways. *Research Policy, 36* (3), 399–417.

Geissdoerfer, M., Savaget, P., Bocken, N. M. P., & Hultink, E. J. (2017). The circular economy – A new sustainability paradigm?. *Journal of Cleaner Production, 143*, 757–768.

Locke, K., & Golden-Biddle, K. (1997). Constructing opportunities for contribution: Structuring intertextual coherence and "problematizing" in organizational studies. *Academy of Management Journal, 40*(5), 1023–1062.

Quinn, Robert E., and Kim S. Cameron. Paradox and transformation: Toward a theory of change in organization and management. Ballinger Publishing Co/Harper & Row Publishers, 1988.

Ritala, P., Schneider, S., & Michailova, S. (2020). Innovation management research methods: embracing rigor and diversity. *R&D Management, 50*(3), 297–308.

Svensson, O., & Nikoleris, A. (2018). Structure reconsidered: Towards new foundations of explanatory transitions theory. *Research Policy, 47*(2), 462–473.

Webster, K. (2017). *The circular economy: A wealth of flows.* Isle of Wight, UK: Ellen MacArthur Foundation Publishing.

1 Walter Stahel on envisioning circular systems, lifecycles and products

About the contributor

Walter Stahel is widely credited as one of the originators of our modern understanding of the concept of a Circular Economy. The founder of the Product-Life Institute in Geneva, Walter has been researching, publishing and consulting on Circular Economy for more than 45 years. His award-winning papers have brought forward concepts that shape our understanding of Circular Economy and its guiding principles, such as Product-Life systems and Cradle-to-Cradle design philosophies. His recent publications include *The Circular Economy: A Users Guide*, and he is Senior Research Fellow at the Circular Economy Research Centre, Ecole des Ponts Business School and Visiting Professor in the Department of Engineering and Physical Sciences, University of Surrey. He is also a full member of the Club of Rome. He was awarded degrees of Doctor Honoris Causa by the University of Surrey (2013) and l'Université de Montréal (2016), and the 2020 Thornton Medal of the Institute of Materials, Minerals and Mining.

Sharing his perspective on a mature and transformed circular economy, from an industrial perspective, Walter Stahel states,

> at the moment the discussion is mostly on material things, and in my opinion, the immaterial ones, especially the full producer liability and innovative new materials, components, systems solutions, are the key change makers [. . .] in addition to my old warhorse stop taxing labour wages. [However, he remarks that] not taxing labour and taxing something else, instead, seems to be an absolute brick wall (or concrete wall) that nobody wants to know about.

He suggests, to do this

> would really mean changing the parameters and the framework conditions – [for example,] even the International Energy Agency has now suggested that governments should stop subsidising fossil fuel production and consumption,

Acknowledgements: We, the editors, are immensely grateful to the contributors, as they have taken the bold step to let their relatively unfettered 'conversations' with us be printed. Each transcript therefore, whilst lightly edited, represent only their thoughts and opinion on the day of the interview and in the context of the questions and discussion, as posed. As 'personal opinions,' they must be treated accordingly and not replicated, repeated or taken to infer anything 'outside' of the interview context.

Note: This interview was conducted by Fiona Charnley on 2 May 2021. This transcript represents the interviewee's thoughts and opinions on the day of the interview only and in the context of the questions and discussion, as posed. As 'personal opinions,' they must be treated accordingly and not replicated, repeated or taken to infer anything 'outside' of the interview context.

https://doi.org/10.1515/9783110723373-003

which is a $5 trillion business. [Pointing to a present crucial debate] how do you . . . stop carbon emissions if you're subsidising it in most countries? Of course, politics is . . . a compromise . . . but, I think the future is possible, but not by keeping the old framework conditions.

I think globalisation is on the way out or has been, for quite a few years. But it would certainly be strengthened through not taxing labour but taxing long distance transport, and among other things, by not subsidising fossil fuels. But it seems these relate to things nobody wants to be the first to do. Now, all the countries more or less agreed that they want to go, by 2050, to a net carbon country. But I cannot see how they do it without radically changing these framework conditions. If these are changed, the right to repair can still be circumvented by the producers by renting products instead of selling them and keeping all the business by producing really long-life products.

Moving away from a material focus, and considering

the immaterial or non-material challenges, then you can also see 'full producer liability.' A Performance Economy almost twins [this concept], in the sense that the performance economy has already closed these liability loops. Once you force all traditional producers to do it, they will probably find out that they might as well start selling their goods or materials as a service rather than a good and get it back, so they can profit from it.

When asked, about the wealth of schools of thought related to Circular Economy, how he thinks CE differs from a sustainability-driven future, Walter replied:

[T]he definition of sustainability dates from 1712, something like this, 300 years ago . . . It came from mining, moved to forestry, and then it very clearly meant maintaining the wealth of stocks, and making a living not from demolishing the stock but from caring for it. [. . .] Recently, I saw a list of 250 definitions of sustainability, which means that everybody has now adapted his own definition and so, we have completely lost sense. I think the same thing is happening, or may happen, with Circular Economy, because there is a similar confusion with the Sustainable Development Goals, completed by up to 250 sub-goals. Nobody really has a clue. It's like a shopping mall, you find everything, but it's not compatible with some of the other things. So, I think first we have to really define a vision of the future.

Referring to Walter's prize-winning paper:

> It's been almost 40 years [1982] since the product-life factor . . . and the concept of the circular economy is still basically the same, but there is also something that is probably a circular society, which concerns the natural, human and cultural capital, which are not really profit-making domains. But the common denominator is caring. And caring is a word that is completely missing in today's sustainability and economic discussion.
>
> If we accept that there is a circular society, which is non-monetary, and the circular economy, which is monetary . . . the reason why I'm focusing on manufactured capital or synthetic materials . . . is that we have no real positive influence on natural capital.
>
> In my opinion, we really have something to say there, simply because agriculture will always depend on droughts and floods and all these weather hazards. Regenerating something would imply that we know what nature wants. The best thing is to leave nature alone, but we haven't got the time or the space anymore. So, we are trying to create almost artificial natures, and then we get many kinds of pests, or floods, weather problems that throw all our plans overboard. In Australia or California, they are fighting droughts and wildfires and everything is blamed on climate change, but I don't think we really understand how the climate works.
>
> So that's why I focus on the manufactured capital, because not only do we have control over it, but we also have a liability. There are all these synthetic materials [that] are incompatible with nature. Nature basically does not produce any waste, so if we just accept that statement, it means that all waste is man-made. And that means we have to take responsibility for it.
>
> Then we have two challenges: to deal with the legacy waste of the Anthropocene[1], technically and liability wise.
>
> We have to make sure the legacy waste from the Anthropocene, such as plastic in the oceans and nano or micro-plastics everywhere on water and land will not continue to be created in the future, so that's why we need circular sciences. We need, and this is the research programme that has started in the UK, . . . we need circular metallurgy, circular chemistry, circular energy, circular whatever.
>
> That also means that the manufacturer has to take control of what he's doing. Because if you want to keep control over materials, you have to make sure that

---

[1] According to Walter "the Anthropocene started on 6 August 1945 with the nuclear bomb. To create a . . . new era, it has to be traceable, worldwide, so that everybody can refer to it. Nuclear bomb . . . fallout was worldwide, which means that wherever you dig, you will always find this nuclear fallout on which you can build."

you get these materials back. Or the objects that contain these materials. the easiest way to do that would be to rent the material. Because then the manufacturer who embodies these materials into his objects is forced to give the object of the material back, in order to stop paying rent. But of course, that idea seems less popular with manufacturers.

If manufacturers become liable for the older plastic in the ocean and micro nano plastics, what do they do? They are bankrupt. Here, we have the UK Research & Innovation National Interdisciplinary Circular Economy Research Programme, and the circular water programme in Singapore, where they actually produce drinking water from wastewater. And you have projects like Subcoal by the Japanese steel industry using paper and cardboard waste mixed with organic waste to substitute coal in steelmaking.

Returning to the idea of non-material loops, what is the role of data and digital technology in the future of the Circular Economy?

One thing which certainly helps the area of operation and maintenance of infrastructure and vehicles and other objects, is digital twins. That starts to be quite common for new buildings. I've just read the report about the prestigious rolling stock manufacturer that is now introducing this for trains with the German railway. Another thing is the Internet of Things, but I think we'll have a problem with data security because you basically become the physical twin of the digital twin.

The other problem is in steelmaking. We discussed this recently; you cannot use 'Internet of Things' [technologies] because all these components don't support temperatures above 80 degrees, and in process technology, very often you get much higher temperatures, and therefore you cannot use digital [technologies].

The main objective of the circular economy is still extending the service life of products in the use phase. There, the calculation is very simply: in a saturated market, if you double the service life of a product, you half the input from mining and manufacturing, and you half the output into discarded products and materials – waste. So, the linear industrial economy cannot be a fan of supporting Project Life Extension.

Blockchain could be useful for that, but it's certainly not the individual user that will develop and apply it and [it is hard] to know who else will do it.

We have manufacturing strategies for every country, why haven't we got remanufacturing strategies?

Well, I'll tell you a secret, China, for the last (I think even) ten years, has had five pillars of its industrial strategy. And one of these pillars is remanufacturing. And there are a number of European manufacturers that do remanufacturing in China but not in Europe. Why is that? Well, because in China they have to, otherwise, they don't get access to the market. And in Europe, they don't want to cannibalise their production, hence the importance of the framework conditions.

Either you really change to integrate remanufacturing, as a substitute for manufacturing, like Caterpillar, or you have this other problem that you are cannibalising your sales through remanufacturing, and then most companies don't want to know about it.

It's like BMW now has a small electric car, and from the beginning the strategy team working on it said that as these cars will last forever, at least electric motor, maybe not the battery, we should not sell the cars, we should rent them or lease them, and then, the top management said forget it. Because then the quarterly results for the company [would] change radically, so you sell them. If you lease them or rent them over 20 years, you get 5% of the turnover beforehand, and of course the one thing managers are always concerned with is their quarterly results, so that could be something similar, also with remanufacturing.

But it's very difficult to discuss topics like this with insiders; if you're not an insider, people don't really want to admit that they simply don't want to do it.

So, what influence do you think new policies such as extended producer responsibility will have? Could that help?

The 'right to repair' extended producer responsibility in the US and EU, has existed for cars for quite a while and has been introduced now for electric and electronic equipment. We will have to see who will react. The 'right to repair' is a great idea from Kyle Wiens. But who is going to do it? You know, most people I think are no longer 'handy man' enough to do it themselves, so they will need a service person to do it, and then if we have these high taxes on wages, it simply becomes difficult to pay for repairs, as long as you compete with cheap imports from China or elsewhere. But if these conditions are changed, then the right to repair can be circumvented by producers by renting the product and keeping all the business in-house and producing really long-life products.

I don't know if you have seen the study on Tesla and the first Tesla leasing fleets that have done 500,000 miles in the US. And they're as good as new. There's hardly any wear as long as they don't have accidents.

I think, sooner or later the car market for E-cars will be near saturation. And then, the manufacturers will really have to think about selling or renting, because this substitution rate will drop. At the moment, it's very high, but once all or most cars are electric, then the substitution will be very small. I don't know anybody who has changed his Tesla, or his Toyota Prius, for a new model only because a new model has come out.

Can you explain a bit about how you see the link between the circular economy and job creation?

You have to be careful, it's mainly the era of R – so it's the refill, repair, reuse, remanufacture – of the circular economy that has to be locally where the products are, especially immobile objects, such as sewers and infrastructure, buildings, big equipment etc. But even repairing other objects, washing machines that you will not send back to the manufacturer. So the manufacturer must have a service system to stay in control; this is what I call the 'intelligent trend' of decentralisation.

Decentralisation is slowly creeping in, in very different areas such as food production, 3D Print in manufacturing and local energy production. The electricity companies' networks try to monopolise photovoltaic power; for example, in France, where EDF has a monopoly on electricity production. If you install photovoltaic panels, you have to pay EDF for using them. And that is an absolute violation of several EU directives, but who cares? What's an EU directive?

Some high-tech products, such as carbon nanotubes, cannot be transported, because if they get mixed up they cannot be separated, so you have to produce them locally. It's the same with some short-lived medical tracer elements. crowd finance, crowd mapping and all micro production (breweries, bakeries) – I think it's becoming increasingly accepted that this future may see a much more decentralised economic activity

And problems with supply chains – when a cargo ship is blocking the Suez Canal for a week, creating absolute havoc in Germany – are increasing. Car factories had to close down because they couldn't get the chips, just one chip that was on one of these blocked vessels. This is becoming not only inefficient but ridiculous . . . we depend on these globalised systems just to get our local products.

I think . . . some countries' governments have realised that remanufacturing allows you to re-industrialise old industrial regions where you still have the skills. The Midwest in the US used to be the factory for the world (which is now China), but they still have the skills, and they still have stocks of older cars and buildings in need of remanufacture and technological upgrading. Governments could actually re-industrialise some regions by using the circular economy.

I hope you remember *Das Kapital* by Karl Marx, where he clearly defined that the only thing that counts is productive labour, and productive labour is defined as 'producing something that can be sold on.' This is the belief that only productive labour creates wealth and welfare, but the circular economy is not productive labour; it's essentially services maintaining existing stocks, which means that economists have to rethink what happens to their theory if you replace productive labour by essential services, because that is the way that the future modern economic theory needs to be built.

Think about nurses and doctors, firemen and delivery people and all other essential services; why don't we pay them decent salaries? At the moment, you pay decent salaries only to productive labour, because those are 'in the interest of society.' If you accept that the essential services based on caring are maybe even more in the interest of a sustainable society than productive labour, then you can imagine that your house of cards starts to crumble.

A bold reminder of this came to light at the beginning of the COVID-19 pandemic:

these people are working overtime, unpaid overtime. And the latest figures I've seen from health workers, especially those working in COVID departments, show that 40% have already quit. We can educate and train [a] fantastic number of new people, but if we cannot solve the problem of motivation and remuneration, these people will be completely discouraged after a few years. Pay is one way to reward them, but not only the (UK) NHS, no health system in the world is prepared to really pay the health workers the same as workers in car manufacturing, including holidays, and bonuses and all these lovely things.

When asked to talk a bit about the link between circular economy and some of the recent terms, for example, net zero, and certainly climate challenges, Stahel responded,

Okay, what is missing here is taking into account embodied resources, embodied water and embodied CO_2 emissions.

Resources spent from 'the mine' to the 'point of sale' is something that politicians simply have never looked at or tried to figure out. Tell anybody that is not in the textile industry, that it takes 10 tonnes of water to produce the cotton for one pair of jeans and so, by doubling the service-life by repairing jeans, you save an incredible amount of water.

But we have never measured this systematically, so we don't know. If we want to convince politicians, they always want data, so we first would have to measure the CO_2 emissions caused by, for instance, erecting a building. [In] the UKRI research programme, there is one on mineral-based construction material that is looking into getting the materials back. Of course, the problem with CO_2 emissions is that the most valuable material in concrete is the cement not the aggregate, the rarest material is marine sand. If we break down concrete constructions, we can basically only make new aggregate out of it. But the big CO_2, the big energy input, is in the cement, so the only way to preserve these CO_2 emissions is to extend the service life of buildings and infrastructure.

In France, the Paris region has now started to put pressure on people to not replace buildings but they renovate, modernise, upgrade them, anything you want to maintain building stocks. But for another reason, because France has run out of landfills to put construction waste.

As Walter begins to summarise, he speculates on policy and international governmental collaborations:

There are fortunately several topics that end up in the same conclusion [product-life extension]. But to convince nations, I think that governments in the G7 are too far away from reality. Like the weather forecast, based on satellites, they are incapable of seeing what the weather will be tomorrow. But they can tell you in two month's time what it will be and then nobody remembers, so that's an easy one. Will the G7 be willing to change the framework conditions such as not taxing labour but taxing non-renewable resources instead?

But we have to be careful because less developed countries first have to build up their stocks of infrastructure. I was on a UNESCO panel a couple of days ago with people from India and Kenya, and they don't quite know what to do with the idea of a Circular Economy. I said, 'well, it's not your problem – you first have to build up your infrastructure and maintain your agriculture. Once you have motorways and all these roads and railways, energy pipelines, then Circular Economy becomes important for your country'.

Some international bodies like the G7 would demand for different strategies for different countries, and then you enter into discussion: is China an LDC? Or is it an industrialised nation? In 2005, I was on the Great Wall near Beijing. There is a Swiss cable car going up, so you don't have to climb up, which is very nice. When you get to the top and come out of the station, you see a nicely made monument where Henkel German, petrochemical and cleaning stuff manufacturer, is thanked for having paid for the renovation of this part of Chinese wall in, I think, 1998 or 2000. When China was still supported because it was a less developed country. Of course, already then they had fooled Henkel, but if somebody wants to give you a gift, why should you say no? And so, if LDCs can get money, you would be astonished how many countries or regions suddenly declare themselves as LDCs. But we should distinguish between countries that have infrastructure, and those who don't have.

Another issue is COP26. I was at COP15 in Copenhagen and was so disgusted that I never followed the other COPs. Have you ever attended a COP? You should, it's great fun, because it's actually two COPs. One is where the politicians meet and then, normally about 20 miles away from that, is the COP for the NGOs. And there's absolutely no contact between the two. The NGOs make all their speeches and the politicians make their speeches and decisions. Then, everybody pretends to have been able to influence the outcome, because they don't have time to really discuss issues, and if they did, they couldn't cope with it because they are not the experts to judge if the economic or industrial strategy is compatible with their own country's wishes. The tragedy of COP meetings is that politicians meet politicians, but they have no clue how real-life functions. The NGOs unfortunately only talk to each other. But, I mean, they are all convinced of what they're saying, but there is no interchange.

With regards to the UKRI research programme, it's such a big step forward for the UK and for research and innovation, particularly around Circular Economy. What are your hopes for the programme?

I think, independent of what comes out, the incentive to look into these opportunities is good. I think it's mainly universities with industry partners, but just looking at these issues, they will discover opportunities that nobody thought about before. In that sense, the universities are really the right frame. Because it's young people that will change the world; some of the recent inventions I've seen are all multidisciplinary by multinational teams – people from China working with Brazilian and British and Italian – coming up with out-of-the-box solutions that you will never get if you follow scientific disciplines.

In that sense, I am very optimistic that we will be astonished by what comes out of the UKRI Interdisciplinary Circular Economy research programme, but we have to give it time, and we have to find ways to motivate and encourage the young researchers and the professors. The horizon 2020 programme is a black box, and you're not allowed to talk about if you work for it, which frustrates me . . . whereas I think the UKRI Interdisciplinary research programme will be much more open. It will give other people other ideas, new ideas. In a sense, I don't know if IS4CE arises as a consequence of Brexit or not, but it certainly is a programme to take control of your own future and sciences and develop the Circular Economy. So, if it's a success, the solutions that come out of it don't need an explanation as to why this makes sense. Because it will be obvious.

Everything is interlinked. That's one of the terrible things about complexity, but the structure behind the circular economy is quite simple. The complexity arises due to people thinking in silos, and the solution is getting them to look at and understand the other silos.

2 Ken Webster on framing potential circular economies

About the contributor

Ken Webster is a visiting fellow at Cranfield University. He was Head of Innovation for the Ellen MacArthur Foundation, where he worked between 2010 and 2018. Ken has led many international symposiums about how the Circular Economy offers significant systems-level change. His book *The Circular Economy: A Wealth of Flows* (second edition 2017) relates the connections between systems thinking, economic and business opportunity and the transition to a circular economy. His most recent book (with Craig Johnson) is *ABC&D Creating a Regenerative Circular Economy for All.*

As an influential voice for many years in Circular Economy discourse, Ken Webster observes:

> [T]hese things (the CE) generate a life of their own, particularly when you're dealing with a heuristic.

He goes on to tell us:

> 'Circular Economy' was reframed at the end of the first decade of this century, partly because of the influence of the Ellen MacArthur Foundation, and the thinking that went behind what they wanted to do. Now what they wanted to do – and I know because I was there doing it – was to create a heuristic around Circular Economy, which had a number of features.

> The role of a narrative or heuristic is to make it understandable and, in a way, a supportive framework. It's not meant to tell the exact truth. It's supposed to be a rule of thumb: we'll do it this way.

> So that has got advantages, but also disadvantages. [He first explains] the advantages [of a] heuristic, and the one that was created, is that people can buy into it. In other words, it's framed or it's positioned to address some of their values. [With respect to this, Webster points out] George Lakoff famously said, "people vote their values, not their interests." I think that sounds counterintuitive. But I think it's been borne out by political changes in more recent years. Does it fit my [or] our values?

Note: This interview was conducted by Stefano Pascucci on 22 July 2021. This transcript represents the interviewee's thoughts and opinions on the day of the interview only and in the context of the questions and discussion, as posed. As 'personal opinions,' they must be treated accordingly and not replicated, repeated or taken to infer anything 'outside' of the interview context.

https://doi.org/10.1515/9783110723373-004

Now, the Circular Economy proposes to do two key things.

> One . . . we can work from the economy outwards. I recognised, or many people would recognise, that sustainability tended to work from outside the economy in[wards]. In other words, the economy needs 'adjusting and regulating' . . . came out very frequently. It needed 'moderating' in what it was doing. So, it was all about social and environmental pressures from the outside. Which is reasonable and that's fine.
>
> But it also came with a lot of this sort of personal guilt thing, you know, you're the consumer, you ought to behave differently. So, it felt a lot like behaviour change, regulation, and as the comedian in America, George Carlin said, in the late 90s, people wanted to make the world safe for their Volvos, that is, it became to look like a very middleclass concern.
>
> And, you know, if you are belonging to certain groups, you would be a sensitive consumer: you would think twice about how you travelled around; you choose things; buy things carefully. Well, there's nothing wrong with that, except that it really doesn't extend to a lot of the poor, who don't have those choices. [Most importantly], it doesn't really appeal to business. It doesn't appeal to blocks like the European Union which wanted jobs and growth. And it doesn't appeal to the creative types, the designers, the should I call them 'the progressives' within business and entrepreneurship, who want to be doing something different, but, at the same time, do something good.

He recalls from his experience at the Ellen Macarthur Foundation, that

> the way to adapt to a changing world is not to say, 'Well, you've got to do with even less,' it is to take that design approach, redesign the economy, if you can. So, if you're redesigning an economy – a very bold statement – you have to build it on existing ideas. And, of course, that's what I was keen on doing. Synthesising what was there, and that meant if we're working from the inside out, we have to have a positive cycle.
>
> Braungart and McDonough [have] done a whole lot of work on that with the Cradle-to-Cradle initiatives. They've been very keen to say that it's not about doing with less, it's about creating an economy which acts almost like a living system and doesn't use material so much as nutrients, and everything is food for the system. This was a bedrock idea and it had been tried out in a number of businesses and elsewhere. Its main appeal is that . . . it's about adjusting the way that we see design and what follows from that. You know, [McDonough] famously said, "let's have buildings like trees and cities like forests." Now, he was teasing us with that. But also, he was talking about positive footprints. We

don't regret being on the planet, we think it's positive because you can rebuild natural and social capital.

Discussing this reframing further, Webster points out:

more, but more of what, was really the important question? You improve the stock of solutions available to humankind, you know, so that was with rebuilt capital, you have more choices, you've got more opportunities for taking something off the flows of materials and products, and improving human welfare. So that was one end of it, it's got to be a positive cycle, not to do less harm.

Addressing the second element that the Circular Economy proposed to change

It was a way to frame it, so that one didn't use the word sustainability, as far as possible. Because one of the clues about framing is that all words have unconscious associations. And so, if you said sustainable economy, people go 'Oh, yeah, fine. We're into that again, are we?' But if you said Circular Economy, circular gives that sense of feedback . . . it's not linear. And it says economy. Oh, that's serious. Now that is very powerful. If you put that together, it's serious and it's not like it was before. [Delving into the nuances of sustainability,] what are we talking about? The SDGs? We've got a list of really good things, but it doesn't tell us where we're coming in or out from, really. It also doesn't tell us what to focus on. It's too big a concept and [sustainability] had worn down in a way, in my view, it had worn down.

With this came a realisation:

Oh, we can redesign industrial systems? And there were several players in that arena: Walter Stahel, with the performance economy or the functional economy; you had people also who were working in architecture. Well, it's interesting that Walter was an architect originally, these are design people, McDononough's an architect and designer.

You have people like Michael Pawlyn looking at biomimetic architecture; we have Janine Benyus in biomimicry. The commonality of all these people, I would say, possibly with the exception of Walter, is they're all taking insights from living systems. [Notably,] it's framed positively and the support for the framing is insights from living systems, which also rests on something more fundamental, which is the idea that the real world works on complex adaptive systems. It's a systems world.

Bringing us up-to-date, Webster reflects:

> We have been using a sort of throughput model. This is a sort of mechanistic, reductionist, practical, helpful – the industrial revolution of machines and throughput whilst degrading natural and social capital. That felt like something in the past, it didn't reflect how real-world systems worked;, it neglected most of the feedback, or focused just on feedback, say, through the monetary feedback loop.

> Now, you have a very different starting point. You're starting with complex systems, you're taking insights from living systems, and you're framing it as a positive cycle, not a 'do less harm' one. And you're also focusing on only three target groups. One is the policymakers, the second one is business, and the third one, if you like, is what I call the progressive designers. People who wanted to do it differently.

With this approach,

> What we left out quite deliberately, was we weren't going to play the social and environmental thing, except as by-products of the process. Now, that sounds ridiculous. How can you leave that out? It's a heuristic you see, we want people to buy in, and to buy in on the basis of their values, which is that the economy matters most. And if we redesign the economy, make it a positive cycle, we can spin off these benefits anyway. Particularly if we're closing the loop, there might be more employment, there might be more growth, but in a different way. So, it downplayed quite deliberately.

Poignantly,

> I remember the Ellen MacArthur Foundation discussed, should they have lay members? Should people be able to join the foundation as a citizen? And we said no, to begin with, at least, this is not playing to citizens, it's playing to elites. It's playing to business groups, it's not a grassroots organisation. It's about being persuasive at a different level, not rejecting those other things. And I think in a way that was very successful. I think, evidentially, it was pretty successful in the beginning because you had jobs and growth, positive cycles led by design, and if you add in the real world of strained resource availability, around the end of 2009/2010, coming out of the Great Recession, as well, but you had rising resource prices, it all looks like something that ought to be done. You know, the need to do another opportunity?

To be clear,

> I want to emphasise [that] I'm not putting down, in any way, social environ-
> mental things. But the question was, how do we shift people a little bit?
>
> Hunter Lovin (of Natural Capital fame) noted recently, "Hypocrisy is the first
> step to real change." In other words, there are a lot of large agribusiness firms
> now getting involved in regenerative agriculture. Now, they don't know what
> they're doing, says Hunter, but the point is, because they feel that they must
> get in on it, they're being in a way, she would say, hypocritical. But the point
> is, they're having to attend to it.

Highlighting a key takeaway:

> I think one of the messages about framing and Circular Economy is: it's not a
> magic bullet for anything. It's a heuristic that takes the thinking further. In
> modesty, it had holes in it from the beginning. And some of those holes were
> deliberate because it allowed us to focus on particular things and not others.
> You know, the old mediaeval idea that the priests would sit around discussing
> how many angels you could get on the head of a pin. We spent a lot of time
> with partners discussing what words to use and what words not to use.

Addressing the question of how Circular Economy has evolved:

> Well, one, you're bound to get some pushback on a heuristic. But the thing is,
> if you're getting pushback on it, at least, it succeeded. You know, it's out there
> operating, so yeah, we need to work on the detail. My reflection is people need
> to get into the detail now, and it won't come up entirely pretty. You know,
> there are plenty of, well, it's inherent in the term heuristic, there are plenty of
> detail things that need to be worked out, right?
>
> Some people are also getting a bit, I would say, uncomfortable with the idea that
> in a way, it's caught so much attention, that it's losing its transformative poten-
> tial. All it is, is okay, this is how the economy works, can we fix the bits that don't
> work very well with materials and energy? The criticism might be, well you're tak-
> ing the world in the way it's structured, and who controls it, you're allowing that
> to continue, but you're trying to fix some aspects of it around the material cycle.
>
> Until they say it's transitional (is transitioning), but not changing, where it
> needs to be transformative is to carry people further, because one of the attrac-
> tions of the heuristic was that people thought, "Oh, this is a way in, we can get
> a very different society this way," or as one of the tools. But it may turn out that
> they're disappointed now. They might be increasingly disappointed because

they go well "What about the social bit? What about the scale? What about operating at different scales? What does it mean about resilience? Does it have anything to say about resilience?" and so on. [In response to these concerns] I think, [there's] a refocusing at the moment around, well, in reality, it was looking at what they call the technical cycle. That's giving you the durable market stuff. But it wasn't doing much about the bio cycle, anyway, it was certainly not doing much about scale. It was all about the big players, so it didn't do that. On an organisational note, there's a difference between centralised global operations and more local open source, open network, distributed Circular Economy.

Essentially, currently,

you've got to an uncomfortable point where the thinking hasn't developed far enough to call it a 360-degree Circular Economy, if you forgive that sort of notion, but it isn't 360-degrees at the moment. And one of the big dangers now is that that process is not completed. Because if it's left just as a heuristic that suits policy and business, it doesn't carry at this very urgent time, it doesn't carry enough of the tools that people will say this can be helped to make us help transform the world.

Reflecting on his own research,

I think the monetary cycle should be included. I've been talking about materials almost exclusively, materials and energy. [Webster reflects] a colleague of mine, Caspar Jorna who was at the Ellen MacArthur Foundation, says "what point is there in fixing the material cycle, if you've still got an extractive money cycle?" Because what you're doing is you're enabling . . . what Brett Christophers calls "rentier capitalism." It's enabling the increased concentration of business, and the owning of economic rents, through having an extractive, rather than circulatory (there's a clue), money system. So, some people like me are interested in tying-in monetary and material cycles, but not many.

Concluding these thoughts:

I say that my perspective is after ten years, we need to do two things: one is to consolidate the big picture, and the second thing is to get into the detail to see what does work and what doesn't. [. . .] In the end, it's all going to be about system conditions, because if we're talking about a systems approach, it's very much like saying, "Well, at the moment, we're playing soccer. Let's allow them to pick up the ball and run around a bit with the ball." I don't know what you might call it, it's halfway to rugby, but it isn't soccer anymore. So, a simple change in some of the rules, not so simple perhaps, but a change in the system conditions would either enable or disable this change. At the moment, system conditions are not changing, but they could do, if we have a clear idea of what it is we are trying to achieve.

Being clear on the mistakes that Cradle-to-Cradle has made by focusing too much on product and too less on system change: Was that part of the initial conversation when you created this narrative that we now call Circular Economy? Or has it emerged step by step or at a later stage?

You're quite right, although it's driven by insights from living systems, and really, McDonough and Braungart deserve a lot of praise for all of the work they did in that, you know, the [eco-]effective, [eco-]efficient, and all of the rest of it. It did look, at that time, as though it was some sharp ideas from a couple of people who were trying to monetise it in a way through consultancies around design and product. I mean, I'm not saying that that's exactly what was happening. But it felt that way.

At the time, the perspective on this:

felt like something that was in design and manufacturing, and perceptions really matter. So, it didn't leak out, in a sense in the way that might be expected, into other arenas. So, what I thought was I'll just paint a broader brush over that and draw things together, which does have overlaps.

In fact, that was one of the big successes for me of the early Ellen MacArthur Foundation days; we actually put these folks in the same room. Some of them for the very first time, they hadn't been there and they were discovering or accepting that the potential of what became called the Circular Economy approach. I think they were pleased in a way that it expanded the sort of work they could do, or their role.

[The Circular Economy] it's an emerging set of ideas, which had got stuck in sort of almost consultancy way, it was very much for business. [Critically,] I felt it could be much broader, more broadly applied. Also, Cradle-to-Cradle sort of felt like product design. How do you design these things? It didn't feel systems enough. And I think there's also reluctance in [certain] writings to engage in the bigger economy. You know, certainly not the economy of monetary systems, that that's been difficult for a number of us. It became a lot looser, they established Institutes that would help share Cradle-to-Cradle ideas. But at the time, it felt a little bit closed.

Moving forward with the discussion,

> What most people do is they borrow ideas, and I've borrowed ideas widely. One interesting thing was I also borrowed from Gunter Pauli.
>
> He wrote something, the blue economy, it's up to its third edition. Now, one of the things I noticed straight away about Gunter Pauli's work is that people were ignoring it. And I looked into that a little bit because I was asking people who are supporting us in EMF, why don't you look more at what Gunter is doing, it looks interesting, and part of it was it's very downscaled. It's very, not just community driven, but it's also a local economy.
>
> It's about circulating cash in the local economy, working with what you have, creating multiple benefits. I think the question was, where's the scale to this for big firms to get a hold of? I think there isn't something for big firms to get hold of, it's far too practical for that, it's far too devolved. And in a way, if you can't make a lot of money out of it, or you can't create a dominant business within that arrangement, some people lose interest. You know, oh, well let's make mushrooms out of coffee waste. That's a reasonable thing that's been going on, a little bit. But it's not meant to be at huge scale. It's meant to be dealt with locally or in different parts of a city. So, it's super small business. One of the things I felt was that Gunter was ignored because there wasn't enough of the big business in it.
>
> I wanted to include it because it was about business and enterprise, but it fell out that when presented with it, most people just turned away from it. They thought, well, that's okay but you know, it just it didn't catch people's attention in the way that the more mainstream folks did. So, I put it in there and this is the same, with biomimicry, oddly. Biomimicry is right at the heart of a Circular Economy, but it doesn't get talked about, very much. It seems to be all about big durables, big business going to scale and digital. You know the magic sauce is digital, which is something which I think we ought to perhaps discuss at some stage, because how do you loosen up these business models? Well, digital's there to do that for you.

When asked specifically, what is next for Circular Economy:

> Well, the dystopian future, I would call it that, is that through digital and the use of digital platforms, we go to this future where you 'own nothing and you will be happy.' [. . .] Now, you won't be happy. That's the point. Because if you don't own things, who does own them? And if you can't repair things, who does this? At what cost? It's this idea that the Circular Economy could reinforce rentier capitalism by providing new asset classes, products components to

materials, which will be controlled by very few businesses. But using digital to track and trace and measure performance, I think that's very anti-democratic, it's not putting resources in the hands of people. It's tidying up, if you like, the resource cycle to the advantage of a limited number of businesses, which some people go, "Hey, well, you cut down the carbon burden," "hey, it stopped doing this," you know, "let's see mining companies doing the recovery of the metals and bringing it back into use."

Webster warns:

There is a huge danger of it if you do not control it, or you haven't got access to it. There's already a great deal of discussion around tractors and other big equipment in America, farmers are complaining; they can't fix their own tractor because they can't get the software plugin, to enable them to do the diagnosis, you know, the analysis. So that's the dark side of pervasive platform because they use network effects. So, we should emphasise the role of scale, they will tend to dominate specific materials, valuable materials, we're not talking about single use plastics or anything like that, which is just a badly designed material.

On a positive note,

what will happen on the other side is that because digital also means people can connect, you can get much more social production, you could do much more localised closing of the loop, people will know where things are, and are accessible. So you get to devolved, resilient, thriving, economies based on access to materials and also access to things like land or machinery. This is part of a revolution, or if you like encouraging different ownership patterns, so that it's not dominated by companies where ownership and control is separated. They're much more integrated cooperatives, sharing platforms and stuff like that.

I think the question is, how do we adjust the balance between these two forces? Because digital and the state of the world – the both of those are on the table. And it's going to be a political decision about how those two balanced up, if they do. If they don't balance up, I think we risk two things: one, losing much of what we call democracy through the increased control of objects that we need or, two, the revival of democracy by enabling people to get hold of tools, equipment, resources, to generate a part of their own living, [and] participate in the economy.

With some final remarks,

> the Circular Economy is something to look at because it talks about circulation, not extraction. I think that's one of the last heuristic jobs, to say, "Well, folks, you know, [Ken believes] even Adam Smith said, it's about circulation, not economic rents." So, let's apply that in the current world. How do we circulate money back into the system to enable people to create or meet their own needs and wants? How do we do that alongside the fact that we have a globalised system? And that makes that interesting because you can say I'm on the side of circularity not extraction, because we've had enough of that extraction thing.

> We want the story to be about circulation in monetary and material cycles. So that's how I would see the dystopia and also everybody has to have a utopia. That's the other thing about narrative. It's a three-act play; the problem is the start with the linear economy is trashing the world, then there is the struggle, let's try some recycling bits and pieces like that, and then there's a utopia, we've got a Circular Economy now.

> When everything's thriving, and capital is being rebuilt. See, that's the cheesy part of a Circular Economy. But you do need a utopia for all the social narratives. And George Monbiot describes it as a restoration narrative. We're in loaded trouble, these are the struggles we've got to do, but these people will see us positively into a restored and balanced world. So, Circular Economy is playing in those waters, and I'm very pleased that it is because I think it's adding something to the quality of the debate.

3 Kate Raworth on creating regenerative and distributive economies by design

About the contributor
Kate Raworth is author of the international best-selling book *Doughnut Economics* and co-founded Doughnut Economics Action Lab. She teaches at the University of Oxford's Environmental Change Institute, based in the UK, and she is a Professor of Practice at Amsterdam University of Applied Sciences.

According to Kate,

> the basics of the concepts of circularity, where waste from one process becomes food for the next and where we use Earth's resources far more collectively, carefully, creatively and slowly go back in many ways to economies that our grandparents and great grandparents knew very well, in the sense of biological materials that were used again and again. One major thing that's different, this time, is the introduction since say, the 1950s, of mass plastics and other human-made materials from persistent organic pollutants to nuclear waste. How do we eliminate those things that can't be circular? And how do we create circular loops that mimic nature's regenerative powers, with the human-made or technical materials that we inherit? So that's completely unprecedented. But a lot of it, I think, we can learn and relearn from the past.

In terms of what a transformed Circular Economy will consist of:

> I'm delighted to say I don't know. I don't know what it's going to look like what it's going to feel to walk down the street. I think we can impose our inherited twentieth-century imagination of industry just with a few loops on it, but there's no way it's going to look like that! And that's what I think is so exciting: it's going to be incredibly creative in terms of how we transform not just where resources flow and how they are used, but all the questions around ownership, to service use. In my family, we recently got rid of our car in January, and I now use a car scheme. To me, that's part of being part of a Circular Economy, one of

Note: This interview was conducted by Allen Alexander on 22 June 2021. It formed part of the ISPIM Virtual Conference in Berlin, 2021. This transcript represents the interviewee's thoughts and opinions on the day of the interview only and in the context of the questions and discussion, as posed. As 'personal opinions,' they must be treated accordingly and not replicated, repeated or taken to infer anything 'outside' of the interview context.

https://doi.org/10.1515/9783110723373-005

the ways of shifting, even though it might not look like circularity. I think these shifts are only the beginning of how a Circular Economy is going to transform the way we have a sense of how we belong and what we share, and you know, how objects belong in our spaces.

So, I don't know what it's going to look like. And I love it every time a new innovation comes along, and you're taken aback. I remember, the first time I saw an electric car plugged into a lamppost in my neighbourhood: it's that moment of, that's a first and now it's going to become part of the new normal. I can't wait to see more of these firsts.

But when comparing "Circular Economy to Sustainability or Eco-Innovations,"

there are so many different ways people use language or labels to distinguish what their group is doing, and sometimes that can be really empowering and exciting and energising.

And actually, Doughnut Economics falls exactly into that . . . what I call Doughnut Economics in a playfully serious way other people might call well-being economies, regenerative economy, circular economies.

I think it's valuable to use this range of language because I think we're searching. And I think as we're evolving our concepts, our language will evolve. As things evolve, the first thing that nature does is to diversify and try out lots of different variations, and these different variations pop up. And then, some of them get amplified. So, I think we're at a stage of lots of language coming out around cycles.

I think it's also really, really important not to get embedded in camps. I'm circular economy, you're regenerative economy, you're wellbeing economics, and I'm community wealth building and you're deep ecology – and these camps set off against one another because they don't use the right nuance of language. And that is the way that this evolving edge and the huge, bigger team and the bigger movement that I think we're all part of can split itself and destroy itself from within. So, I really like embracing a lot of language!

But if you ask me how do I see things nested within each other, to me, the word 'regenerative' is at the top layer. So, when I talk about doing Doughnut Economics, we need to move from degenerative to regenerative industrial systems, and indeed, agricultural systems. And I see Circular Economy as a particular way of thinking that is very much part of regenerative design and essential. I don't think you could do regenerative design unless you included within it the principles of Circular Economy. So, I celebrate all those different languages . . . I just don't put my energy into slicing them up and distinguishing between them, because I think this is 'big team' work. And it's really important to recognise ourselves as part of that 'big team'. Different people, bringing forward different concepts and aspects of it, but it's all interconnected.

But before we go onto another thing, its really important to say that when it comes to sustainability or regenerative, I will go for regenerative, and one of the reasons is because sustainability can be interpreted as sustaining, keeping things, making sure we don't run things down from where they are.

But of course, we are living on a planet that has been profoundly rundown, and so I think the power of using the language of regenerative is recognising the starting point, which is: we're massively overshooting many planetary boundaries and local ecological boundaries; we have degraded, and we've inherited a degraded system. And therefore, the first act is of restoration and repair and healing and bringing something back, whereas 'sustaining' risks making it sound like the baseline is what it is, and we just need to sustain it, at its current reduced capacity.

Another reason why the language keeps shifting is because companies try to capture it. Valuable words get coopted and people get frustrated. Many people associate the term sustainable development with corporate and political capture, while others say, no no, this is the overarching concept that I'm aiming for – and when people feel the language has been coopted then they won't use it anymore. And I think that's also partly why people keep moving on.

I like the word regenerate because it fixes in our minds the fundamental metrics and cycles that we're thinking about. If we start with the metrics of the living world, then we root our aspirations and goals and standards in living systems. And then we invite business to make itself compatible with that, Rather than starting with what shareholders have come to expect as the normal returns and double-digit growth and taking finance and extraction as a norm, it roots us back in the living world.

Regenerative by design is working with and within the cycles of the living world, and it requires us, first, to understand them. We must all learn the basics of ecology. Economics means the art of household management so we should all be required to understand the nature of our planetary household before we even start to claim to be able to manage it.

When asked what three wishes she would like to have to shift our current global perspectives on economics, she suggested firstly,

> every economics tutor in the world, when you're teaching your first class of economics, don't start where every class seems to have started over the last 50 years. Economic theory, supply and demand, which is just the absolute standard thing, and it puts the market at the centre of our vision as if, suddenly, the economy is the market. That's a massive assumption. And it puts price as the metric of concern, which means that everything that falls outside the price contract, by definition, is called an externality.

> Economists accept this definition and say, Well, of course, yes externalities matter Well! If they matter, why would we go around talking about the near death of the living world as an 'environmental externality'? That alone tells us this framework does not serve our times, and the mental framework is completely inadequate for the reality that we face.

> So, number one: don't start with the market; start with the embedded economy. The economy is embedded in society and embedded in the living world. That means the economy is a social construct, and we can change it and reinvent those relationships, interactions and mechanisms. We can reinvent markets, the state and the commons. So that's number one.

> Number two: the goal of the economy! The goal of the economy, again, it's so deeply embedded in twentieth century thinking that it's never drawn, it's never explicitly discussed. You hear it in every politician's speech, in every macroeconomics class: endless economic growth, and it is presumed to be endless, no matter how rich a nation is, already. I'm sitting in the UK, one of the richest countries in the world; richer than many countries have ever been before. And yet, all of our governmental policy, all of our economics teaching, is based on the assumption that the solution to the UK's problems lies in yet more growth. And there's something absurd about that. At what point will economists say 'we have gone through the growth phase, and now it's time to grow up'?

> Think of everything in the living world: nothing in nature thrives by trying to grow endlessly: it would either destroys itself or the system on which it depends. So why would we think our economy is going to be the one outlier? Instead of having endless growth as an implicit – and what I call, a cuckoo goal (that creeps into the nest and kicks out other things) – what is the goal?

> Let's make it explicit . . . how about this as a goal: to meet the needs of all people within the means of the living planet? And, if you don't agree, fantastic, because now we're actually having a conversation about it. And we can reshape this. But let's have a goal. Because if we don't have a known goal, how on earth

do we talk about success of the economy? How on earth do we measure the success of any one policy, and if we're going to get to this goal, here's my third wish: we need to transform our idea of the dynamics.

Simon Kuznets, a brilliant 1950s economist, looked at American, UK and German data on what was happening to inequality over time and it seemed to show that, as these economies grew, they first got more unequal but then things evened up. He was really surprised because he had expected the gap to widen, not narrow. The resulting shape – an upsidedown U curve – came to be known as the 'Kuznets Curve' and it gave the (false) impression that, if you care about inequality, you should be wary of redistributive policies because they may slow down growth, and growth itself promises to even things up again. Then in the 1990s some environmental economists picked this idea up again. This time they looked at local air and water pollution data across many countries and said the same thing's happening with environment: as the economy grows, pollution will first increase but don't worry, as growth continues, it will clean things up again. This Environmental Kuznets Curve, I think, has been phenomenally damaging because the picture, and hence the narrative, is so simple. It gets into our minds, and it has influenced decades of environmental training: as economies grow first expect pollution to increase, but later it will decrease. But today when we move from looking at local to global environmental impacts, that is not at all true. Nations' ecological footprints and carbon footprints tend to rise with GDP unless governments take very concerted action to start trying to pursue decoupling and start to bend that curve. So instead of mistakenly relying on growth, we need to create economies that are, by design, regenerative and distributive. So that was my third wish.

But these perspectives are all rather macro; what about real changes we can see and the 'how' of all of this? Adopting a national perspective and citing New Zealand as an example, Kate is

not at all surprised that Circular Economy, Doughnut Economics and mission-led strategies are known and gaining traction in New Zealand and, of course, I'm delighted by that. I think it's in good part due to the Prime Minister, who I've heard saying in public that she's read Doughnut Economics and it reinforced a lot of what she already thought. She's one of the few national leaders who has said, 'We're going to create a wellbeing economy, so let's produce a wellbeing budget.'

In my mind, she's one of the strong examples of a national leader who has said, 'let us give an intentional goal to our economy – and it is not endless growth – and let's try and put this new thinking into practice.'

But how do we do this at the national level? Well, I think you have to do it by starting to do it, for example, just as Jacinda Ardern is doing: let's have a well-being budget, let's have the leaders speak to this vision and then create budgets that actually start to reflect this, in how we spend public money.

But also, I think that it's crucial to put in place legislation and regulation that stands behind the message we're giving. For example the Netherlands has said 'we're going to be 100% circular by 2050'. That's a wild ambition. And everyone must be thinking, what does 100% circular even mean? To me that's like Kennedy's 'let's go to the moon'. We don't know how to get there. But we're only going to figure out how to get there if we try.

What I really like is that the Netherlands also said, let's be 50% circular by 2030: that's a really serious ambition within less than a decade. And I know the city of Amsterdam have said 'let's have all built environment tenders be circular by 2023'; and from 2022, 10% of all government procurement will have include circular commitments. This is impactful because it takes this long-term vision of transformation, turns it into serious decadal ambition, and then brings real contracts near for the pioneers. Amsterdam are saying, hey, companies, you are welcome to keep doing business in our city, but if you want to stick around, you've got to get circular. It sends a powerful long, loud, legal message to all business. That's the new boundaries of doing business in Amsterdam over the next decade. You want to stay? You know what you're going to have to do! As a result, I sense a level of energy amongst students, amongst architects and urban planners in Amsterdam that I just don't see in many, many other places, because there's a clarity about what's coming. Now, many students graduating from university there say 'you mean, I studied circularity, and now I actually get to put it into practice here?'. Whereas elsewhere, it's like, yep, put those dreams away in your back pocket and go and get a 'proper' job!

Now considering the role that companies and industry might play in a Circular Economy transition:

Moving towards circularity is in some places being initiated by people organising circular cafés at the weekend; the fact that we still call these 'circular cafés at the weekend' shows that we don't yet take this transition seriously – shouldn't it just be a permanent venue in the city centre, open every day for ongoing repairs?

Some companies starting to engage with the concept of circularity have been taking a very narrow closed-loop company-centric view of it, designed around the idea that customers will return products to the company for reuse by them. Well, the average household in the global North owns about 10,000 products.

There's no way we're going to send things back to every single manufacturer. What's more, Nature herself would never try to do it this way. Nature doesn't turn a daffodil back into a daffodil or a peacock back into a peacock. Nature creates an ecosystem of resource building blocks – like chitin, keratin, lignin and cellulose – that are then passed through an ecosystem of use and reuse.

To me, humanity's circular challenge is how to make a series of ecosystems, say, of metals use, of ceramics use, of plastics use, of textiles use. Individual companies are never going to manage to come together to do that. In fact, current anti-monopoly law often says they're not allowed to: if you're talking to each other, you must be colluding. So we need the state, at the industrial level, to step in and say, 'we need you to come together in order to create a circular industrial ecosystem'. That's why I think the work that some organisations do of mapping the resource flows in a place crucially helps to take a bigger-than-one-company, bigger-than-one-industry view of those resources and show that we collectively need to come together and work this out.

Finally, to her thoughts on current industrial adoption:

Greenwashing is a huge threat to what we do through Doughnut Economics. Plenty of companies would love to stick the Doughnut on their website, talk about it now in their PR. I've been approached by banks and fossil fuel companies – do you want to do a little video for us? Thank you, but no thank you. You've got a lot of work to do in transforming and until you start to work on that then you can't use the Doughnut concept in your work. It's a huge risk, because companies have whole teams of people figuring out how do we show that we do enough work to show we're part of the transformation, but while still getting the maximum shareholder returns that we want to provide. It creates huge cynicism amongst the public who say, I can't tell the difference between these things, one of them is truly transformed, the other is complete cover. I also think that greenwash understandably creates cynicism among activists and people who dedicate their lives to trying to create real change. So far, what we've done with Doughnut Economics, is say, no companies can use the concept of the Doughnut publicly at the moment, but we are going to be opening that up into a set of tools, inviting ambitious companies into this space. So again, coming back to what I was saying about eco-linguistics and framing around 'regenerative': frame the space around the living world, and how the living world actually works, and then invite last century's corporations into that space. Remember that everything in the human economy is a humanly designed construct, and its design may well be out of date, and utterly incompatible with what we now understand we need to do, and it can be redesigned. So then invite business in to explain how it is going to transform to be part of the regenerative and distributive dynamics we know we need to create, because business so often is pulled back because of the way it's owned and financed.

Here's the key Doughnut Economics questions for any company. We say be a detective about that company, whether you work for it, love it or hate it, are leaving it or joining it, are the CEO or the newest employee. Always ask:

> What is the company's **purpose**?
>
> How does it **network** with its suppliers, its customers and its industrial allies?
>
> How is it **governed**? And who has voice in that decision making? What are the metrics of success? And what are the incentives given to middle management?
>
> Crucially, how is it **owned**? Because how it's owned – whether it's by venture capital, or shareholders, or employees, or by its customers, or by family – how it's owned profoundly shapes the final design trait:
>
> How it's **financed**, and what that finance is expecting or demanding. So many companies are owned and financed in ways – whether through shareholding or venture capital – that say, 'Well, I'm just here for the high financial returns, right? I mean, why else would I be here, I want my double-digit returns year on year, on year, on year'. That is going to totally undermine any true purpose.

Too often the word 'circularity' or 'regenerate' gets stuck on the front of that purpose, but it's not real – the company is doing the minimum it needs to do to get away with using that language. To me, it's crucial to look at these traits and ask yourself, what is the deep design of this enterprise? And can it truly be part of transformation or not? This will tell us which businesses belong in the 21st century and which do not.

4 Frances Wall on virgin resources, scarcity and circularity

About the contributor

Frances Wall is Professor of Applied Mineralogy, Camborne School of Mines, University of Exeter and the Principal Investigator for the UKRI Interdisciplinary Circular Economy Centre for Technology Metals (Met4Tech). This centre brings together world-leading researchers to maximise opportunities around the provision of technology metals from primary and secondary sources, and lead materials stewardship to create a technology metals circular economy roadmap for the UK. It is part of the UKRI National Interdisciplinary Circular Economy Research Programme.

As a thought leader in minerals and mining resources, Professor Frances Wall takes us on a journey to visualise a transformed circular economy:

> I'm going to start on my idea about what a transformed circular economy might be with mining, because often that's kind of missed off, once we start with products, and then go round in circles from the products in our ideal world. I want to go right back to the beginning and say that I think that it's a basic thing that humans do like farming – we use materials from the earth.

She suggests that

> in a transformed circular economy, we will still be [mining], for a long time to come. Because of the number of people on the planet and because of the way that technology is changing, we need new materials. And because the circle isn't perfect, and what we use, even in my ideal, I think it'd be a long, long time before we're so clever that we can recover everything at end-of-life.

To Professor Wall, an ideal transformed circular economy is

> when we've mined some minerals and made metals, then we will never waste them. [Critically, she does acknowledge that this is] a very idealistic view to take [and perhaps,] we won't be perfect, but in my mind, there will be really good circular economy principles [required] too.

Note: This interview was conducted by Allen Alexander on 8 July 2021. This transcript represents the interviewee's thoughts and opinions on the day of the interview only and in the context of the questions and discussion, as posed. As 'personal opinions,' they must be treated accordingly and not replicated, repeated or taken to infer anything 'outside' of the interview context.

https://doi.org/10.1515/9783110723373-006

In terms of how to get there,

> things like designing out waste and pollution is absolutely something that all extractive industries are trying to do already and can get much, much better at, so there should be much more efficient processes and really new ways of extracting and following the mineral veins down. You don't have to dig up loads and loads of material when you're going for some of these very specialist technology metals. They may be dissolving out with solutions; [also we would be] much cleverer at using by-products, so mines are much more efficient operations.

> [Also] when they come to processing, and you come to make the metals or mineral products that go on into the manufacturing supply chain, it's seamless; between things that are coming around for the first time and things that are coming around that have already been used. In the same way a smelter . . . might take things in from recycling and also raw ore, but [in the transformed CE] that will be absolutely second nature. Nobody, in my ideal world, will have a second thought about that.

> [In her visualisation] they'll all be raw materials providers, so you won't have miners on one side and recyclers somewhere else. It will be a much more integrated industry providing raw materials for manufacturing.

> Then, as we go on through the story, everything will be tracked. And its information, including its environmental footprint, will be held with it, so that everybody knows where things have come from.

> That's very important for responsible sourcing as well as environmental stewardship. And very important as it goes into manufacturing, they will be thinking about making their products for a long and circular lifetime – that's clever things, so they can be taken apart and recycled at the very end of life.

Further to this discussion, she questions:

> [W]ouldn't it be great if you also had some more modular thinking in product [design]? Imagine the car comes out as a new car, and then it goes back to the factory after a year or two, because they're going to change, maybe you want a different something in there, cleverer battery, three different seats, and it just keeps going backwards and forwards instead of straight down the line towards the scrapyard. [. . .] So, more modular manufacturing, designed for recycling right from the start, then off go these products with the metals, including the specialist metals that I work on mostly, inside them. And when eventually they come to end-of-life, we have much cleverer ways, so we know exactly what's there. And we know which things can come out.

In summary,

> I foresee in my – not entirely perfect – transformed circular economy, that [we] will be really clever at, say, recycling things where there's fearsome regulations already today. Like car batteries, maybe electric car motors, certain things will be 100% efficient on recycling and sending them back round, in one way or another. And then, there'll be, in technology metals, they might only be there in parts per million in some uses, really tiny, and that's why I say that I don't think my transformed economy is still a perfect circle, because you get these elements sitting in products in very small amounts. And although as academics, we can talk about recycling everything, I think, in the pragmatic business world, it'll be a long time before we can close the circle.

When reflecting on whether the circular economy differs from the sustainability movement,

> I hope that it doesn't differ very much from this sustainability movement. But when I talked a little bit about mining, when you extract metals from the ground, they only come out once, if you see what I mean. And then, off they go into the manufacturing supply chain. [Concluding that] metals are inherently very sustainable materials, and that they will survive and can be recycled.

> [However,] they've probably left the place where they started. So in terms of sustainability, we understand now how important it is when you do that first stage of extraction, that the wealth and the advantages that can come from that really go into that community or into that country, where they're coming from. That's what creates your sustainability, whether it's manufacturing growth, or education, or better health services, skills etc. that can go on into other industries. So the point of sustainability is not necessarily part of circular. But it's really important in the way that we think.

> And then, I guess there's all the other things along the supply chain, so you could recycle perfectly. But if it's being done in a way that's not healthy and safe, that creates all kinds of pollution problems and things, then that's not going to hit the sustainability agenda either.

> So maybe I'm making a point then, that it's very important, not only that we do things that are circular but that we do those things in the right way, as well. So, it's a kind of responsible sourcing agenda, which is really big in the extractive industries. But we need to be very careful about responsible recycling, and obviously, all of manufacturing, the corporate social responsibility and all the sustainability issues that go alongside, too.

When asked whether there is a possibility of a regenerative dimension in mining, considering regenerative ideas predominantly stem from natural and biological systems, she responded:

> I like that question, because metals don't need to be regenerated, do they?
>
> Because they're there, they don't disappear in the first place. We grow food, we eat it. So you have to grow back and grow the cabbage again, but a metal will be there. It is the ultimate sustainable material.
>
> It's a really different way of thinking from the regenerative kind of biological cycles, we're very much into the biological cycles of things, aren't we, and actually, the metals [and] material world isn't quite like that.
>
> So, you're regenerating products if you like with your metals, aren't you, rather than you don't have to regenerate the metals themselves, that's not going to work, they don't disappear.

Acknowledging that when considering the extraction of a particular metal, there's no way of refilling mines, or regenerating any of that content other than collecting it in service, the conversation moves to repurposing, remanufacturing and extending product lifespans:

> I think that raises a really important question about who owns that metal. So, maybe we'll see this [resolved] but I don't know what the answer is going to be, in the future. At the moment, the miners will mine it, process it and sell it on to somebody else who does the next stage. [The problem] is, it's somebody else who sells it to somebody else, who may sell it to a consumer or company that's using it, and then, off it goes, maybe into recycling or whatever. There are models, of course, where somebody along the chain owns it and then hires it out to somebody else. I think that's something interesting to watch; those [new] ownership models. It might be miners [who] own their metal and just hire those metals out. I think there are problems with that, but I also think it's possible.

Have you seen any examples that that would suggest that that's possible, where mining communities could retain the ownership of their minerals?

> No, I can't think of any. You know, this [is an] interesting thought process, isn't it, when we're sitting in Cornwall, which is a very famous mining region, and the iconic old engine houses are very much part of the scenery. And if you think of the tin, and the copper that has come out of there. Where's that gone in the world? [Reflecting on this:] I would say that most of the tin is probably in rubbish dumps all over the place, because it was used for domestic cans, and so on.

There are tin mining projects looking to reopen; will we be any better? And the answer is, maybe, but it's not a given, because the tin now goes into electronics, but it's not the gold in electronics, so it's often not economically worth, at the moment, recycling.

But it certainly could and should be recycled, so we should know in the future. But should a mine in Cornwall keep hold of it? I don't know, that's risky, isn't it? What if no one suddenly wants tin in 20 or 50 years' time, and all this tin starts coming back to Cornwall? What would we do with it?

On a more optimistic note,

> there are models where, for example, in catalysts where you have platinum and the chemical company owns them; they just send them back to the manufacturer to recondition them . . . so they're not owned probably by the mining companies, but they're owned by the people who are using them, and then they'll go around, and they'll be recycled. So there certainly are different models of ownership coming already. But the ultimate would be for the miners to retain ownership of their metals. And I think that's a very nice thought experiment to do and see if that might happen.

Can you see a situation where we start to treat landfills like mines? And we develop a stream of mining that's not first using mineral resources, but instead, it's starting to sift through landfill or other waste piles to extract essential resources?

> Well, yes, I suppose, is the first answer. Because if you think of waste piles from mining waste, they are already sifted through and reworked, and have been, for a long time. Because if you have put something on one side as waste at one time, then if technology is changed . . . or simply the price of the metals has changed, it makes it worthwhile to go back and look, then that's been . . . an obvious thing for people to do. So, there are many examples where those kind of real big bulk wastes are being looked at, where we go and start to dig through domestic waste.

> I imagine that's a sign that we've failed totally, to look after what we've got. It is good jobs for geologists, like me – we know about how lithium and rare earths and cobalt and copper and all these metals [and] how they occur in the rocks to begin with . . . Suppose you have to go and learn about how people put stuff into landfill. I don't think there's any system at all, is there? So how would you explore it? It is a complete nightmare, terribly inefficient, and I

really hope we don't have to do it. I'm sure people are doing it to small extents, but I think because it's so difficult, because it's such a jumble and because it's so badly organized, it's actually still much easier to go back and find new resources. Then, of course, we've done sensible things with landfill, like do other things with the land, so that we don't necessarily want to go and dig it all up again.

One of the key themes, or one of the key research themes from 20/30 years ago, was the resource-based view of the firm, which allowed you to add knowledge to resources. An emergent model that has arisen from this is the scarce resource-based view. Have you seen those pressures on mineral and mining resources from a scarcity dimension and if so, which ones do you think are the most critical in the way that we consume at the moment?

Each country, recently, such as the USA, and the European Union as a group of countries, has actually looked very carefully at which they regard as the most critical resources and produced their own lists. So, the European Union was one of the first, and it produced in 2010, our list; I think it was of 14 different metals and minerals that they regarded as critical.

That is a very specific term, it means things that are really important for the economy, but which have supply risk, potentially, that is, they might all come from one country. So, if there was a problem in that country, then there will be a high supply risk.

Often, that is done at country level, but it can be done equally well at company level. And companies, for example, Rolls Royce, that are making aero engines that they want to last for many decades, if they put a real specialist metal like rhenium into that, they want to be sure that they can get any future supplies of rhenium that they need to, safely, and for a long time.

Those kinds of companies have looked very carefully at supplies, and about the potential restricted resources. But it's all about supply and demand, and remember, there's nothing that we're actually going to run out of, at all. So sometimes you see big headlines, you know, we're about to run out of rare earth metals, or we're going to peak copper or something like that . . . It's very dynamic, it's not worth going to explore for new metal resources, unless the price is high enough and people actually want it and then, investors will spend money for exploration companies to go and find some more resources.

You never see a full picture, if you just look at the published information from the British Geological Survey, or the United States Geological Survey about what the resources there are or what reserves there may be, in the world. That only tells you what we know about and have defined, with some kind of quantitative degree of certainty, that we know about, but of course, there's the rest of the earth, and we've only really scratched the surface.

In highlighting the reality of resource scarcity, Frances points out:

However, it's all about supply and demand and the cost of producing, so there's the financial cost. But then, there's also the environmental cost, and we're much choosier now about that. Quite rightly, we are much more particular about how things are being mined. We don't want the pollution that's gone with it in the past, and we want much better community and societal benefits than have been produced from some mines, in the past.

With her concluding sentiment suggesting,

it's really about the constraints of energy and water and all the resources that you need to produce. So maybe the scarcity is on that side, we can't just carry on digging low-grade resources in ginormous mines and then sending them into the linear economy and stuffing them into landfill. That's the thing to try and get away from, to get to a much more efficient economy and more circular economy.

To provide a further example of this:

Think about aluminium, it comes from a rock called bauxite, and that that's the most economical and, I would say, environmentally friendly. It's actually quite an environment- and energy-intense process, but that's, by far, the best way to produce aluminium. But in rock, like granite, there's aluminium silicate there, the feldspars. If anyone looks at what a granite rock is, [the] big white crystals in there, they've all got loads of aluminium. We could produce aluminium from there if we wanted to, but that's really difficult and expensive. There's loads of aluminium, it's one of the most common metals in the crust, so it's about the economics of it.

I suppose coming back to . . . my original vision for my transformed circular economy . . . a company . . . if it wants new aluminium, it will . . . just go and buy aluminium and it wouldn't worry about whether it's coming around again. Aluminium is really easy to make nice and ready to go again, from a recycled [perspective]. I guess, then, what we'd hope is that they're also bringing round products, in more innovative business models, that keep . . . higher value, before they have to remake their own aluminium or buy it again as new metal.

So, I guess that would be the trick with the circular economy . . . not only thinking about "where's my original resource coming from?" but they're thinking: What can I do to be clever with my products to keep them at higher value, before they need to go back to the start into the component raw materials?

Translating this from practice to research, what are the critical research questions that you are facing currently with these challenges?

We have a centre called the Met4Tech, a UKRI Interdisciplinary Circular Economy Centre for Technology Metals – the real specialist metals like lithium, or neodymium or cobalt, indium, and rhenium. Things that . . . are mined in quite small amounts, probably just 10s or 100 thousand tonnes per year. Orders of magnitude less than the big industrial metals like copper, or iron, or aluminium.

In comparison, these technology metals are relatively small amounts that we need, and therefore, not many mines are actually mining them in the first place. But, they're disseminated hugely in use, whether that's through consumer products or particularly in the products that we now need for decarbonisation. Whether that's electric cars, or wind turbines and solar panels.

These specialist metals are fundamentally important to what we want to do and to decarbonize and combat climate change, but they're relatively new (some of them) into the economy; things like lithium, we've not used in anything like the same amounts as we need now, over the next 20–30 years.

[At] the Met4Tech Centre, we're going back to the beginning and thinking about the original primary extraction and how to apply circular economy principles there. We [are] calling them circular economy geo models. And we're looking at Cornwall as one of our case studies there. We're looking at what happens if we have a couple of lithium mines, some tin mines, some geothermal energy, which is all under development, at the moment; what does that ecosystem look like? Should we bring in the value chain and have some smelter technology? What kind of technologies would be most environmentally friendly, there? How could these businesses really link together? You know, to put them together in a circular economy ecosystem, and do some lifecycle assessment alongside that to actually look at the best environmentally friendly solution. [Referring to an opportunity discussed earlier, with regard to treating landfills as mines,] there is mine waste in Cornwall, maybe alongside that we should be reusing some of the waste or cleaning up some of the waste. That's one thing that we're doing in one of our themes.

With regard to critical research questions,

we have a theme . . . where we're using some chemistry and materials science and looking at how we can improve the manufacturing, so that [products] can actually be taken apart much more easily and recycled or reused, later on; [for example], some really clever ways to de-laminate things, so they'll come apart nicely. We are working directly with some of the manufacturers on that. That's our team at Leicester and Birmingham looking at how we recycle better. We know how to recycle the major metals but things like batteries, of course, are a whole new challenge for the recycling teams.

The key takeaway Frances wished to highlight was

the fact that we can get everybody together in Met4Tech. We've got business specialists, geologists, chemists, engineers, material scientists, social scientists, and we can look along the whole value chain. Really looking at the thing together rather than a geologist researching in one area, and a material scientist looking at recycling over here and social scientists doing something different. The fact that we can all get together and look at issues of responsible innovation, and, overall, how and what the flows look like, in the UK. We're talking internationally to people, we think technology metals will be an international circular economy, rather than just in the UK; we don't think we'll ever be self-contained here . . .. We're going to make a roadmap for the UK, so to really talk to companies, learn from our own research, and from other people in this area and think about how we produce that transformed circular economy and technology metals for the UK.

So that's our killer question, if you like, and our solution is to get all of us together and that's been really nice in the first six months of the project, listening to other people's points of view and learning each other's language and thinking about how we're going to be able to make these models that put all of our thoughts together, for the ultimate roadmap. And I think that is absolutely essential, isn't it; even if you're some 'techie' person, like me, used to studying minerals at one end of the chain there, when you want to get the circular economy right, you have to all go and talk to each other, and really work together to produce those solutions. So that's really the thing that we can do that, you know, other projects that we're doing can't; really excited to see what we come up with.

5 Hugo Spowers on transforming established industrial trajectories with a circular business strategy

About the contributor

Chief Engineer and founder of Riversimple, Hugo Spowers is an Oxford University-trained engineer and entrepreneur. His first business was in motorsport, designing and building racing cars and restoring historic racing cars. He left motorsport for environmental reasons and set up OScar Automotive in 2001 (which became Riversimple in 2007) on the basis that a step change in automotive technology is both essential and possible.

The first fuel cell car to emerge was the LIFECar, developed by a consortium Hugo brought together with the Morgan Motor Company and presented at the Geneva Motorshow in 2008. The small Hyrban technology demonstrator followed in 2009. The prototype of the Riversimple Rasa (the 'Alpha') was launched in 2016, followed by a much-improved customer-ready version (the 'Beta') in 2018. A pilot trial of 20 Beta Rasas is running in Monmouthshire.

Hugo is responsible for all technical aspects of the cars and for the architecture of the business itself. He is considered something of a thought leader on the Circular Economy and has been invited to give talks on entrepreneurship at Imperial College, London and Cranfield University, among others. At the Real Innovation Awards in October 2019, hosted by London Business School, Hugo was awarded the George Bernard Shaw Unreasonable Person Award "for someone who has shown enormous tenacity and stubbornness in pursuing an idea despite the difficulties encountered along the way."

When asked about a vision of a transformed Circular Economy, Hugo felt:

Rather than talk about The Circular Economy, can I talk about circular economics? The principles are so particularly exciting because not only do they fill in so much of the jigsaw of a sustainable industrial society, but there is also a clear line of sight to their implementation. Even in our mad economic system, a truly circular business is inherently more profitable than a linear one. If the recipe makes more money, it is reasonable to expect entrepreneurs to adopt it because that's what they're after!

I think the two key features that are essential to realising the potential of the circular economy are that products stay on a single balance sheet from clean sheet of paper through to end-of-life and that all operating costs are internalised on that same P&L. Reductionism rules OK and the temptation is to adopt

Note: This interview was conducted by Allen Alexander on 7 January 2022. This transcript represents the interviewee's thoughts and opinions on the day of the interview only and in the context of the questions and discussion, as posed. As 'personal opinions,' they must be treated accordingly and not replicated, repeated or taken to infer anything 'outside' of the interview context.

https://doi.org/10.1515/9783110723373-007

individual elements, but the benefits of all the features of circular economics are not additive – they compound and adopting half of the principles won't deliver half the benefits; they could even be negative! Consultants will always try to tease any subject apart; everything is more saleable to clients piecemeal, and it probably earns bigger fees. However, the benefits won't be realised, early adopters might catch a cold and be put off, and the end point that an incremental strategy will reach will be different and sub-optimal.

It's important to understand that circularity is a business model rather than an 'economy.' Recently, I read an attack on the Circular Economy from the world's leading proponent of the Steady-State Economy, Herman Daly. A growth-based economy is at odds with a steady-state economy, and linearity with circularity, but circularity is complementary to, rather than competitive with, steady-state economics.

Commenting on the potential confusion between sustainability goals and Circular Economy:

Riversimple's purpose, specifically, is the elimination of environmental impact; I don't believe that reduction of environmental impact is an adequate goal, because being less unsustainable is still not sustainable. Circularity is a necessary element of a sustainable industrial society, but it is not sufficient.

A circular business model fundamentally changes product and service design, unlocks greater profits for less resource consumption, but it doesn't, on its own, get us to maintaining the natural capital balance sheet. Systematically increasing the natural capital that is leveraged, or the energy demanded, is still not sustainable. Just as product and service design are transformed by the business model (circularity), we need to step up to the next level of design as well – ideology, purpose or governance, whichever term you prefer. As Einstein famously said, "No problem can be solved from the same consciousness that created it," and a circular business driven by maximising shareholder value will still grow inexorably, externalise, minimise labour standards, perpetuate global inequities, foster geopolitical tensions and erode the bioproductive capacity of the planet; we also need to address governance of limited liability corporations to align their interests with the common good.

Moving toward a more focused discussion, exploring more specifically the challenges of trying to create a shift change in a highly embedded industrial trajectory:

The cars we've been building over the last 15 years benefit from a whole system design approach. Starting from a clean sheet of paper allows us to build a completely different sort of vehicle, rather than trying to shoehorn hydrogen technology into a vehicle model and architecture and a manufacturing model that has been optimised around combustion engines for 100 years.

And when I talk about whole system design, what I mean is that we never examine an element of the design and compare it with alternatives at a subsystem level; we build a comprehensive model of the whole system, the whole car, and we then try various different solutions for each component. Quite frequently, you will end up choosing a component that is heavier, more expensive or less efficient, because it gives you a car that's lighter, cheaper or more efficient.

It's very counterintuitive, but we're trying to optimise the whole system rather than to optimise the elements of the system, because we believe that by doing so, you actually 'pessimise' the whole. This system-level innovation is really what yields the breakthroughs. It's what allows us to build a commercially viable hydrogen fuel cell car, here and now, with existing available technology.

As I say, whole system design is, in some respects, counterintuitive. One of the things that we constantly are conscious of is the unquestioned assumption, particularly in the Western world, that changing one thing at a time is the prudent thing to do. That's absolutely true if you're refining a mature system but, when you're going through a step change, it's an absolutely catastrophic strategy.

The fact that we're changing multiple elements simultaneously – the car, the materials, the architecture of the car, the business model and so on – reduces all the stress on this new fuel cell technology. Focusing at a system level has enabled us to make a breakthrough, both in performance and commercial viability, using existing technology.

The fuel cell is a 10 kilowatt fuel cell made for forklifts for Walmart warehouses, and a 10 kilowatt fuel cell is just about enough to power three domestic kettles. But this car does nought to sixty in nine and a half seconds, and it is enormously good fun to drive, I can assure you.

This contrasts with the Mirai, the hydrogen car from Toyota. It's a brilliant bit of engineering – I don't want to be seen in any way to be critical of the industry, because I think it is extraordinary what they've achieved here. But it does demonstrate the limitations of trying to make a step change incrementally.

As Lloyd George, great Welsh Prime Minister, once said, "You can't cross a chasm in two leaps." If you're going to build a conventional car, with a fuel cell, you have to persuade the fuel cell to behave like a petrol engine, which it really doesn't do very well. Consequently, Toyota have poured billions, literally, into pushing the state-of-the-art of fuel cell technology way beyond its comfort zone, and as I say, the engineering is brilliant. But to our mind, it's solving problems entirely of their own creation, necessarily, and understandably; the motor industry is the most refined and mature industry we've ever had in the world, but the more technology-specific you become, the harder it is for you to do something fundamentally different.

Now our car isn't an equivalent to the Toyota, it's only two seats, not four seats, we're not pretending otherwise. But the Toyota has over ten times the power from its fuel cell, and uses over three times as much hydrogen per mile, and the differences should not be so stark. That just demonstrates the opportunities available by designing at a whole system level, particularly in an area with such immature technology, where the opportunities are richest for reinventing the concept.

We're often asked about how we can ever hope to compete with major auto-manufacturers, but I don't think we could have done this from within the auto industry actually. Now, obviously, as a start-up business, the one constraint is always a lack of capital, which is not the case for a major manufacturer, but the great luxury that we have is a clean sheet of paper, we have no legacy constraints. And this is a luxury that should not be underestimated.

I'm from a motorsport background, and so I often like to use an analogy from motorsport. We now dominate motorsport globally in the UK, but 60 years ago, we were absolutely nowhere. In 1959, Ferrari dominated, with hundreds of people building really beautiful engines; they were fantastic bits of work, and they focused on that subsystem as they regarded that as the heart of the beast.

Then, five men in a shed in Kingston went and built a different sort of car, the Cooper. They bought an engine off the shelf that anybody else could buy. It was nowhere near as powerful as Ferrari's engine, but they turned up at the Monaco Grand Prix, and Jack Brabham won. He won eight out of nine Grand Prix that year, they completely demolished Ferrari. And then, Lotus came along and did an even better job, quite frankly. Between them, Cooper and Lotus built the foundations of the modern motorsport industry.

Their breakthrough was at a system level, a different sort of car with the engine in the other end, but the extraordinary thing about the story is that neither Cooper nor Lotus ever built an engine. Not only did they get to the sharp end of motorsport, but they stayed there through system-level innovation. This is analogous to where we are now in the auto industry; the process of electrification is the most profound change that the industry has seen in the last 100 years.

Taking a wider, economic, environmental and social focus, Hugo acknowledges,

[W]e're also going through a step change in the environment within which we operate. By that, I mean principally the natural environment, but also the business environment and the regulatory constraints introduced as a damage limitation exercise in response to environmental degradation.

We talk of a funnel as a generic illustration of the problems that we all face, every country, every company. The principal constraints today (carbon emissions, resource depletion, energy security etc.) were simply not on the radar in the twentieth century when the existing industrial model developed.

Technology is accelerating all the time, population is growing, the affluence of that population is growing, and our technological ability to consume is accelerating, creating a converging funnel and trying to stay within the walls of the funnel is expensive and painful. These are costs on the economic bottom line, and they're only increasing, but it's not surprising because the profit motive of the standard, take-make-throwaway business model lies way outside the funnel.

If you sell cars, you make more money by selling more cars; you're rewarded directly for maximising resource consumption. I've wondered how we can ever have a sustainable industrial society based on rewarding industry for the opposite of what we're trying to achieve. But it's also not very smart business because, although we can argue about the timescales for various peak resource issues, what we can't argue about is the direction of travel. Without putting a timescale to it, these constraints are only going to bite deeper, going to lead to more and more commodity price shocks, more and more regulatory pressures. And our view is, as with cars designed for hydrogen fuel cells, that it is much easier to design a business model and strategy to suit the conditions of the twenty-first century, at the system-level, than it is to tweak a business model that was designed to do something fundamentally different.

Anchoring this back to the company he started,

> so what we're trying to do at Riversimple, and through the Circular Economy [principles] more widely, is to build a business model that profits from staying away from these [resource, commodity, regulatory etc.] constraints.

We essentially achieve that through a combination of two themes: One, by aligning interests with all the actors in the system, in every relationship we build and, two, by turning all these costs into sources of competitive advantage. I want to say that this whole system design is applied not just at the level of the product and the business model, but there's also a third level of design; design at the level of ideology or purpose. For us, this is sustainability and is manifested through our corporate governance.

Talking of how the company is structured, Hugo reflects:

> We have an unusual corporate governance model; it is unequivocally a 'For-Profit' model, and all the money flows in pretty much the same way as it does in a normal business, but the control of the company isn't solely in the hands of the equity shareholders.
>
> We have six custodians representing the six key critical stakeholder groups with whom we interact, and the board's fiduciary responsibility is to balance and protect those six benefit streams – Investors, Environment, Staff, Customers, Community and Commercial Partners – rather than to maximise one. We argue that, because the business depends on all these groups, we want to maximise the goodwill that we engender from all those stakeholder groups. You can't hope to maximise goodwill from five of them if their interests are subordinated to the sixth. The guiding principle for the board, the fiduciary responsibility, is to balance and protect these six benefit streams.
>
> This also aligns very, very closely with the Welsh Wellbeing of Future Generations Act, which I really think is still, to this day, a really globally ground-breaking bit of legislation. I am sure it's a foretaste of the future, and other countries are going to have to implement it.

Returning to the systems-level challenge,

> I talk also about whole system design and trying to optimise the system, rather than to optimise the elements of the system. The three levels of design of the organisation (the governance, business model and the product) are all highly intertwined and interdependent. We're trying to optimise the gamut of all of them, rather than to look at each one in isolation.
>
> The business model, in particular, profoundly changes the design of the car. We're often told that we're trying to change too much at once, but we argue that actually, when you're going through a step change, changing multiple things simultaneously reduces risks and barriers; the cost of the Toyota Mirai development programme compared to Riversimple's is a good example of this.
>
> There's no doubt that whenever a new idea comes along, all the conversation, like a lightning rod, goes straight to the reasons why it can't be done, and I think that generally those reasons are true. But they assume that the new idea will replace the old idea, and unfortunately, the old idea is embedded in a network of relationships. It co-evolved with this network, and all the players it connects, but these relationships are all the reasons why the new idea won't work. If you have the luxury to do so, being able to redesign the new context

around the new idea, all those barriers disappear, and suddenly the new idea looks an awful lot more appealing.

That is the best explanation I can give as to why when you're going through a step change, changing multiple elements simultaneously reduces risks and barriers.

Reflecting on where he lives and the strategy for locating the company in Wales:

System-level change also creates opportunities for Wales that we have not enjoyed, of late. The reason I say that is that in a mature sector, like the auto industry, scale is incredibly important.

For example, we make almost every bit of a car in Wales, but we don't actually make a car. If you're going to build a new Ford Fiesta, you need scale, because it's such a mature and highly refined model; each generation may only be 1% better but that 1% is very, very difficult to achieve when the technology is so mature. However, when you're going through a step change, a system-level change, the opportunities, the bites of the cherry that you get are much, much bigger. The real opportunities are in system-level innovation, and this doesn't need scale; it needs agility and this plays into the hands of smaller companies and countries compared to eras of incremental improvement.

It's not just the auto sector; the energy sector as a whole is going through the same sort of transition. When you look at the Circular Economy and resource constraints that we all face, the Circular Economy changes virtually everything. I think it creates opportunities, particularly at this moment in time for Wales, to really establish a position at a global level in industries that are immature, because there is much less competition than there would be if we waited until the incumbents have been replaced by mature new models.

Hugo now focusses on one of the key economic tenets of circularity – the role of sale-of-services.

The business model of the industry we start with, is all about the sale of product, it's what we've been doing since the Industrial Revolution. If you sell a car, optimistically about 40% of the revenues generated by that car through its life ends up going to the manufacturer, and the other 60% of the revenues go elsewhere. There's very little you can do about this if you sell cars.

If, on the other hand, instead of selling product you sell service, 100% of the revenues for the life of the car go to the manufacturer. Typically, we would expect a customer to take a car for a three-year contract; it can effectively be their car, as it would be, if they 'bought' a car on a PCP or a lease, and it sits on their drive. But with a finance deal, the asset actually remains on the balance sheet of the finance company.

In our case, it remains with Riversimple, and the customer makes a monthly direct debit payment, with a fixed monthly rate plus a mileage rate, like a usage rate on a phone. Critically, this covers all costs; it includes not just the maintenance, but insurance and fuel. At the end of the contract, the car comes back to us, but we don't sell it into the second-hand trade; we provide it to a second-, third-, a fourth-hand customer.

If, for instance, a fourth-hand customer takes a ten-year-old car, the cost to the customer of motoring remains entirely predictable. If they do 10,000 miles a year, they will know exactly what it's going to cost them and they don't face the risk of a major component failure, economically taking the car off the road. We're responsible for that. The car that's ten years old will be as reliable as a new one; otherwise, we wouldn't supply it to you because we're responsible.

The car is of course designed for this model, so there are no moving parts in the car other than the wheels. There's no other metal-to-metal wear, no lubricants, no oil changes. All the structural materials are inert, so there's no corrosion, and components that do degrade are designed to be service items. So there's no component that can economically take the car off the road if it fails.

This aligns our interests with those of our customers because, instead of obsolescence and high running costs, our interests are longevity and low running costs. It also aligns our interests with those of the environment, policymakers and impending regulations, because instead of being rewarded for maximising resource consumption, we're rewarded for resource conservation and energy efficiency.

This alignment of interests creates a self-regulating mechanism which, we believe, is much more effective and much more efficient than active regulation. As long as resource and energy efficiency reduce profits, we need to rely on regulation, but there is no reason to go beyond compliance with regulations, which are weaker and later than they should be because industry inevitably lobbies against them, and we need to police compliance with those inadequate regulations. In contrast, self-regulation through aligning industry's interests with the outcomes that we're all after is more effective because we have a reason, profit, to make every car as energy- and resource-efficient as possible – a driver to excel on these metrics rather than reluctantly comply. It is also more efficient because you don't have to sample and test products – nobody is going to cheat if it reduces their profit – and you don't have to pay anybody to do it.

Finally, it changes all the economic barriers to bringing new low-carbon technologies to market. If you sell cars, the 'price to the customer' depends absolutely on the build cost of the car, and the automotive market is incredibly cost-sensitive. It doesn't matter if it's a battery car or a hydrogen car; to bring it to market, you have to match the extraordinarily low supply-chain costs of petrol engine cars today – and I don't believe there's a product on the planet that's remotely as good value for money as a modern motorcar with all that complexity and refinement. Inevitably, any zero-emission car, in the early days, is going to carry a cost premium in the marketplace. To get the cost down, you need to get the volume up and to get the volume up, you need to get the cost down. So it's a chicken and an egg.

If, on the other hand, you sell the service of the car, the price to the customer is not based on the build cost of the car, but on the lifetime cost. The two key features of this model are, first, that the car must stay on the same balance sheet from a clean sheet of paper when you're designing it right the way through to end-of-life, and secondly, we must internalise all operating costs on that same 'profit and loss' sheet.

Now that of course gives the convenience to the customer of the single transaction, but it also completely changes the design drivers because we're not building a product to sell but a revenue-generating asset to sit on our balance sheet. We're more comparable to a property company building offices (e.g., we want the revenue to last as long as possible, and we can amortise it over the longest period possible). In this case, the lifetime cost is the driver for the pricing to the customer. That does of course include the build cost, but it also includes all the operating costs and end-of-life costs too. Because we know it's going to be our vehicle at the end-of-life, we design for maximum recovery of value – that's not just raw materials but components that can be refurbished and used again – which of course are much higher value than just raw materials.

We also design to minimise the operating costs, because we're paying for all the spare parts. And we're also paying for all the fuel, so it's worth our while investing in efficiency. Finally, of course, we want to maximise the length of the revenue stream because it's a revenue-generating asset on our balance sheet.

Those three elements of longer revenue stream, lower operating costs and end-of-life value recovery can all offset a higher build cost, so that we can come to market at the same price to the customer, long before the build cost is anything like as low as that of a petrol engine car.

There's one other interesting outcome of this and that is a shift in focus, I believe, from a strategic driver towards minimum acceptable quality to a driver for highest possible quality. For example, if you look at the life of white goods, it keeps on going down because future savings are discounted, and the purchase price is absolutely the key driver in sales. You obviously can't have a fridge that's less reliable and with a much shorter life than all your rivals, but there's no point in a much longer life, because it's not as profitable – and all your competitors have these same drivers.

On the other hand, if we sell the service, the longer that we can make the revenue stream last, the more reliable and more efficient – that is, a higher-quality product – the more profitable it is, at the same price to the customer. When we hear the expression 'quality pays,' we normally mean that customers will pay more for a higher quality product. But that's not what I mean; I mean that actually, a higher quality product is more profitable at the same price to the customer.

This alignment of interests is also true in all our supply-chain relationships. Currently, if you buy components from suppliers, interests fundamentally are opposed, because we want the lowest possible price for the longest possible life as a manufacturer of a car, whereas the supplier wants the highest possible price with the shortest possible life, so they can sell us another one.

If we move the service model upstream into our supply chain, the same thing can happen in that we can align our interests with those of our suppliers. Take the example of our fuel cell. It's called a fuel cell stack because there's a stack of cells, each with a membrane covered in very finely divided platinum to catalyse a chemical reaction on it.

The membrane electrode assemblies are made by separate companies, who supply them to the fuel cell company, and all these components sit in our car. As I explained earlier, the car remains on our balance sheet, but is in the hands of the customer, who pays a monthly fee for having the car, plus a mileage rate. If we extend our 'sale-of-service' upstream into the supply chain, we don't want to buy fuel cells from our fuel cell supplier, we want to pay them for the electricity; we want to pay them a monthly figure for having the inventory in our hands, pay them for the runtime, because it degrades the fuel cell and also pay them for the efficiency with which hydrogen is converted into electricity to bring their business interests in alignment with ours.

Furthermore, rather than buy the membrane electrode assemblies, the fuel cell manufacturer can lease them on exactly the same basis. Finally, the membrane manufacturers can lease the platinum from the mining company. At the end-of-life of the fuel cell, the platinum (as a catalyst) can be 100% recovered. Effectively, the mining company gradually decouples its revenue from new material, new virgin resource extraction, because their revenues increasingly are coming from the stock of platinum on their balance sheet.

It is also a good example of how the business model changes technology development; a leading supplier of membrane electrode assembly has explained to us that they would double the platinum loading . . . for two reasons. First, they know they will get all the platinum back, and secondly, it would make the fuel cell more efficient, and they're going to be paid more for that.

Essentially, this brings all the suppliers and manufacturers onto the same side of the fence. It's a better basis for a collaborative working relationship, because we have complete alignment of interests over the cheapest way of delivering mobility over the whole lifecycle of the car. This is true not just for the fuel cell, but the tyres, multiple other components, electric motors and so on. And this is, in an ideal world, where we want to end up, but we don't need all our suppliers to embrace this straight away; we want suppliers to do so only because it is better business for them.

Finally commenting on a key enabler,

we don't believe that this would be possible, it would just absolutely overwhelm the system, were it not for blockchain. Suppliers are using data from the car to bill us, but we are providing that data, so we are developing a blockchain solution to automate trust, right the way through the value network, so that all parties to this value network can have complete trust in the fidelity of the data.

Hugo is pursuing these 'Circular Value Network' concepts with Exeter University and Swansea University through a research centre at Riversimple, Circular Revolution:

Long live the revolution.

6 Andy Rees OBE on waste, reuse, recycling and the power of positive governance

About the contributor

Andy Rees OBE has 41 years' experience of working in the environmental field, including spending the last 25 years in the area of resource efficiency and a Circular Economy. He carries numerous accreditations and memberships. Andy has been Head of Waste Strategy at the Welsh Government for the last 21 years and is responsible for waste and Circular Economy strategy policy. In 2017, he received the *Material Recycling World* magazine's 'Editor's Choice Award,' and the Chartered Institution of Wastes Management's 'Waste and Resources Leader of the year' award, in recognition of Wales' success in recycling. Andy was awarded an OBE by the Her Majesty the Queen of the United Kingdom for "services to the environment and recycling in Wales" in the New Year Honours 2019.

In considering a transformed Circular Economy (CE), Andy suggests,

> One of the points that we frequently make is that a Circular Economy itself is a process and not actually the end outcome. We've got to be very careful not to get into a situation where we have unconstrained economic growth that leads us to use more and more materials. So, we could have a Circular Economy that keeps those materials in use for as long as possible, but we would end up using more materials overall.
>
> We don't want to trash the rest of the planet because of our use of resources; it's about doing what's fair. It is also what is going to have to happen, as the rest of the world's economies grow like ours. We can't then all have a three-planet resource-use economy!
>
> So, we have got to focus on dematerialising our economy, and our ways of life, and perhaps not relying on an economy that requires us to buy an endless mountain of stuff that doesn't last very long. Trying to fundamentally shift that economic model will be a massive challenge, but that's where we need to get to at the end. We need a dematerialised economy.

Note: This interview was conducted by Allen Alexander on 19 November 2021. This transcript represents the interviewee's thoughts and opinions on the day of the interview only and in the context of the questions and discussion, as posed. As 'personal opinions,' they must be treated accordingly and not replicated, repeated or taken to infer anything 'outside' of the interview context

https://doi.org/10.1515/9783110723373-008

Andy then reflects at a more individual level, suggesting CE needs to be

> based upon different forms of happiness, other than buying perpetual amounts of stuff, so the stuff that we do need, needs to be valued and last a lot longer and use fewer materials.
>
> Food supply and demand and our dietary habits are also another key part of a Circular Economy. Especially in terms of reducing our exploitation of natural products from land in all parts of the world – trying to avoid deforestation, for example. And, thinking of the adverse impacts on biodiversity, how we use our food and what we eat – they've all got to be a key part of this as well.

When pushed on our individual responsibility and how we explain parity to individuals in developing nations, he notes:

> [W]e need to lead the way and reduce our own resource use and then show how that works and then support the developing nations. Not constrain what they're doing, but help provide them with a better, more sustainable pathway.
>
> So, they can neatly leapfrog the bad things that we've done in the past into a much better way of doing things. It mustn't be seen as them being constrained, but rather being given the opportunity to improve on what we have done. Using resources badly can cause horrendous pollution and harm to human health. When we went through our own industrial revolution, we spoiled our landscapes. I live in Swansea, which was once the cradle of world metal refining back in the eighteenth and nineteenth centuries, and it completely destroyed the landscape and probably harmed people's health. I think a Circular Economy will help avoid that in the future by finding a far better way to manufacture and use things and avoid those really bad health impacts. So, I see a Circular Economy helping developing nations avoid the pitfalls and traps that we ourselves went into. We have to lead the way and show how it can be done. But we can also learn from the developing nations. They have some great ideas as well, so it's not a one-way street at all; for example, their use of natural materials as medicines and as food and bio chemicals. There's some clever stuff that's going on all around the world.

And reflecting on his professional role:

> Our overarching goal within the Welsh Government's Circular Economy Strategy, Beyond Recycling, is to have one planet use of resources, using only our fair share . . . Not only is that a moral imperative, but it is also linked to our 'A

Globally Responsible Wales' goal under the Well-being of Future Generations (Wales) Act.

We see the bio economy and using more bio-based materials as being really important here, particularly moving away from the use of fossil fuels. But, of course, that has to be done carefully, and we've got to make sure we still have enough land to grow food, etc. Growing non-food crops obviously has to compete with growing food crops, so that side of things needs to be resolved.

Reflecting on a holistic or a systems perspective, he confirms:

it needs that strategic holistic approach. We need to look holistically at whole supply chains, from start to finish. We also need to make sure, from a government perspective, that all the relevant pieces of the jigsaw are in place – and that's how the Circular Economy Strategy was put together. A Circular Economy and resource-efficient approach need to be embedded throughout everything that everybody does. It's not just an add-on or a tick in the box. A Circular Economy needs to be mainstreamed as the new business as usual, for sure.

Following on the theme of a shift away from materiality and considering how people can be influenced by policy, and in particular considering the different generations and also some of the diverse socio-economic areas in Wales, Andy goes on to suggest:

I think there will be a shift by some people, but not everybody. For example, my dad managed to buy a washing machine eventually in the late 1960s, and then he kept repairing it. He wouldn't have dreamt of throwing it away, but that was a time when those products were quite expensive. But they were designed to last, and you were incentivised to repair, because buying a new one would cost too much money. They were higher quality; when my parents passed away and I cleared out their house, they still had the same fridge 50 years on that they had when they moved in, because it was designed to last. My parents never saw the reason to replace it. And, to be frank, I think it's trying to get people to realise that; OK, now the upfront cost maybe a bit more expensive, but actually over the whole life of the product if it lasts a long time, it can be repaired and actually save you an awful lot of money.

The challenge there is we have marketers and advertisers who are very adept at getting us to buy products and the latest or the newest thing. But, maybe, we could harness the power of that marketing and advertising expertise to appeal to people along the lines of "you know why it'd be better to buy a longer life product." The product might need to be funkier and more iconic, and it might also cost a little bit more up front, but overall, it will cost a lot less, over its whole lifespan.

There are countries in Europe, such as Germany, who've excelled in providing high-quality goods that last a long time, and there are British manufacturers as well, who've done the same. But they've been outcompeted by the 'sell it cheap, it doesn't matter if it doesn't last very long, you can buy a new one' approach, and we need to move away from that. The new Right to Repair legislation is progress in that direction. It's a whole cultural shift in mindset really.

When asked to contrast CE with a sustainability-driven vision of the future,

[W]hen I talked about a holistic approach earlier, what sets us apart in Wales is probably the 'Well-being of Future Generations (Wales) Act'.

Our Circular Economy Strategy, Beyond Recycling, is totally built around a sustainable development approach. To us, a Circular Economy is a key part of that sustainable development journey and inextricably linked with it. I talked earlier about the 'A Globally Responsible Wales' goal that's one of our sustainability goals; the 'A Prosperous Wales' goal is not just about making people a bit wealthier and creating more jobs, it intrinsically has a low-carbon, resource-efficient economy as part of it. There is also the 'A Healthier Wales' goal; I mentioned earlier that changing the way we manufacture products will reduce emissions that will help on health. And a Circular Economy will help deliver a more socially responsible Wales through creating more jobs, better jobs, different jobs etc., with the skills associated with that.

To us, there is no disconnect between the Circular Economy and sustainable approach; the two are absolutely interlocked. Our statutory duty under the Well-being of Future Generations Act is to maximise our contribution against all of the seven goals. That drives our thinking, and so it's not just "are we doing this?," therefore, we tick that goal; no, it is "what more can we do to maximize our contribution against each of the goals?"

Actually, there isn't an environmental goal specifically, but there is the 'A Resilient Wales' goal, which is about having resilient ecosystems. But a resilient Wales is also about having resilient businesses, so it's not just about nature resilience. To us, a Circular Economy will help maximize a contribution towards most of those goals, to be honest. It's not about doing something less bad; it's about doing something good across the piece.

Another angle, which we are also very concerned about, is responsible sourcing. You've got your Scope Three carbon emissions to address, and you need to look at the supply chain. But those Scope Three impacts also have impacts on nature and biodiversity or on social issues such as modern slavery, etc. We're starting to have really useful conversations about trying to help our businesses become more resilient . . . including if they're sourcing raw materials from conflict zones, or from areas with modern slavery or damaging the environment, which is not good for their reputation, and they may find that customers are not buying their product because of that. But also, morally it's the right thing to do, but probably for a business it's about being really sure where your raw materials come from.

Giving you an example, within the food supply chain, there was that experience of what was being reported to be beef wasn't beef, it was something else. Look at the bad reaction that happened there. There are people who have always worried about fair trade, which is good, but the impact on deforestation of soy and palm oil used in animal feed and in processed food needs to be addressed too . . . I think George Monbiot once said something like "we care about the environment, but we never worry when what we buy might have killed an albatross." He was quite rightly pointing out that we merrily buy all these products without a moment's thought about the environmental impact, or the social impact and what might be happening in other parts of the world.

A Circular Economy helps address that because it looks critically at the flows of materials, the whole life cycle, and tries to make sure that we don't over exploit those natural resources. It wins on so many different levels really, but it's just trying to persuade people to make that transition and to make these arguments. The other day, I had a really productive discussion with colleagues where I gave a presentation on Circular Economy and mentioned responsible sourcing, and I was surprised how much that sparked debate afterwards because I think it struck home – "there's a risk . . . for businesses if they get it wrong." And it's all about education and highlighting those points to people to sell it to them – do this right and your business will be a lot better and more resilient in the future.

When challenged about the role of policy-led CE transitions (contrast against grass roots movements, or industrially led activity), he goes on to note:

I've been in this job for 20 years, working for government within the Circular Economy, and so, of course, naturally there's an element of top-down that we've employed. For example, we set statutory recycling targets, and we set other legislation. But we're lucky in Wales that we're relatively small, we can speak to stakeholders really easily. We can pop down the corridor and speak to somebody in a different department or a different director quite easily. And within the Wellbeing for Future Generations Act, we apply the five ways working, for example, through co-production and an integrated approach. We've learned that trying to work throughout the whole life cycle of something with whoever's involved in it, from start to finish, is really important. That includes listening to and working with grass roots movements led by people and communities.

The classic disconnect between designers of products and those who have to manage waste at the end shows you've got to address everything at every level. Getting innovations from business, getting the ideas is really important, and we can't just create legislation and targets without thinking of the impacts and discussing it with businesses, including as to what would help them. That's what we're doing with our new Non-Domestic Premises Recycling Regulations. We're having informal discussions with key stakeholders to say, "Look, how can this work for you?" "How can this legislation be improved and tailored so you'll understand it so it's enforceable, so it's realistic?"

We may need to phase things, not require everything to be separated on day one. Also, we've said "can you come up with good case studies?" – good examples, particularly when talking to trade bodies, are invaluable. There are some businesses already innovating in this space, and we want to encourage that sharing of ideas. You've got to do it at all levels, to identify where there are any market failures that need addressing by government. And only then address them if there is a clear market failure and everyone agrees that's where the intervention needs to come in. Then, if there's a space where a government can operate, then that's what it needs to do, step up and do it.

Returning to the subject of how the public have embraced this combination of an educating and legislating approach, particularly considering some of the socioeconomic polarisation that exists in Wales, Andy indicates:

When we started, we had a 5% recycling rate, so we're now on 65%, and the public have by and large done it entirely voluntarily.

We haven't done pay-as-you-throw and . . ., and, apart from one or two local authorities, recycling hasn't been mandated on residents. But we tapped into quite a latent desire, as every survey that I've ever seen of the Welsh public shows them to be ahead of the rest of the UK in terms of environmental awareness. I think there's a strong sense of community in Wales that helps, so I think we've tapped into a latent desire to do the right thing. But the key thing has been to phase everything in; provide the recycling collection right service; make it relatively easy . . . and then do a lot of awareness-raising.

We have engaged children on recycling through Eco-Schools, for example. I think we've got the highest proportion of Eco-Schools in Wales, out of the whole of the UK. We have a very strong education campaign, strong awareness-raising campaign, and we had a £3million TV advertising campaign over three years in the mid-to-late 2000s, which was really tapping into children. There were child actors with a strap line of 'it's our future don't throw it away' and saying 'don't throw it away to landfill,' so there was quite a lot of subtle and not-so-subtle behavioural insights and awareness to help raise recycling levels.

Gradually, over time, it became normalised for people to do the right thing. Our recycling campaign last year, 'Be Mighty,' really taps into being proud of Wales doing so well. It was 'most people are doing this, so come on the rest of you, let's beat everyone else and become number one in the world.' At the last count, we had the third highest household waste recycling rate in the world, behind Germany at number one, and Taiwan at number two.

Surveys frequently show that the Welsh public are more aware of environmental issues and more convinced that recycling works than those in other parts of the UK. A recent survey by Incpen . . . showed a significantly more positive attitude in Wales to recycling, compared to the rest of the UK. It's a mixture of providing the service, providing education and a little bit of nudging as well. Things like restricting the residual waste collections have worked. We had long discussions with local authorities about whether that was the right thing to do.

They deliberately only collect black bag waste every two weeks, at most, but they provide a weekly food waste recycling collection service. If you get annoyed that your black bag smells, it is because you're not putting your food in the food waste bin, which is actually collected every week. So, there's a lot of 'nudging' that went on as well.

Then, finally, there is a school of thought that you've got to make recycling easy, so you chuck it all co-mingled into one bag or bin. Whereas we deliberately, for a number of reasons, asked "please can you separate it into several containers because you'll get a better quality." We want to feed high-quality recyclate to our manufacturers. Pembrokeshire Council, now our highest recycling authority, moved to multi-stream kerbside sort just over a year ago. We're achieving incredibly high recycling levels with making it a little bit more difficult for people to recycle, if you see what I mean. What we want is for people to think about the waste that they're producing and not make it too easy to throw it away. That's how, with our separate food waste collections, we have managed to reduce food waste more than any other part of the UK. We think that's because people are separating out the food waste and they can see it, demonstrating how much money they're wasting or how much food they're wasting; avoiding that old expression "it's somebody else's problem," just bin it and it goes out of sight, out of mind.

In a nutshell, people are taking more responsibility in Wales for the waste they produce, and hopefully more of them think about it.

Once again, putting Andy on the spot, the conversation moves to how 'front of the pipe' solutions, rather than 'end of pipe' waste-focused solutions could be implemented. The response is:

[F]unnily enough, that is the $64,000 question, which I keep putting to behavioural insight academics. "What can we do to influence people to get them to that one planet resource use," because that is about shifting people's aspirations. What makes us happy by purchasing stuff that isn't instantly replaceable? How can we buy less stuff but get equal or more happiness from other experiences?! And the stuff that we do buy needs to last a lot longer. And then, when you no longer want it, it needs to be reused, repaired, repurposed or re-manufactured where at all possible.

We have by no means cracked this yet. We're at the foot of Everest on this one, I would say, and we need a lot of help – from academics, but also from designers and advertisers, as I mentioned earlier, and marketing experts. But how are we actually going to do this? As a start, we are moving further up the waste hierarchy in terms of promoting reuse, remanufacture, repair. And we are also working with communities on zero-waste projects.

I'm lucky that my shop that I can walk to down the road in my village is a zero-waste shop, and since the pandemic started, I have bought most of my food there. The amount of plastic waste I produce has absolutely plummeted and now I value the food that I buy a lot more. But the problem is that I'm very lucky I can afford to pay a bit more and that's another part of the challenge here, to provide an 'affordable' attractive solution.

For everybody to achieve this needs a real cultural shift, and it needs, thought leaders, community champions, to drive it; but if that doesn't work then that's where government may have to come in.

There are new powers under the Environment Act, which just came in, for environmental product labelling and environmental product standards. I know there's been a long academic debate about how effective labelling is. If you have a carbon or materials footprint on the label, how many people are really going to pay any attention, but if government set an embedded carbon or materials standard for a product, then that would change the game.

It makes sense to, first of all, encourage a voluntary approach by business, and we'll try every way possible to educate. But if that doesn't work, then some form of government intervention (Wales, UK or OECD or UN level) eventually may be needed to shift things. The 'Right to Repair' requirement is the start on that journey, for example.

So, focusing on 'Right to Repair' Andy reflects:

I belong to the generation that used to hire TV sets and VHS video recorders, and I've once had a lease car. The Riversimple idea [see Hugo Spower's Transcript and Industrial Case Vignette] with their new hydrogen fuel cell car is a potentially very good one. The concept is the driver does not actually own the car, and it'll still be owned by Riversimple. There's another concept where the critical raw materials in electronics could still be owned by the original manufacturer, so they could be sure of getting them back. I also know Celsa have looked at whether they should lease their reinforcing steel bars in construction because of the long lifespan.

For some type of products it would work, and for others it probably wouldn't, depending upon the mindset and the values that people hold. The concept is known as 'servitisation,' where it's what you get out of a product as a service, not the actual product itself that counts. Cars are a classic example with leasing. And in terms of reuse and repair as core parts of a Circular Economy, cars can be endlessly repaired, and people are happy to buy second-hand, third-hand. People do that because new they're, expensive they will be. But they are designed to last and there's the legal requirement for spares to be provided for repairs.

But to delve a little deeper, if we mirror nature in aims of a zero-waste ecosystem, won't issues of retained ownership cause more problems? Andy agrees:

You're right to raise that. The issue of ownership is one of the key blockers.

There are examples of where it's worked perfectly well in the past. I mean how many of us actually own our house? In reality, they are owned by the building society or probably the bank. Abroad, most people rent their properties, and they don't own them. So again, it's how the marketers have influenced us so that ownership is seen as the best thing. And we need to change that mindset because it's what you get out of that product, not owning it, that's the important thing.

There's another example, libraries. You don't have to own the book, you just need to read it. So, with the library, you borrow the book, you read it, you've enjoyed it, you give it back. You don't get anything extra by actually owning the book. If you want to read it again, you borrow it again!

Finally, on reflection around individual purchasing habits and how they might hamper CE transitions:

That's another thing that's resonated with me over the years – we have far too much choice. And that choice leads to inefficiency and the inability to do that modular approach. Why do we need so many different types of everything . . . it really leads to inherent inefficiency in the whole system. We need to re-evaluate what really makes us happy.

7 Ladeja Godina Kosir on transformation of people and the power of networks

About the contributor

Ladeja Godina Kosir is Founder and Executive Director of Circular Change and is also an internationally renowned expert on the Circular Economy. Ladeja was a finalist for the Circular Leadership Award 2018 (Davos WEF) and named in the EU 'women4future' campaign, which features women active in research, innovation, education, culture and sport. She is recognised as the regional 'engine for circular economy transition.' She is the co-author of the Roadmap towards the Circular Economy in Slovenia, Serbia, Montenegro and Chile, author of several articles on circular economy and a regular speaker at international circular economy events.

According to Ladeja, when visualizing a transformed circular economy, she notes,

> if we start with . . . sustainability, which used to be a very popular term, and then compare it to Circular Economy; where I see the biggest difference is that sustainability is offering us the playground where we are looking at how to preserve resources, and really how to behave in a way so as to maintain everything that is aligned with the nature of society, and so on. When we are talking about Circular Economy, what I see as the major difference is that [CE] is much more grounded, in a way – we have a concrete business model behind it.

> Yes, we are thinking about . . . resources again, but [also] how . . . to design things in a way that we can maintain value for as long as possible. It is much more focused, and it is offering different opportunities for different stakeholders. One is, of course, business, but the other is . . . citizens. The third, of course, is decision-makers. [CE is therefore] really based on . . . interconnectedness . . . and interdependency of all stakeholders . . . [and] very often . . . network governance. It is a radical corporate [and societal] collaboration needed to make this happen in practice.

> It has [also] become increasingly important to consider the relationship between [the adoption of CE and] the natural environment. Maybe, what we have faced now, during this pandemic, during this health crisis, we [have] realised how [CE] is connected to the climate crisis as well. We [have] realised, hopefully, that we are part of nature, not the masters of nature. When we think

Note: This interview was conducted by Allen Alexander on 22 June 2021. It formed part of the ISPIM Virtual Conference in Berlin, 2021. This transcript represents the interviewee's thoughts and opinions on the day of the interview only and in the context of the questions and discussion, as posed. As 'personal opinions,' they must be treated accordingly and not replicated, repeated or taken to infer anything 'outside' of the interview context.

https://doi.org/10.1515/9783110723373-009

about basic values around what we need, this is very much related to nature. We need water . . . fresh air, food, shelter and we as humans need relations and love . . . the purpose in our life. But there are cycles in nature and . . . I would say that we forgot or [tried] to cheat nature in a way; to make something grow forever and to make our economies flourish forever, which is not possible, not by the way we are producing and consuming right now, particularly in the developed world, it is not sustainable and it is not aligned with nature.

There are limits and within these limits, we have to find a way to survive as humans, because nature has the answer . . . things are flourishing . . . they go away and then they wake up again. We, as humans, we don't have the option to 'do over'. . . but as we [live] now, we really have to rethink our way of living. . . producing and consuming.

When reflecting on how to rationalise unsustainable and continued increase in GDP within the networks that Ladeja leads,

[W]e talk a lot about: What is a circular economy model bringing? Is it about sacrifice and the whole 'Green Deal' and the story behind that? And now green recovery? Or is this a kind of a renaissance in a way?

For sure, this continuous growth in a way that it has been measured, through GDP . . . is not the only way of measuring prosperity. Here comes the question 'What does it mean to live a good life?' And what does it mean to [create] the conditions where we can do it in a different way?

If I take an example of cars; we like to own cars, but those cars are parked most of the time, 93% is the average that a car is parked in Europe [for its lifetime]. So, it is not a resource that we are using wisely, what we can do? We can share the car with someone, or not own it, but use a platform where [cars] are available to be used. Or when we are driving, not to drive alone, but to take someone.

When we are thinking about the circular transition and transformation, what we see is how different topics are related. I touched on mobility; then we have food systems, health system, education, innovation and so on. Only by merging and collaborating within this environment; only through co-creation, can we find new solutions. It is not just about now having the slogan 'build back better'; it is not about building back; it is about reinventing the wheel. Here, the need for fresh ideas, for asking questions, for trying and failing and everything. . . connected strongly to this innovation process. This is a must.

But this is something that also takes us out of our comfort zones, so we come to [perspectives of] 'yes, we should look, before [making a] change' and for transformation; or 'but not starting with me.' We should start, each of us; and then discover the language for co-creation.

Reflecting on the notion of 'reinventing the wheel,' some argue it may be useful way to draw some parallels from the past when considering the future of Circular Economy. She notes:

> often we are asked: Are we moving forward? Or are we moving backwards? Are we doing something or promoting something that used to be normal? [where society] forgot about that? Or is that something new?
>
> [Ladeja thinks] the answer should be, yes, we started with nature, and nature has the answer. [She goes on to reinforce this by saying:] We, as humans, let's say in the last 50 years, with consumerism, we forgot how to value things. If we look back, for example, my grandma used to have a refrigerator that worked for 50 years and it was repairable. It was normal to change parts and so on.
>
> Now, we are so proud of the smart appliances, and everything is smart and digital. But on the other hand, we see that we are not repairing these things; we are not maintaining them, so we have to find a new balance.
>
> I would say it's not going backwards but using all the solutions that are available to us nowadays. If we compare it again with the health crisis, luckily we have vaccinations or medicines . . . and luckily, we have new knowledge . . . but it is about how we produce and maintain that for as long as possible. With Circular Economy, we are starting with design; it is not just about recycling and saying, 'okay, at the end of the cycle, we recycle;' it is how we produce, how we design, how we maintain and also how we organise jobs around that or the services that are needed, and then at the very end, of course, recycling.
>
> Just one reflection more, because we are working a lot on the national Circular Economy roadmaps, and you see that in different countries, there is different heritage and different culture. If I compare [what] we did in Slovenia, for example, and we contributed to one in Chile and Latin America. What do they have in common? Yes, we can have a similar approach towards the process, but within the process, we have to be very sensitive [to] what is unique in that particular environment and what are their values?
>
> What is the culture, for example. In our part of work, what we have noticed in X socialist countries, is that ownership is highly valued, because owning something means you gain respect, as you can afford it; that you're successful. When you go to Scandinavia, this is not the case.
>
> Most likely, it will be much easier to talk about sharing cars in Sweden, than sharing cars, let's say in Serbia or Chile, for example. What we realized in Chile,

because they are a leaner country regarding their form, but very circular and vivid in culture, and perhaps also when comparing between cities and rural life. To integrate and empower those people, who are not living in cities, and for us to start to become aware of their approach is valuable because they're very much circular. They're not urbanized; they're not using the solutions we are using in the cities. By being sensitive to culture, to people and to the knowledge and practices that exist . . . we can create solutions that are applicable in a country or in a region. It is not just copy, paste [sensitivity, cultural awareness and valuing local and embedded knowledge], if we want to make this global transformation.

When considering what a circular lifestyle means to her, Ladeja says:

I'm very much in favour of demystifying the word 'Circular Economy' and 'Circular Economy Lifestyle,' which is something detached from us.

Starting from myself, for example, in my life – what I think, okay, do I need something new or can I repair, or can I share? How I move around? Can I walk? Or can I take a bike or can I share a car, and so on. But what you ask is related to that which we are working on; the scale of the city as well as on the scale of the networks in which I'm engaged, really bringing this circular lifestyle to people.

Encouraging them to start with small steps, such as carrying your own shopping bag with you instead of using plastic bags or, as I said before, walking instead of driving the car for one kilometre and so on. Also addressing things like fashion; because we are sensitive now we have fast fashion, which is one of the biggest polluters – what can we do? Can we at least start with recycling and then, of course, thinking about not having a fast fashion style, but maybe going to second-hand shops in the city, instead.

To [enable change], I think there are two things that are important to implement this lifestyle: the solutions are accessible and affordable, so I don't have to go an extra mile or pay more. This lifestyle is not something that should be considered only for second-class citizens, but something that is trendy and something you decide to do because it is . . . beautiful and fun at the end of the day. We put a lot of effort in this direction.

As the power of a network is greater than the sum of its individual actors, in terms of ensuring that those networks are successful and create this power that they need, what are the key lessons to take away?

European Circular Economy platform that I chair (now for the third mandate) has the ambition to become the network of networks. Lessons learned are that we are bringing the [content] from the ground, to the structures on top. Here, we are talking about the network governance and about us as transition brokers . . . Within this platform, in the Coordination Group, we are 24 organisations of different sizes and different backgrounds. It is really something very vivid, and we are nourishing the dialogue among us.

What are the takeaways? What we have realised is very often we do not know who is working on what and by having this joint platform, we recognise the different projects as well as the ambitions and goals . . . The collaboration is forced on us; let's say, through that. The second is that we discuss what the burdens are or what are the topics that we would like to address. We have leadership groups within the platform, addressing those particular questions. The one I'm leading right now is particularly on network governance because we have realised that there is not enough collaboration between the governments and the other stakeholders.

The third thing I would like to stress is how important it is to maintain the identity of each organisation under this umbrella. When we are networking, often . . . another structure is born and the identity of this umbrella organisation is somehow sheltering others. What we like to do is, first, identify the stakeholders and encourage them, engage them and then nourish them.

In leading an international network that's based around systems thinking and systems approaches, what struggles or different barriers have you faced in leading that network? And how have you learned to overcome them?

I got the idea to give a title like 'Innovation Lost in Translation,' because . . . what very often happened within different stakeholders is, we get things lost in translation. We need the skills and competencies, very nicely described in the book by Jacqueline Cramer about Transition Brokers.[1] We see our role is to have capability to listen, and also to hear because you're working with different stakeholders: one scientist has one language, a CEO with another language, NGOs with a third language and so on.

In our process, we [focus on] transition . . . and . . . a systemic approach. So, you can't say, 'Okay, this is my interest, and I go for my goal;' [you have to] put the interest on the table, and then [we] orchestrate all those interests and goals – [focusing] the final outcome that it is win-win in the long term. But the challenge is, in the short term, someone will feel that, 'wow, I'm the winner' and someone else will feel, 'oh, at this very moment, I feel like a loser.' But we

---

1 Jacqueline Cramer, How Network Governance Powers the Circular Economy, 2020, https://holland circularhotspot.nl/wp-content/uploads/2020/12/How-Network-Governance-Powers-the-Circular-Economy-Ten-Guiding-Principles-for-a-Circular-Economy-Jacqueline-Cramer.pdf.

[focus] on . . . trust and transparency, these two words are of significant importance that we go through this process with deep trust.

On the other hand, with transparency of achievements and metrics as we go through this circular transformation, we often lack indicators because we knew what to measure for the linear economy. When it comes to circular, it is more tricky.

To have [effective] monitoring of where we are . . . and what we have jointly achieved [helps to stop], one stakeholder feeling 'Oh, I'm only contributing but not gaining anything.' It is not only about financial capital; within this network, we are managing social capital, and of course, taking in account, natural capital.

In attempting to generate indicators of success, Kosir candidly notes:

I will be very subjective and personal at this point. Because as you said, it is easy to play the game with big players; with big organisations; with those who are visible. But what I see as the achievement of the European network as well as of other networks I'm engaged in, is really to find those who are small and invisible, to engage them, and then to empower them. At the very end of the process, they may represent . . . 3% of the whole pie, but for me, that is the most important seed because from this a lot can grow. If they wouldn't be recognised and empowered, they would maybe disappear or, or say, 'Oh, we should quit.'

So my personal purpose is to give some light, to focus some light on those small but very, very valuable stakeholders and encourage them to keep on going and collaborate with vigour and then grow in a positive way.

Part II: **The state of transition**

Stefano Pascucci
Introduction

Will Circular Economy ensure a rapid and effective transition of our globalised market economy at the brink of a climate and socio-ecological crises? Alternatively, should we expect a series of (incremental) transitions connected with (or in preparation for) a 'great transformation'?

In this part of the handbook, we have invited scholars and experts to reflect and share their thoughts on these scenarios. Often, extant scholarship positions Circular Economy (CE) at the crossroad of business and industry strategies, on one hand, and public policymaking or regulation, on the other hand, suggesting the emergence of a transitional agenda (Fischer & Pascucci, 2017; Lieder & Rashid, 2016). CE seems to trigger narratives of *change, creativity, design, innovation, progress, growth,* and *development,* all framed as uncontentious fields compatible with a collective image of sustainable futures, with a still up-and-running market economy, full-speed globalisation, and well-looked-after planetary boundaries. Can the CE agenda really offer all this in 'one go'?

In terms of business strategies, policymaking and institutional change, CE is often considered the only 'world view' left to humanity for stimulating more sustainable production, distribution and consumption, while reducing material use, stimulating recycling and waste management (closing loop of resources), responsible innovation and the overall corporate environmental strategies (Borrello et al., 2020; Cardoso, 2018). Industrial ecology, performance economy, Cradle-to-Cradle™ design, complex adaptive system thinking, and biomimicry, are often identified as the intellectual and conceptual foundations of CE, particularly in its definition and principles (Borrello et al., 2020; Geissdoerfer et al., 2017; Stahel, 2016; Webster, 2013). While the discourse around 'circularity' has been floating around the academic and practitioner arenas for a few decades (Kirchherr et al., 2017), it is the publication in 2012 of the report 'Towards the Circular Economy' by the Ellen MacArthur Foundation (EMF, 2012) that catalysed the move of CE into a recognised 'world view,' where 'changing market economy from within' could inspire and stimulate wider transitional changes (Borrello et al., 2020; Cardoso, 2018; Geissdoerfer et al., 2017; MacArthur, 2013). The subsequent engagement of the EMF with the World Economic Forum and the European Commission, as well as the set-up of networks and partnerships like the CE100, have quickly crystallised CE as both a business and a policy-oriented agenda (EMF, 2015). From that foundational event, and in less than a decade, think tanks, NGOs, business networks and national and regional initiatives around the world have all contributed to develop a CE agenda for change, innovation and a *sustainable transition.*

As highlighted in the chapters contributing to this part of the Handbook, the key frames of this agenda have pivoted around a few key narratives, such as *designing out waste, keeping resources at the highest possible value, use of renewable energy systems,*

https://doi.org/10.1515/9783110723373-010

closing loops of nutrients, cascading, being *regenerative and restorative.* These frames, narratives and discourses have all been identified as 'circular principles and practices,' sort of shared institutional and organisational logics, in the emergent field of CE (Konietzko et al., 2020; Morseletto, 2020). In parallel to the definition of more formalised policy and regulatory frameworks, novel and distinct business practices have been reported, for example, in the relatively abundant literature on circular business innovation models (Bocken et al., 2014, 2016), or so-called circular economy transitions (Geissdoerfer et al., 2018b; Ghisellini et al., 2016; Homrich et al., 2018). In this literature, CE principles and practices have been investigated in all relevant fields of production, distribution and consumption, from finance to logistics, from product design to food systems, from regulations to labour markets. However, while these strands of knowledge have been influencing the debate on CE for almost a decade, a thorough reflection on the capacities of CE to trigger a sustainable transition seems to lack a common understanding. For instance, why is a clear definition of what is considered a 'circular practice' still lacking? How can CE trigger changes in relevant organisational fields if we do not even have shared industry standards? What can be identified as a circular job? Who is representing the interest of 'circular businesses' and do we have even circular businesses as such? And beyond the economic and transactional spheres of CE, how can we envision a transition for a market economy without mobilising narratives on a social dimension of CE? For instance, why are we so reluctant to discuss a 'circular society'? Instead, the CE debate seems still characterised by a high level of ambiguity and complexity (Cainelli et al., 2020; Corvellec et al., 2020). As highlighted by the authors of the contributed chapters, CE is influenced by multiple and contrasting views, often dominated by practitioners experimenting and introducing new practices and norms at business- and supply-chain levels in institutional voids (Fischer & Pascucci, 2017). In order to reflect on these views, as well as the frames, narratives and discourses informing them, we have asked eleven teams of CE experts to give their contribution and asked them to support a process of co-production of knowledge to better understand the 'state of transition' from a CE perspective.

Based on this ongoing debate, this part of the book presents a collection of academic perspectives on 'hot topics' related the state of the CE transition, organised through 11 dynamic and heterogeneous views on 'what's going on' in the CE arena. With this purpose in mind, we have invited teams of colleagues from different countries and backgrounds to offer their own perspectives on these topics. With a co-design approach in mind, we have selected 11 topics and teams that set the stage for what we believe is an intellectually challenging and stimulating journey. Merryn Haines-Gadd, Fiona Charnley and Conny Bakker start this journey with the question of how to move from design thinking into practice to make CE transitions possible. Saskia van den Muijsenberg's contribution looks at design and innovation from a different perspective: what can we learn from nature? How can we design transitions in Circular Economy connected to Biomimicry and an ecological world view? Both chapters build on the extraordinary experience of the colleagues in the field of design and creativity, and offer unique views on biomimetic and ecology-based design thinking processes. The second block of contributions address one of the core topic of the CE transition debate. How can we innovate and change businesses through CE principles and rules? And what is circular innovation? Daniel Guzzo, Janaina Mascarenhas and Allen Alexander tackle these questions in Chapter 10 and focus on 'The transformational power of Circular Innovation'. Our colleagues Paavo Ritala, Nancy Bocken and Jan Konietzko look at the relations between business models and innovation in the CE transitions, and offer to our readers 'Three lenses on circular business model innovation'. Chapters 12 and 13 move the debate into the core of what needs to be changed at organisational and system levels: finance, accounting and regulation. Aglaia Fischer, Diane Zandee and Marleen Janssen Groesbeek present a state-of-the-art discussion on the role of finance and accounting, a place where innovation and change are necessary to enable any form of CE transition. This is a rather unique contribution looking at the current literature and we suspect will set the stage for more conversations in the future. Similarly, our colleagues David Monciardini, Eléonore Maitre-Ekern, Carl Dalhammar and Rosalind Malcolm present a cutting-hedge multidisciplinary chapter on the role of regulation, and the related emerging research agenda. We are sure this chapter will not disappoint and again stimulate debate and possibly a research strand on CE transitions on its own.

One source of tensions and main critique in the CE debate has been the lack of understanding and perhaps acknowledgement of the relevance of social dynamics in the transition. We may say that the social dimension of CE is rather neglected or limitedly debated. We have asked two teams of scholars and experts to start filling the gap. Steffen Böhm, Chia-Hao Ho, Helen Holmes, Constantine Manolchev, Malte Rödl and Wouter Spekkink have tackled the topic of social activities and community-based transformations, and how they are connected with CE actions. Their chapter is thought-provoking and agenda-setting at the same time, so it will not disappoint the reader looking for a critical and stimulating view. In Chapter 15, Esther Goodwin Brown, Marijana Novak, Constantine Manolchev, Sharon Gil and Esteban Munoz look into the implication of CE transitions for the future of work and the job market more in general. In a digital and automated era, how do we reconcile employment with innovation and technological change, so deeply connected with CE? And what are the implications of CE transitions and job creation? The answers are discussed in this outstanding contribution.

The final three chapters of this part of the handbook look at critical resources, materials and material practices as the concluding section. In their chapter, Fenna Blomsma and Geraldine Brennan focus on 'Resources, waste and a systemic approach to Circular Economy' and offer an interesting perspective on the relationship between CE, resources and systemic changes. Marta Ferri, Alison Stowell and Gail Whiteman address one of the most detrimental crises of our time, and discuss how to tackle the plastic crisis through CE practices and strategies. In Chapter 18, Kim Poldner and Domenico Dentoni elegantly bring the reader into how 'Aesthetic engagement' informs our transition into sustainable futures.

References

Bocken, N. M., De Pauw, I., Bakker, C., & van der Grinten, B. (2016). Product design and business model strategies for a circular economy. *Journal of Industrial and Production Engineering, 33* (5), 308–320.

Bocken, N. M., Short, S. W., Rana, P., & Evans, S. (2014). A literature and practice review to develop sustainable business model archetypes. *Journal of Cleaner Production, 65*, 42–56.

Borrello, M., Pascucci, S., & Cembalo, L. (2020). Three propositions to unify circular economy research: A review. *Sustainability, 12*(10), 4069.

Cainelli, G., D'Amato, A., & Mazzanti, M. (2020). Resource efficient eco-innovations for a circular economy: Evidence from EU firms. *Research Policy, 49*(1), 103827.

Cardoso, J. L. (2018). The circular economy: Historical grounds. *The Diverse Worlds of Sustainability, 3*, 115–127.

Corvellec, H., Böhm, S., Stowell, A., & Valenzuela, F. (2020). Introduction to the special issue on the contested realities of the circular economy. *Cultural Organization, 26*(2), 97–102.

EMF (Ellen MacArthur Foundation). (2012). *Towards the circular economy, report vol. 1 – Economic and business rationale for an accelerated transition.* Cowes, UK: Ellen MacArthur Foundation.

EMF (Ellen MacArthur Foundation). (2015). *Growth within: A circular economy vision for a competitive Europe.* Cowes, UK: Ellen MacArthur Foundation.

Fischer, A., & Pascucci, S. (2017). Institutional incentives in circular economy transition: The case of material use in the Dutch textile industry. *Journal of Cleaner Production, 155*, 17–32.

Geissdoerfer, M., Morioka, S. N., de Carvalho, M. M., & Evans, S. (2018b). Business models and supply chains for the circular economy. *Journal of Cleaner Production, 190*, 712–721.

Geissdoerfer, M., Savaget, P., Bocken, N. M., & Hultink, E. J. (2017). The circular economy – A new sustainability paradigm?. *Journal of Cleaner Production, 143*, 757–768.

Ghisellini, P., Cialani, C., & Ulgiati, S. (2016). A review on circular economy: The expected transition to a balanced interplay of environmental and economic systems. *Journal of Cleaner Production, 114*, 11–32.

Homrich, Aline Sacchi, et al. "The circular economy umbrella: Trends and gaps on integrating pathways." Journal of Cleaner Production 175 (2018): 525–543.

Kirchherr, J., Reike, D., & Hekkert, M. (2017). Conceptualizing the circular economy: An analysis of 114 definitions. *Resources, Conservation & Recycling, 127*, 221–232.

Konietzko, J., Bocken, N., & Hultink, E. J. (2020). Circular ecosystem innovation: An initial set of principles. *Journal of Cleaner Production, 253*, 119942.

Lieder, Michael, and Amir Rashid. "Towards circular economy implementation: a comprehensive review in context of manufacturing industry." Journal of cleaner production 115 (2016): 36–51.

MacArthur, E. (2013). Towards the circular economy. *Journal of Industrial and Engineering Chemistry, 2*, 23–44.

Morseletto, P. (2020). Targets for a circular economy. *Resources, Conservation & Recycling, 153*, 104553.

Stahel, W. R. (2016). The circular economy. *Nature News, 531*(7595), 435.

Webster, K. (2013). What might we say about a circular economy? Some temptations to avoid if possible. *World Future, 69*(7–8), 542–554.

Merryn Haines-Gadd, Conny Bakker, Fiona Charnley

8 Circular design in practice: eight levers for change

Abstract: Circular Design, as a practice and approach, has grown in popularity in the last decade, with academics and industry alike proposing many strategies and methods that facilitate this in products. Yet, very few day-to-day products are actually circular. Therefore, in this chapter, we sought to analyse and reflect upon what progress Circular Design has made within industry, uncovering the key barriers and opportunities for how it is implemented. Through interviews with industry experts applying Circular Design in practice, this chapter identified the 'classic drivers and barriers' influencing the sustainability of products but also identified several new insights or 'levers for change' that are impacting the advancement as well. It is proposed that if these levers are ignored, they could potentially continue to hinder advancement, but if addressed, could help to unlock activity within this area and help speed up the transition to a fully circular product system.

Keywords: Circular Design, implementation, products, industry perspectives

Introduction, review and approach

Evolution of circular design

Over the last 30 years, many design frameworks have been proposed to improve the environmental profile of a product (Sheldrick & Rahimifard, 2013). Generally referred to as Design for Environment approaches, they are concerned with implementing environmental considerations within the design process (Boks & Stevels, 2007). However, knowledge and understanding within this field have evolved over time, as has the terminology and classifications used. Below summarises some of the known approaches:

Green Design – focuses on a single issue or aspect of a design's ecological impact, such as the recyclability of the materials used

Eco Design – considers the whole life cycle of the product, aiming to anticipate and minimise environmental factors, from design to manufacturing and end-of-life (ISO/TR 14062:2002, 2002).

Sustainable Design – integrates the triple bottom-line perspective, aiming to synthesise ecological, economic and social considerations (Bhamra & Lofthouse, 2007)

https://doi.org/10.1515/9783110723373-011

Within academic discussions, the nuances and varied benefits of these approaches are widely debated. Green design, at least within product discussions, tended to focus more on sustainable or green materials, and now is predominately associated with architecture and the built environment practices (Tseng et al., 2013).

Eco Design, on the other hand, primarily contributes to carrying out environmental assessments and strategic decision-making (Vallet et al., 2013), but can often be time-consuming and requires significant knowledge and expertise (Bhamra & Lofthouse, 2007; Sheldrick & Rahimifard, 2013). Lastly, Sustainable Design integrates the multi-stakeholder perspective, expanding our understanding of the implications of design, which challenges designers to embrace concepts of meaning and integrity of their products that goes beyond the obvious ecological benefits (Chapman & Gant, 2007).

However, considering the complexity of today's global product systems, traditional approaches that focus on optimisation, such as Green and Eco design are no longer fit for purpose and require a radical rethinking of product and system innovation to ensure that higher environmental gains can be made (Keskin et al., 2013).

Thus, while the Circular Economy concept has emerged as a promising framework for preserving and regenerating the environmental and economic value of resources within a system (Geissdoerfer et al., 2017), so too has Circular Design, as a method that can facilitate this transition by enabling materials, components and products to retain their value across multiple lifetimes (Den Hollander et al., 2017; Moreno et al., 2016).

Considered as part of taking a **Design for Sustainability** approach that aims to facilitate a systems shift through the radical redesign of products and services (Chick & Micklethwaite, 2011; Sumter et al., 2018), Circular Design requires designers and product developers to be systems-level thinkers (Moreno et al., 2016), able to actively integrate lifetime extension strategies while also being mindful of the implicit business model that surrounds the product (Bakker et al., 2014).

With the rise in popularity of Circular Design, researchers and practitioners from academia and industry alike sought to develop tools and strategies to facilitate circularity in products.

Examples from industry and third sector initiatives include: the *"Four Design Models" by* The Great Recovery Project (2013), *"Design on Demand"* by Forum for the Future (2016) and the *"Circular Design Guide"* developed in collaboration between IDEO and the Ellen MacArthur Foundation (2017).

Examples from academia include: *"Products that last"* by Bakker et al. (2014), *"Circular Design Framework"* by Moreno et al. (2016), *"Circular Design Tool"* by Moreno et al. (2017), *"Typology for Circular Product Design"* by Den Hollander et al. (2017), *"Cards for Circularity"* by Dokter et al. (2020) and *"Circularity Deck"* by Konietzko et al. (2020).

Though each suggests different procedures for enabling circular product systems, such as tools or frameworks, all offer insights regarding both design strategies and business model archetypes, the design strategies proposed, such as Design for Material Recovery or Design for Remanufacturing, address the more product-level factors, whereas the business model archetypes such as circular supplies or access-based models speak of how the function and value of the product is delivered and recaptured. This, perhaps, is what sets Circular Design apart from other Design for Environment approaches. Not only does it acknowledge and account for the entire product eco system through business model innovation, it integrates multi-lifetime thinking through product lifetime extension strategies, facilitating the maximising of the value of the products and materials from both an economic and environmental perspective.

Yet, it is important to note that while Circularity and Circular Design might offer some new insights, its philosophies and principles are built upon its predecessors. As these approaches evolved, regardless of the 'new' terms and practices that have been defined, the core ambition and driving principles of each, from eco design to circular design, remain the same: to reduce the environmental impact of products, through tactics such as reduce, reuse, repair and recycle.

Within each of these approaches, many *design for x strategies*, such as design for durability, reparability, serviceability, disassembly, upgradeability or emotional durability, have been framed and reframed over time by many authors and practitioners (Bakker et al., 2014; Haines-Gadd et al., 2018; Den Hollander et al., 2017; Ljungberg, 2007; Moreno et al., 2016; Mugge et al., 2005; The Great Recovery Project, 2013; Van Nes & Cramer, 2005). Nonetheless, to facilitate the environmental factors, the central values were carried forward with each iteration. Therefore, rather than defining and debating what design strategies might be classified as "eco," "sustainable" or "circular," it is perhaps enough to contemplate any and all tactics that fundamentally enable products to have extended use, reuse, be repaired, refurbished, remanufactured and recycled, as these ultimately are the key pathways that facilitate circularity and sustainability in products (Bocken et al., 2016; Nubholz, 2018).

Implementation of circular design in industry

In 2015, the European commission released a report, *"Closing the loop – An EU action plan for the Circular Economy,"* which outlined both legislative and non-legislative initiatives for accelerating the transition to a circular system. Due to the policies and regulations outlined, the initiative has not only helped to build new infrastructures and enabled a technology push for the region, but has also been a driving force for increasing momentum for research and the implementation of Circular practices across many industries (Mhatre et al., 2021).

Regarding Circular Design, evidence of how these practices have subsequently been implemented in industry is varied (Dokter et al., 2021). At a design level, several manufacturers have defined their own strategies and guidelines regarding circular product design (Dell, 2020; H&M, 2019; Inter IKEA Group, 2017). At a process level, some are exploring circular business models or services (Caterpillar, 2020; Signify, 2020), while at an organisational level, circularity targets are beginning to be integrated as part of their key performance indicators (KPIs) and commitments (BAM, 2021; Inter IKEA Group, 2019; Philips, 2020; Signify, 2020).

Though these examples indicate that we are moving in the right direction, most products on the market today still operate and cater to linear systems. Furthermore, considering that only 8.6% of the human-made world is classified as actually being circular (Circle Economy, 2020), the question is, what progress are we making towards a fully circular product system and what is hindering our advancement? Therefore, to address these questions, this chapter seeks to understand not only where industry is in its journey towards developing circular products and what headway is being made in circular design implementation, but also what is both positively and negatively impacting this progression.

Approach of the study

To address the key question of this chapter, knowledge and insights were collected from industry experts who are currently involved or were previously involved in the development of products, services and business models that utilise Circular Design principles and strategies.

This enabled the research to gather case study examples and evidence of how circular design is currently being put into practice. A qualitative approach was employed, as it allows for the gathering of rich, deep insights into the experience and context of participants and how this can influence practice and outcomes of a particular process (Ritchie & Ormston, 2013). This qualitative data was collected utilising semi-structured interviews conducted online using Zoom and Teams, over a two-month period. Each interview lasted 60–80 min and was recorded.

To gather a holistic perspective of Circular Design implementation, the key topics this inquiry sought to understand were related to: which Circular Design strategies are currently being considered, how and where these have materialised within the CE product initiatives, what are the challenges and success factors that lead to these being implemented and how Circular Design as a practice is being considered within the organisation. Table 8.1 shows what information was gathered from each participant as well as the list of questions that were asked.

Table 8.1: Questions asked in interviews.

Participant/company information	– Organisation – What the company produces – Their role within the company
General CE questions	– Can you describe how you work with the Circular Economy in your company? – What type of activities or projects do you undertake that are related to the CE
Circular strategy focused questions	– Can you describe a project that you consider has been successful in terms of implementing circularity at a product level? What key factors were important to its success? – What CE or circular design strategies did you consider? Why did you pick these over others? – What was your approach to developing the circular solution? Did you use any tools or methods? – When you considered adopting Circular design strategies, did you also contemplate the business model around the product as well? If so, in what way? – From your experience, which strategies are proving to be the most challenging to put into practice? And why? – Which strategies are proving to be the easiest to put into practice? And why? – Have you observed an improvement in the circularity or sustainability of your product as a result of adopting Circular design principles? If so, how have you measured that?
Diffusion of Circular Design questions	– When did you start using Circular Design in your products and what triggered this exploration? – Where do you think your organisation is on its journey into implementing Circular design? – In your opinion, has the way circular design been used and considered within the organisation changed over time? (i.e. Is this a niche concept or something that is becoming more mainstream?) – If more mainstream – what factors led to this occurring? – If still niche – what factors are preventing it from progressing?

Participant selection

It was decided that data would be gathered from organisations that have been exploring Circular Design within their products for over five years, as these participants would have more in-depth experience and knowledge to share. Within each organisation also, only industry experts who are directly involved in the Circular Design process were approached to ensure that they have the appropriate expertise.

Seven participants from six organisations were approached and interviewed. Their information has been anonymised, but following is a description of the organisations and what they produce.

List of organisations interviewed

–   Organization A: A multinational home appliance manufacturer who sells products such as white goods, ovens and vacuums cleaners under several different brands
–   Organization B: A multinational manufacturer of consumer electronics and healthcare products, such as large medical equipment
–   Organization C: A multinational manufacturer of lighting products
–   Organization D: A multinational manufacturer of optical and imaging products such as lenses, cameras, scanners and printers
–   Organization E: A manufacturer of bamboo clothing
–   Organization F: A manufacturer of parental products such as push chairs

Thematic analysis was employed, as, aside from being the most widely used method for analysing qualitative data, it is also the most useful for "capturing the complexities of meaning within textual data" (Guest et al., 2011, p. 11). The interview data was transcribed, collated and then analysed by all authors in order to identify the key implicit and explicit ideas or themes within the data. The results of this process are shown in the second section of this chapter.

Results

The progression of circular design

So, what progress have we made towards circular products and what is impacting the implementation of circular design within industrial contexts? All the organisations we spoke to are actively integrating circular design principles into their products and had at least one product or service offering that could be classified as contributing towards Circularity. Table 8.2 presents a summary of the main strategies they are considering:

Several of the organisations have been engaging in these circular activities longer than others, but it is also understood that lead times between sectors can differ, affecting the rate at which the outcomes of Circular Design can be realised. For example, within consumer electronics, it was suggested that *"We can design things now, but the evidence won't be seen in the marketplace for four-six years, in some cases: we are a slower industry"* (Organisation A). Regarding how they view themselves in their

Table 8.2: Summary of strategies and circular products on market.

	Strategies being considered	Circular offerings on the market
Organisation A	Refurbishment, Reuse, Repair, Design for Recycling, Recycled Content, Leasing	Subscription Robot vacuum in Stockholm (B2C)
Organisation B	Design for Serviceability, Repair, Refurbishment, Recycled Content, take-back, subscription models for consumer products, predictive maintenance	Refurbished MRI machines (B2B), Coffee machine 75% recycled plastic (B2C)
Organisation C	Serviceability, Longevity, Reuse, Upgradability, Refurbishment, Recycling, Take-back scheme Leasing	On-demand 3D printed lighting (B2C & B2B), Light as a service (B2B & B2G)
Organisation D	Durability and Longevity, Design for Assembly, Repair, Refurbishment, Adaptability and Upgradability	Refurbished high-value printers (B2B)
Organisation E	Material Health, Durability, Recyclability, Disassembly, Recycled Content, Emotional Durability, Take-back schemes	Sustainable denim, Circular Technical Jackets (B2C)
Organisation F	Durability, Longevity, Serviceability, Recycled Content, Leasing	Repairable and serviceable stroller, which is often resold on second-hand market; sells recalled or minor-defect strollers (B2C)

journey towards circularity, most organisations described themselves as either 'in the thick of it,' 'an intermediate phase,' or 'making progress', while Organisations C and E consider themselves to be ahead within their respective sectors.

However, regardless of the sector, there were some common challenges and successes observed within the interview data and two points became evident. Firstly, that many of the insights or factors identified could be categorised as 'classic drivers and barriers' within this context, and that they are still present and actively influencing the sustainability of products. Secondly, in addition to these typical factors, there are also several new insights or 'levers for change' that are affecting progress that, if ignored, could slow down progress; but if addressed, could help to unlock the activity within this space and speed up the transition. Therefore, these two points of reflection will make up the structure of this section and will be explored and discussed using examples from the interviews.

Classic drivers and barriers to change

The factors listed in this section were categorised as 'classic' due to the fact they have been either observed or explored as key barriers or drivers by previous studies and they cut across sectors.

Classic drivers

Legislation – Most organisations stated that the introduction of policy commitments and/or regulations actively influenced or triggered them to consider circularity and sustainability factors. *"Legislation is always an enabler; the momentum is really growing in society. But there is an urgency for action and the green deal is helping"* (Organisation C). Moreover, with new standards around 'Design for repair' being introduced, some organisations are not only wanting to be compliant, but excel and *"have a good score"* (Organisation B).

Customer demand – *"The Greta effect is real!"* (Organisation A). The growing societal awareness of environmental factors has not only impacted employee perceptions inside their companies, but has also resulted in increased pressure on organisations to engage with sustainable products. *"Overall, society is more sensitive to the topic, and we are seeing this with [our] people too"* (Organisation B).

For Organisation C, this is affecting demand and they are *"seeing a pull for circular products; our portfolio for professional luminous is growing for different purposes."* Ultimately, this trend of environmentalism is something that Organisation C is very much aware of and how it impacts the market; so believe *"if there is more pull, than it gets easier to implement."* However, while this is a positive step for circularity and sustainability, arguably though, the burden for change is still being placed on the consumer and not led by the organisation.

Investment in research partnership and exploring new resources – All organisations are currently, or have previously, engaged in research partnerships with academic institutions and/or other organisations to carry out research and development (R&D) towards circular offerings. Organisation E, in particular, identified a gap in technologies available for recycling textiles *"R&D is where the solutions are going to come from, and so we are very happy to get involved with this process ourselves."*

Sustainability Leadership – Organisations A, B, C and E attributed the successful implementation of circular design as a result of having leaders and managers who are actively driving this type of agenda. In particular, Organisation A remarked that *"Our CEO is a really strong Advocate [. . .] there's so much passion in the company."* Moreover with *"the trickle-down effect, pretty much everybody in the company has to have something about sustainability in their yearly review, which has never happened*

before. There are people who don't have it in their title, trying to show how they're con-
tributing to sustainability."

This, in turn, is then impacting company perception, ultimately helping to attract talent to the organisation, as, *"for many employees, it is a strong reason why*
they joined the company."

Classic barriers

Cost of circularity – Many previous studies have offered that often economic investment is needed to transition from linear to circular offerings (Ellen MacArthur
Foundation, 2013; Ghisellini et al., 2016; Rizos et al., 2019). It is likely this will be
required at a product level for activities such as R&D, and at a systems level for activities such as reverse logistics, for the recapture of value at end-of-life. While it is
argued that circulating products will eventually result in a return on investment
(Bocken et al., 2016), this still requires organisations to initially foot the bill. This
was observed more significantly by Organisation F, stating that, within design, *"we*
can do whatever we like, assuming it doesn't cost too much," as ultimately, *"The bot-*
tom line is also always still a sticking point – how much will this cost, how much will
it make us." Though they attempted to implement recycled and circular materials
into their products, they found that *"most of the more circular materials tended to*
be more expensive" and this increase in cost was hard to justify to the organisation.

Mindset shift and proving the business case – Most organisations remarked that
proving the value of the circular business models was a significant challenge, as
not only are they *"still trying to determine where the real and best value is"* (Organisation A) within these systems, but the economic benefits of some activities are simpler to quantify; *"it's a lot easier to convince the business to use a recycled material*
in the product because this can be proved; the supply chain exists, it can be costed.
To change the business model is much harder" (Organisation A).

Investment will also be required to get these new business models implemented
as *"[today's] products are made for one-time selling, so in order to redesign a new prod-*
uct and a new product ecosystem that is optimised and profitable for circularity, requires
huge investment for such a service. This can be a big risk to take." (Organisation F).

This ultimately requires organisations to consider new mindsets as well: *"it's*
relatively easy for design and engineering to propose how they would do something,
quite difficult for the company, to have sufficient confidence to try a different way of
working" (Organisation B). Nevertheless, Organisation A is having some small successes in this area, having successfully launched a performance-based robot vacuum; citing a low barrier to entry cost as a main factor that drove uptake.

Issues of scale – All organisations reported the use of pilots or trials to explore circular products and services. However, Organisation A observed that scalability can

be an issue that hinders progress: *"In certain markets, certain offerings do well and others do not, therefore in terms of the scale of the business model, you can't just do it the same everywhere and think that it will work."*

Environmental trade-offs – Managing environmental trade-offs is a topic widely discussed in both design (Ma & Moultrie, 2017; Prendeville et al., 2014; Salari & Bhuiyan, 2018; Vanegas et al., 2018) and CE contexts (Bracquene et al., 2020; Glogic et al., 2021; Millward-Hopkins et al., 2018). Similarly, within industry, Organisation B remarked that *"if you've worked in sustainable design, you are always trying to understand these trade-offs [. . .] if circularity is the best option."* While Organisation E found balancing circular design strategies (such as emotional and physical durability, disassembly and recycling) a challenge from a technical design perspective, Organisation B found communication of these trade-offs to decision makers more crucial, stating that,

> The businesses are not used to this; usually, you need to be good at presenting these options to people[1] as well. It's hard for people to manage new thinking, then new terms and complex trade-offs. Sometimes they trust you; other times they are sceptical, and it's about making sure the trade-off is communicated properly to your stakeholders.

Organisation B is larger than Organisation E and involves many more stakeholders in decision-making processes, which is why this is probably more salient. Nevertheless, all organisations stated that Life Cycle Assessments (LCA) are central to how they make evaluations on which strategies to implement; however, Organisations B and D specifically stated that the business case or economic assessments 'really' influence what direction is to be considered.

Though positive that environmental assessments are considered as a best practice for product development, Organisations E and F both recognise their limitations when trying to implement new materials or product ecosystems, and are exploring new databases and data collection research to improve their accuracy.

In conclusion, though these points have been determined as 'known' or 'classic' factors within the field of circularity, it demonstrates that, intrinsically, some drivers are working and should continue to be pushed further, such as regulation and sustainability leadership, and that some barriers are yet to be resolved, and require more research and investment. However, it also shows that, although these companies are considered pioneers in Circular design, each organisation had only one or two products within their portfolios that could be considered as circular. Therefore, these typical drivers, while still relevant, are not enough, and there are

---

1 'People' within this context refers to key stakeholders of the product development process such as designers, product managers, engineers, business management etc.

other avenues or 'pressure points' that have been identified that require more attention, if we are to unlock the potential within this field.

Questions to contemplate further from this section

Does being circular really cost more than being linear?
How should we be picking which strategies to use over others?
What data do we need to prove the business case?

Levers for change

The factors identified within this section have been grouped together as they represent points that potentially, if overlooked, will continue to slow the progress of the circular products, but if addressed and developed, could act as linchpins for speeding up the transition.

1. Be a catalyst for change

From the data, it was observed that many of these organisations were acting as catalysts or facilitators for change, both up- and downstream within their supply chain. This involved activities such as educating their suppliers, facilitating new networks and engaging in thought leadership within their respective sectors to help integrate circularity into their products.

In particular, Organisation E wanted to ensure that at the design stage, their product could be fully recyclable at end-of-life. Traditionally, they relied on their upstream supplier to provide knowledge on recyclability of the materials in their product; however, they discovered these topics had not been considered. Therefore, they *"realised we needed to be facilitators of sorts"* and linked up a recycling processor with their supplier. Similarly, Organisation F wanted to use recycled content within their products but found this was not currently available, so are facilitating conversations with other materials suppliers to explore this possibility. Lastly, Organisation A are engaging in similar activities, however found that due to their brand pull in some markets, *"we have the scale for negotiation on some materials, components or processes, but we don't have for others."*

Moreover, beyond facilitation and driving new knowledge within manufacturing, Organisation E also feels that in the fashion and textiles sector, there is a significant gap *"between solutions and those that need solutions"* regarding circular or sustainable materials and appropriate end-of-life processing. They observed that some large brands do not want to invest in change, so they feel they should act on thought leadership on this topic: *"hopefully it will be smaller guys like us who will be able to push the bigger businesses in that direction."*

In conclusion, as more organisations engage in these discussions and attempt to drive change where they have influence (i.e. within their respective supply chains), the more these topics will diffuse into everyday practice. However, looking at the strategies these organisations are driving, they are primarily centred on materials and involve actors upstream more than downstream. So moving forward, while it is important these upstream discussions continue, it is vital that facilitation occurs downstream as well. This will not only assist in the establishment of the collaborative networks required to enable take-back schemes, re-use and the recapture of value, but will also help to demonstrate the business case for circular products.

2. Integrate downstream circularity

For some organisations, addressing upstream innovation (i.e. changing the design, replacing materials for more recyclable or recycled content) is proving to be easier than downstream innovation (i.e. setting up take-back schemes, subscription models etc.). All organisations we spoke to are considering and implementing recyclable and recycled content materials in their products; in particular, Organisations B, E and F have products in the market that are made predominantly out of recycled content. Organisation A remarks, *"Recycled materials have come a long way and there seems to be an increasingly healthy Marketplace. Companies are interested in this and I think the material suppliers realize that."*

However, downstream, ensuring these materials are recovered and recycled is proving to be a more challenging prospect:

> Hardest thing to do is end-of-life solutions [. . .], for fabric and textiles industry where there is no infrastructure for it; there aren't very many technologies that are scaled up and commercially available yet, and there is a massive disconnect between brands, their suppliers and recycling processors" (Organisation E).

Furthermore, Organisation B observed that recycling is not something that manufacturers like them have any direct control over; it is the Waste Electrical and Electronic Equipment (WEEE) system that manages this waste, not them.

While material innovation might be considered 'easier,' is it really the most pressing strategy to facilitate circularity in products? As Organisation E observed: *"to what extent is ensuring that there is recycled content a good idea, if it's not being recaptured?"* The upstream innovation guide developed by the Ellen MacArthur Foundation (2017) believes it helps to address the problem at the source by preventing waste being created in the first place, while most certainly true when considered for packaging. Is this the right strategy for more durable products with longer lifespans?

Extending the physical and emotional lifetime of a durable product at the use phase is the most important strategy for minimising its environmental burden (Bocken et al., 2016; Haines-Gadd et al., 2018) and should be considered first when designing for circular products (Den Hollandar et al., 2017). While only Organisation E stated that they were actively considering the emotional durability, all organisations are considering physical durability factors. In particular, Organisations A, B, C, D and F found implementing strategies such as serviceability or repair at a design level to be relatively straightforward. However, initiating take-back schemes, leasing and the operations required for refurbishment, reuse or recycling are proving to be a much trickier proposition, especially for products in the B2C market, as Organisation C remarked: *"in the reverse logistics, there is still lots to be gained."* Despite the challenge, all the organisations are exploring the implementation of these initiatives, and Organisations B, C and D currently have these networks and operations established to recapture and refurbish high-value products, but only for the B2B market.

Moreover, Organisation C is taking things one step further and designing their products to be Circular-Economy ready.' They have recognised that while the downstream system is not ready to cycle these products just yet, when it catches up and the structures are put in place, their products will be ready, as the best current strategy for design to consider is to 'make the circularity of a product possible' and the rest will fall into place (Organisation F).

Lastly, Organisation E struggled with setting up reverse logistics, as the collaborator they partnered with could not handle the scale of returns and had to close the initiative down due to economic constraints. They felt they had 'done it the hard way' by trying to set up this initiative from scratch, but if were to do it again, they would partner with third-party suppliers to manage returns. This highlights another key concept of circularity – the importance of partnerships and collaboration – which will be explored further in this next point.

3. Be proficient in collaboration

For products to cycle in a functioning circular economy system, networks will need to be established to capture and cycle this value. This inevitably requires collaboration between system actors, and is cited as a key enabler for making circularity work (Organisation C). Yet, as several organisations highlighted, to create a *"working Circular ecosystem, this is the real challenge. This is where the real magic happens; in collaboration, not in the separate functions or disciplines; in collaboration, inside and outside your organisation"* (Organisation B).

To achieve this utopian vision of collaborative circularity, it is often proposed that partnerships need to be built on mutual benefits (Charnley et al., 2011); however, for Organisation A, this has been a point of friction: *"we can all see the benefits of collaboration, but when it comes to viable projects that the business will buy into, it's quite hard to find alignment because the interests are different."* While it could be argued that this is an indication of a mismatched partnership, Organisation B believes issues such as these arise due to economic constraints: *"it is the place we need to invest, we do not budget for collaboration; we reserve budget for product innovation, BM innovation, and not for collaboration."*

While increasing investment in collaboration could help to unlock some of these sticking points, this would also need to be examined in parallel with comprehension of circular economy concepts from a system-wide perspective. It was observed by Organisation C that:

> in the market, we do not have harmonised understanding of circularity; when I am at conferences, recycling companies only think about materials, other companies only think about spare parts harvesting etc. There is such a diffused way of thinking about circular. I think that is a big issue we have.

What was interesting about this point is that all organisations we spoke to are keen and willing to explore collaboration, but are finding the process time-consuming (Organisation F) and, on occasions, unsuccessful (Organisation A, E). Therefore, perhaps we need to be considering not only what methods or processes we need to facilitate for a better collaboration between system actors, but what expertise we might need to push this further. As Organisation B offers, collaboration *"should be looked at as a separate skill set, or competency area for any sustainability or CE professional."*

4. Have collective circular comprehension: pull in the same direction

> Even some people who are engaged in sustainability still don't really understand what Circular Economy or Circular Design really is (Organisation E)

In general, CE is still thought to be a 'niche' concept, both internally and externally. Subsequently, for some organisations, this has resulted in a varied understanding of what CE is and how this contributes to business functions. Organisation A considers CE to be a subset of sustainability, and even though considered to be holistically integrated across many different functions, they are sometimes faced with the question of *"do we want to just make our existing business more sustainable, or do we want it to be circular? In some cases, it's a bit of an either-or."* Moreover, some think circularity is just about recycling materials; for others, it is about business models.

Organisation D, on the other hand, is prioritising the development of a carbon-neutral product over a circular product, but is contemplating how circularity might contribute to this, which prompts the question, is there a tension between creating circular, sustainable and carbon-neutral products? Perhaps, though some believe it is just a matter of better aligning net-zero goals with circularity principles (MI-ROG, 2020). Regardless, what this could indicate is that we need to ensure that not only is everyone on the same page, but are also reading from the same playbook. As Organisation E observed how getting different teams such as both product and marketing 'buy into' the concept, and understand 'how this impacts their job,' really helped in the successful implementation of their circular products.

For a few organisations, if circularity and circular design is also considered as a long-term strategy for the organisation, this will both help to align internal activities and result in a competitive advantage in the future. This final factor was observed most significantly with Organisations B and C and is viewed as making the most headway in this regard in two ways.

Firstly, both organisations integrated 'circular product revenue targets' as part of their KPIs, which has significantly helped to drive change. While Organisation B is creating strategic product roadmaps for how to reach these targets, Organisation C compared CD principles with their existing KPIs to ensure these translate into circular products. And while CE or sustainability was previously seen as part of compliance, it is now viewed as 'becoming core to the business, and employees are considering integrating these ideas into their existing ways of working.

Secondly, both attribute business structures as an enabler for the development and delivery of circularity. Organisation B has a horizontal unit that advises and guides many different departments and teams (from product to R&D to operations) on the best strategies for circularity, from both technical and business-model perspective. By not being tied to a particular department or a product line, this enables them to have a more holistic, systems view of driving the circular product development process.

Organisation C, due to the nature of their focus on offering service-based circular offerings, has created a dedicated *"sales force that understands the proposition, can sell it, and is comfortable enough to explain the benefits to the customer."* And by having a *"good service team,"* this can really help to manage the relationship and provide service through the contract. They highlighted that the ability of the sales force to successfully communicate the benefits of switching from linear to service-based model was crucial for adoption, and so is something that must be implemented.

5. Build up circular capabilities: knowledge, tools and metrics

Most organisations indicated that they are constantly developing their knowledge and capabilities around circular design and circularity principles, which was being facilitated through various activities. Firstly, through continuous learning practices, such as training programmes or looking to external resources and guides: Organisation E had employees attend an academic training programme on the circular economy, while Organisation C runs its own training internally on Circular Design and circularity. Most organisations stated they took inspiration or influence from external guides and tools developed by academia and sustainability consultancies; in particular, Organisation E found the Ellen MacArthur Foundations principles of sustainable denim to be a helpful starting point for their exploration, as *"this gave us something to aim for – a really clear set of guidelines."*

Secondly, through the internal development of tools and methods for both facilitating circular design and measuring circularity: Though Organisations B, C and D have developed their own internal design guidelines for enabling circular products, all the organisations are actively using circular design strategies and frameworks in one capacity or another. Regarding measuring the environmental performance, most tend to use LCAs as the primary methods of assessment. Yet, when considering circularity specifically, some are exploring this in more detail than others. Organisations B and C have defined specific metrics for measuring the circularity of their products, Organisation E has some indication but does not measure these formally, while Organisation F, on occasions, monitor the resale value of its products, which it considers as more of a 'quality indicator' than a circular indicator.

Measuring product circularity – not only its carbon footprint, but also its ability to be circular – is an interesting challenge, one that many in academia are trying to solve (Boyer et al., 2021; Bracquene et al., 2020; Linder et al., 2017; Mesa et al., 2018; Shevchenko & Kronenberg, 2020). However, it is promising that the industry is also developing knowledge in this area as well, which, hopefully, will help speed up the transition in this space, as Lord Kelvin famously offers, *"if you cannot measure it, you cannot not improve it"* and this is certainly something that needs improving.

Lastly, Organisations B and C formalised some of these activities through the development of *'Centres of Excellence'* that have enabled the company to not only to build but also consolidate their knowledge and capabilities in sustainability and circularity, internally. However, it was also reflected that to truly help implement circular design into products, we need to go beyond developing expertise for a select few, and integrate it more widely within the organisation, and that *"Real change happens within the existing functions of an organisation when people become aware, activated, engaged. To execute these new values of CE is the real journey; basically, to make yourself obsolete; to make CE thinking, the norm"* (Organisation B).

6. Tweak and totally transform: Where to start

For most organisations, there was a tendency to servitise existing products and tweak the design over time to improve circularity, rather than start from scratch to create entirely new circular product and service offerings. One reason stated for why this was the case is that *"you have to start somewhere"* (Organisation B), but it is assumed that considering the time it takes to launch a product, servitising existing units requires fewer up-front costs and provides the opportunity to explore the business case. Most improvements being implemented were about increasing serviceability, ease of refurbishment, disassembly and repair; all factors that are necessary if products are to be maintained, recaptured and redeployed for future users or use cycles.

However, this prompts the question, to what extent can Circular Design be retrofitted? As Organisation F argued, most products and product ecosystems are designed for one-time selling and would need to be completely redesigned if they are to be optimised for profitability. So, would there come a point where you might need to start from scratch anyway?

Not necessarily. It was observed that this could be dependent on whether you are discussing a high-value product or a low-value product. Organisation B offers:

> large equipment has inherent value, typically high enough for a business model or service model to be good enough, without the need to contemplate Circular Design strategies as much; these systems have been optimised to work efficiently across a longer period of time, due to the investment needed for manufacture. [. . .]. For smaller equipment, the business model cannot be made unless you design for circularity, from the start. Only when you start integrating Circular Design, can you begin to repurpose, in a profitable way, products into the market.

With this in mind, right now, is circularity only being implemented for high-value products? One limitation of our data set is that we were only able to speak to manufacturers of, arguably, high-value products (mostly electronics, etc.). Although Organisation E produces lower-value items (socks and underwear), their current circular offerings on the market have a higher price point, which they believe contributes to the likelihood that it will fulfil its circular potential and not be discarded at end-of-life: *"For higher-value products, consumers are more likely to research it, donate it, reuse it, and maintain a longer relationship with it; so more likely to care where it will end up."*

At least for high-value equipment and products, it is thought that *"the business model is leading"* (Organisation B) and these types of products are, in general, considered to be *"assets in the field"* (Organisation D) that can be a source of revenue for many years, if they are designed to last. However, arguably, the business model is also leading for low-value items as well, as ultimately this influences how it is designed. Nevertheless, it is also thought that perhaps as servitisation of products becomes more commonplace to consumers, the redesign of products can properly begin to suit circular business models (Organisation, F).

Moreover, Organisation A reflected that in most cases, *"the hardware takes a lot longer to develop than the software,"* so digital products that help enable the servitisation of existing products is faster to roll out, and will demonstrate the value much quicker than physical changes. So perhaps then, it could just be a case of adopting different approaches for different types of products.

At least for Organisations C and E, they decided to go back to the drawing board and develop their product offerings with a circularity intent. For Organisation C, in particular, not only does circular design increase the ease with which the products can be maintained and servitised, it also enables *"an extension of functionality"* over time, such as the inclusion of new features, as, *"there is a lot of freedom for functionality when you do the right design."* So, one could even argue, beyond improving the sustainability that, perhaps, circular design is inherently just good design.

7. Consider the circular consumer

Thus far, research on the circular economy and circular design has tended to focus more on technical challenges of implementation (Lofthouse & Prendeville, 2017) and less on the role and experience of the consumer in this new type of system. A few studies have tried to address this perspective. For B2C products, one study created a tool for new product developers to design opportunities for emotional connection so that consumers want to keep and maintain their products far longer (Haines-Gadd et al., 2018). Others examined the attitudes of consumers towards second-hand or refurbished products (Baxter et al., 2017; Mugge et al., 2017) and the role of the consumer in the circular economy (Camacho-Otero et al., 2020; Chamberlin, 2021).

For the B2B market, these emotional and behavioural issues were observed to be less of an issue (Organisation D), but for the B2C market, understanding how consumers feel and engage with circular products is an important step that must be considered to accelerate the transition as *"we learn so much more by having more people buy into it and really what we're trying to do is learn and figure out how viable a business model is"* (Organisation A).

Within the industry, this is starting to be more comprehensively examined as well, and most organisations indicated they were investigating the consumer experience dimension of circular design as *"it is a bit futile to design entirely circular product that then no-one wants to buy or wear; so placing the needs of the consumer needs at the core is key"* (Organisation E). Moreover, by adopting the principle, *"By thinking with the customer, you can help to further improve the performance"*, it helps to build a product (Organisation C) where users develop more meaningful relationships with it and this approach could contribute towards your organisation being considered a preferred supplier.

But it was also remarked that, especially for circular consumer products, this may require participation from the user to help make this happen. For example, Organisation E is exploring one such possibility and found that:

> in order to make the jacket truly circular, the waterproof level of the jacket is not as high as we'd like, and needs to be refreshed once a year; so once you buy the jacket, you also receive a lifetime subscription service and receive a liquid wash that goes in the washing machine.

Also, if a product needs to be repaired, Organisation A wants to understand consumer attitudes and willingness to engage in this process and distinguish *"how to make that easy and desirable."* Furthermore, to enable repair or returns at end-of-life, continued consumer engagement will likely be required to maintain some visibility on the products, as *"we've done all the work to make sure it can be recycled; why wouldn't we make sure it does get recycled?"*

Lastly, for some organisations, lack of information or data on existing or future consumers is slowing down the progress: *"The biggest challenge was finding a customer for used products"* (Organisation D); they had predicted that these were most likely in developing markets, but found it difficult to propose a solid business case and products to investigate this further. While for Organisation A, due to legislation being different between regions, there is a varying willingness from consumers to share data on product use to help inform circular design improvements: *"Circularity only lives with the flow of data . . . data is the oil, if you can't understand the flow of things, then you can't create a circular economy."*

In summary, it is a positive step that organisations are considering the consumer side of the circular products, but this is still an undeveloped area for circular design in general, and we need that consumer pull and acceptance if these types of products are to be more widely adopted and integrated.

8. Integrate feedback loops to improve circularity

Information feedback loops are important for enabling closed-loop supply chains as they facilitate the gathering and sharing of data that can help improve product design and optimisation of the value within the supply chain (Koppius et al., 2011). If this is missing within companies or systems, this can result in lack of awareness of how upstream changes impact downstream processes. Therefore, in most cases, this means there is no incentive for businesses to consider how circular design strategies might improve the environmental or economic profile of their products or systems. As Organisation F remarked, currently within their organisation, there is *"no business driver or case for making improvements."* That, *"if [they] were to enhance the serviceability of the product,"* they would not be able to measure or offer evidence that *"it would save [them] so and so money"* or not. They argued that this *"feedback loop is missing"* and that *"it could be better, more visible."*

In addition to internal feedback loops, external feedback loops can be useful for improving the circular design of the product as well. As shown with Organisation E, they gather feedback from their consumers regarding any issues that arise as result of circular design changes made to the product. This not only allows them to continue to adapt and improve circularity without affecting satisfaction, but also offers the company the opportunity to provide a rationale behind their design decisions to the consumer. Similarly, Organisations B and D gather feedback information from service engineers to design to improve the ease of repair and refurbishment of their products as well. Therefore, organisations implementing circular design should consider implementing initiatives that allow them to gather, measure and input data upstream, as ultimately this could help companies to make more informed, intelligent decisions regarding the sustainability and circularity of their products.

Questions to contemplate further from this section

How do we encourage consumers to share data to make products more circular?
How can we capture data on our products to improve decision making?
What skills should we learn to improve collaboration?
What does making products 'circular economy ready' entail?

Discussion and conclusion

Pace of change: are we accelerating the right parts of the system?

Climate research is urgently calling for a radical redesign of how we deliver value through products and services, if we are to slow the impact of climate change (IPCC, 2018). Therefore, we must be mindful of which parts of the system we are addressing and to what degree these will impact the pace of change for this transition. It could be argued thus far that most of the circular thinking has been applied to the 'low-hanging fruit' opportunities (i.e. material innovation for recycled, renewable and recyclable content and the servitisation of high value products). While these are important activities towards making the transition, they are arguably incremental steps of innovation and we are still predominantly operating in the business-as-usual mindset, selling one-time products to consumers and relying on downstream actors to manage the problem. These pioneering organisations are taking steps towards mitigating these issues, yet there are few circular products available on the market; many barriers and challenges still to overcome and there are many more stakeholders within the system who are yet to begin their journey. It is important to note, however, that most organisations we spoke to are large multinational organisations and are undergoing a transition from linear to circular offerings, rather than building in this approach from the start, which inevitably will take longer.

Therefore, reflecting on the 'eight levers for change' identified:
1. Be a catalyst for change
2. Integrate downstream circularity
3. Be proficient in collaboration
4. Have collective circular comprehension – pull in the same direction
5. Build up circular capabilities – knowledge, tools and metrics
6. Tweak and totally transform – where to start
7. Consider the circular consumer
8. Integrate feedback loops to improve circularity

It is proposed that if these were to be addressed more significantly, they might help to unlock some of the sticking points that organisations are experiencing, which could not only mitigate the classic barriers such as proving the business and scalability, but also potentially speed up the transition to more circular products. However, it is also recognised that this is not an exhaustive list of suggestions and recommendations for how to enable circular products. The focus of this study was on the implementation and advancement of circular products, from a design and organisational perspective. There are wider system enablers such as tax breaks and financial incentives that also could speed up the pace change within this area, such as those adopted in Sweden (Starritt, 2016).

Internal organisational challenges

As mentioned in the literature review, circular design principles are an evolution from previous environmental strategies and research. As such, this can diffuse understanding and perspectives on what being 'circular' or 'sustainable' might entail. While, to some, this might be seen as a matter of semantics – to create functional, cohesive circular product ecosystems – there needs to be a shared understanding about what all of us are trying to achieve. Though all the participants we spoke to are acting as CE champions inside their respective companies, most stated that circular thinking is still considered to be a niche concept; therefore, getting company-wide buy-in and comprehension is a huge challenge. To address this challenge, organisations need to firstly continue to build capabilities and knowledge around circular economy topics so that each stakeholder understands how their function contributes towards building circular products systems. And secondly, encourage organisational leadership figures to galvanise their organisation towards this direction through measurable targets and internal incentives, which would not only help to inform priorities, but also keep everyone motivated and focused on the north star of what we are aiming for.

Externally managing circular product ecosystems

Value in a circular economy does not flow in one direction; it is something that is exchanged by system partners (Lancelott & Haines-Gadd, 2020). Managing the flow of value between stakeholders is a huge challenge, especially if there is lack of knowledge of sustainability topics amongst suppliers and customers. While some organisations are acting as catalysts for change, this is often only within their respective supply chains; but hopefully over time, as this thinking spreads, these practices will become more commonplace. Furthermore, although upstream innovation amongst some of these organisations is seemingly 'easier' to implement than the downstream processes, as we move more towards the concept of value constellations rather than value chains, perhaps this will become less of an issue.

The evolving role of the circular designer

The implementation of circular design was observed as an activity that goes beyond the designer and is a process that involves many more actors within the organisation.

While the design stage is still an important part of integrating circular principles, to make products truly circular, partnerships, networks, logistics and infrastructures need to be established to support the circular product eco-system. Although the role of the designer has evolved significantly over the last 30 years, as part of implementing circular thinking, it may need to continue to develop even further.

Beyond managing the environmental trade-offs between different strategies, implementing circular design strategies from a technical products design perspective was shown to be a relatively straightforward activity. It was the more system-level challenges such as collaboration between stakeholders or initiating reverse logistics that are causing more friction. As Organisation B observed, collaboration is hard and is a separate skill set that is needed from those implementing circularity. Therefore, perhaps designers and design thinkers, who are often considered to be facilitators, translators, communicators and systems thinkers, could be the actors needed to make this circular transition occur more smoothly, ultimately expanding the role of the designer over time, requiring them to become, in a sense, 'circular design thinkers'.

Limitations

This study was only conducted with six companies, most of which are large-scale organisations. To advance this perspective of circular design within the industry, further interviews with smaller organisations should be conducted.

Conclusion

In conclusion, within this chapter, we sought to understand what progress has circular design made within the industry. Through interviews with circular design practitioners who are implementing these into products, it was observed that there are many 'classic drivers and barriers' that are still present, both enabling and hindering the adoption of circular thinking, such as legislation or sustainability leadership and proving the business case. However, there were also new insights that have not been discussed within the literature and have been proposed as the 'eight levers for change' that, if addressed more significantly, could help accelerate the transition to more circular product ecosystems. We need organisations to be catalysts for change within supply chains and consider how to integrate downstream circularity as well as upstream circularity. They should increase their proficiency in collaboration, ensure there is a collective understanding of what circular means and to continue to develop their capabilities in circular practices. Organisations should both tweak and totally transform their products, depending on what it is, consider the circular consumer experience and, lastly, ensure that feedback loops are integrated to help drive improvements. It is through these actions that we might finally get to a space where being circular is not niche, but the new business-as-usual approach.

References

Bakker, C., Wang, F., Huisman, J., & Den Hollander, M. (2014). Products that go round: Exploring product life extension through design. *Journal of Cleaner Production, 69*, 10–16. https://doi.org/10.1016/j.jclepro.2014.01.028

Bakker, C., & Schuit, C. S. C. (2017). The long view – Exploring product lifetime extension. *UN Environment*. https://doi.org/10.2307/2961837

Bhamra, T., & Lofthouse, V. (2007). *Design for sustainability – A practical approach*. Aldershot: Gower.

BAM. (2021). *BAM Impact Positive report 2021*. Available at: https://bambooclothing.co.uk/wp-content/uploads/BAM-Impact-Positive-Report-2021-v11.pdf. Accessed October 25, 2021.

Baxter, W., Aurisicchio, M., & Childs, P. R. N. (2017). Contaminated interaction: another barrier to circular material flows. Journal of Industrial Ecology, *21*, 507–516. https://doi.org/10.1111/jiec.12612

Bocken, N. M. P., de Pauw, I., Bakker, C., & van der Grinten, B. (2016). Product design and business model strategies for a circular economy. *Journal of Industrial and Production Engineering, 33*(5), 308–320. https://doi.org/10.1080/21681015.2016.1172124

Boks, C., & Stevels, A. (2007). Essential perspectives for design for environment. experiences from the electronics industry. *International Journal of Production Research, 45*(18–19), 4021–4039. https://doi.org/10.1080/00207540701439909

Boyer, R. H. W., Mellquist, A.-C., Williander, M., Fallahi, S., Nyström, T., Linder, M., Algurén, P., Vanacore, E., Hunka, A. D., Rex, E., & Whalen, K. A. (2021). Three-dimensional product circularity. Journal of Industrial Ecology. https://doi.org/10.1111/jiec.13109

Bracquené, E., Dewulf, W., & Duflou, J. R. (2020). Measuring the performance of more circular complex product supply chains. *Resources, Conservation and Recycling, 154* (November 2019), 104608. https://doi.org/10.1016/j.resconrec.2019.104608

Camacho-Otero, J., Boks, C., & Pettersen, I. N. (2019). User acceptance and adoption of circular offerings in the fashion sector: Insights from user-generated online reviews. *Journal of Cleaner Production, 231*, 928–939. https://doi.org/https://doi.org/10.1016/j.jclepro.2019.05.162

Camacho-Otero, J., Tunn, V. S. C., Chamberlin, L., & Boks, C. (2020). Consumers in the circular economy. Handbook of the Circular Economy, November, 74–87. https://doi.org/10.4337/9781788972727.00014

Chamberlin, L. (2021). Transforming Consumption: design for engagement, meaning and action in a circular economy. Norwegian University of Science and Technology.

Caterpillar. (2020). *Circular economy*. Available at: https://www.caterpillar.com/en/company/sustainability/remanufacturing.html. Accessed October 25, 2021.

Jonathan., C., & Gant, N. (2007). *Designers, visionaries and other stories: A collection of sustainable design essays*. London: Sterling, VA: Earthscan.

Charnley, F., Lemon, M., & Evans, S. (2011). Exploring the process of whole system design. *Design Studies, 32*(2), 156–179. https://doi.org/10.1016/j.destud.2010.08.002

Chick, A. E., & Micklethwaite, P. (2011). *Design for sustainable change.* Ava Publishing, Switzerland.

Circle Economy. (2020). *Our world is now only 8.6% circular* Available at: https://www.circle-economy.com/news/our-world-is-now-only-8-6-circular. Accessed October 25, 2021.

DELL. (2020). *Circular design – Revolutionizing how we use resources.* https://corporate.delltech nologies.com/en-gb/social-impact/advancing-sustainability/sustainable-products-and-services/circular-design.htm

Dokter, G., Thuvander, L., & Rahe, U. (2021). How circular is current design practice? Investigating perspectives across industrial design and architecture in the transition towards a circular economy. *Sustainable Production and Consumption, 26,* 692–708. https://doi.org/10.1016/j.spc.2020.12.032

Dokter, G., van Stijn, A., Thuvander, L., & Rahe, U. 2020. Cards for circularity: Towards circular design in practice. *IOP Conference Series: Earth and Environmental Science, 588*(4). https://doi.org/10.1088/1755-1315/588/4/042043

Ellen MacArthur Foundation. (2013). Towards the circular economy: Opportunities for the consumer goods sector vol 2. *Ellen MacArthur Foundation,* 1–112. https://doi.org/10.1162/108819806775545321

Ellen MacArthur Foundation. (2017). *Upstream Innovation a guide to packaging solutions.* Available at: https://plastics.ellenmacarthurfoundation.org/upstream.

European Commission. (2015). *Closing the loop – An EU action plan for the Circular Economy.* https://doi.org/10.1016/0022-4073(67)90036-2

Forum for the Future. (2016). *Design for demand.* Available at: http://designfordemand.forumforthe future.org/. Accessed October 25, 2021.

Geissdoerfer, M., Savaget, P., Bocken, N. M. P., & Hultink, E. J. (2017). The circular economy – A new sustainability paradigm?. *Journal of Cleaner Production, 143,* 757–768. https://doi.org/10.1016/j.jclepro.2016.12.048

Ghisellini, P., Cialani, C., & Ulgiati, S. (2016). A review on circular economy: The expected transition to a balanced interplay of environmental and economic systems. *Journal of Cleaner Production, 114,* 11–32. https://doi.org/10.1016/j.jclepro.2015.09.007

Glogic, E., Sonnemann, G., & Young, S. B. (2021). Environmental trade-offs of downcycling in circular economy: Combining life cycle assessment and material circularity indicator to inform circularity strategies for alkaline batteries. Sustainability (Switzerland), *13*(3), 1–12. https://doi.org/10.3390/su13031040

Guest, G., MacQueen, K. M., & Namey, E. E. (2011). *Applied thematic analysis.* SAGE Publications. https://books.google.co.uk/books?id=Hr11DwAAQBAJ

Haines-Gadd, M., Chapman, J., Lloyd, P., Mason, J., & Aliakseyeu, D. (2018). Emotional durability design nine-a tool for product longevity. *Sustainability (Switzerland), 10*(6), 1–19. https://doi.org/10.3390/su10061948

H&M. (2019). *Sustainability performance report 2019.* Available at: https://hmgroup.com/wp-content/uploads/2020/10/HM-Group-Sustainability-Performance-Report-2019.pdf. Accessed October 25, 2021.

Den Hollander, M. C., Bakker, C. A., & Hultink, E. J. (2017). Product design in a circular economy: Development of a typology of key concepts and terms. *Journal of Industrial Ecology, 21*(3), 517–525. https://doi.org/10.1111/jiec.12610

IDEO and Ellen MacArthur foundation. (2017). *The circular design guide.* Available at: https://www.circulardesignguide.com/. Accessed October 25, 2021.

Inter IKEA Group. (2017). Inter Ikea sustainability summary report FY17. Inter Ikea group. Retrieved at: https://preview.thenewsmarket.com/Previews/IKEA/DocumentAssets/502623.pdf

Inter IKEA Group. (2019). IKEA Sustainability Report FY19. Inter Ikea group. Retrieved at: https://about.ikea.com/en/newsroom/2020/02/27/breaking-the-trend-ikea-reports-a-decrease-in-climate-footprint

IPCC, 2018: Summary for Policymakers. In: Global Warming of 1.5°C. An IPCC Special Report on the impactsof global warming of 1.5°C above pre-industrial levels and related global greenhouse gas emission pathways, in the context of strengthening the global response to the threat of climate change, sustainable development, and efforts to eradicate poverty [Masson-Delmotte, V., P. Zhai, H.-O. Pörtner, D. Roberts, J. Skea, P.R. Shukla, A. Pirani, W. Moufouma-Okia, C. Péan, R. Pidcock, S. Connors, J.B.R. Matthews, Y. Chen, X. Zhou, M.I. Gomis, E. Lonnoy, T. Maycock, M. Tignor, and T. Waterfield (eds.)]. Cambridge University Press, Cambridge, UK and NewYork, NY, USA, pp. 3–24. https://doi.org/10.1017/9781009157940.001.

ISO/TR 14062. (2002). ISO/TR 14062:2002 Environmental management -- Integrating environmental aspects into product design and development. Retrieved at: https://www.iso.org/obp/ui/#iso:std:iso:tr:14062:ed-1:v1:en

Keskin, D., Diehl, J. C., & Molenaar, N. (2013). Innovation process of new ventures driven by sustainability. *Journal of Cleaner Production*, *45*, 50–60. https://doi.org/https://doi.org/10.1016/j.jclepro.2012.05.012

Koppius, O., Ozdemir, O., & van der Laan, E. (2011). Beyond waste reduction: Creating value with information systems in closed-loop supply chains. *ERIM report series research in management*. https://papers.ssrn.com/sol3/papers.cfm?abstract_id=1961468

Konietzko, J., Bocken, N., & Hultink, E. J. (2020). A Tool to Analyze, Ideate and Develop Circular Innovation Ecosystems. *Sustainability*, *12*(1), 417. MDPI AG. Retrieved from http://dx.doi.org/10.3390/su12010417

Lofthouse, V., & Prendeville, S. (2017). Considering the User in the Circular Economy. Product Lifetimes and The Environment - PLATE 2017 Conference Proceedings - 8–10 November.

Lancelott, M., & Haines-Gadd, M. (2020). CIRCULAR BUSINESS MODEL DESIGN GUIDE. PA Consulting.

Linder, M., Sarasini, S., & van Loon, P. (2017). A metric for quantifying product-level circularity. *Journal of Industrial Ecology*, *21*(3), 545–558. https://doi.org/10.1111/jiec.12552

Ljungberg, L. Y. (2007). Materials selection and design for development of sustainable products. *Materials & Design*, *28*(2), 466–479. https://doi.org/https://doi.org/10.1016/j.matdes.2005.09.006

Ma, X., & Moultrie, J. (2017). What stops designers from designing sustainable packaging? – A review of eco-design tools with regard to packaging design. *Smart Innovation, Systems and Technologies, 68*, 127–139. https://doi.org/10.1007/978-3-319-57078-5_13

Mesa, J., Esparragoza, I., & Maury, H. (2018). Developing a set of sustainability indicators for product families based on the circular economy model. *Journal of Cleaner Production, 196*, 1429–1442. https://doi.org/10.1016/j.jclepro.2018.06.131

Mhatre, P., Panchal, R., Singh, A., & Bibyan, S. (2021). A systematic literature review on the circular economy initiatives in the European Union. *Sustainable Production and Consumption, 26*, 187–202. https://doi.org/10.1016/j.spc.2020.09.008

Millward-Hopkins, J., Busch, J., Purnell, P., Zwirner, O., Velis, C. A., Brown, A., . . . Iacovidou, E. (2018a). Fully integrated modelling for sustainability assessment of resource recovery from waste. *The Science of the Total Environment, 612*, 613–624. https://doi.org/10.1016/j.scitotenv.2017.08.211

Millward-Hopkins, J., Busch, J., Purnell, P., Zwirner, O., Velis, C. A., Brown, A., & Iacovidou, E. (2018b). Fully integrated modelling for sustainability assessment of resource recovery from waste. *Science of the Total Environment, 612*, 613–624. https://doi.org/10.1016/j.scitotenv.2017.08.211

Moreno, M. A., Ponte, O., & Charnley, F. (2017). Taxonomy of design strategies for a circular design tool. Product Lifetimes and The Environment - PLATE 2017 Conference Proceedings - 8–10 November, 275–279. https://doi.org/10.3233/978-1-61499-820-4-275

MI-ROG. (2020). The Circular Economy and Net Zero Carbon. Circular Economy approaches to reaching net zero carbon infrastructure | White paper. https://aecom.com/content/wp-con tent/uploads/2020/08/MI-ROG_white-paper-4_circular-economy_2020.pdf

Moreno, M., De Los Rios, C., Rowe, Z., & Charnley, F. (2016). A conceptual framework for circular design. *Sustainability (Switzerland), 8*(9). https://doi.org/10.3390/su8090937

Mugge, R., Schoormans, J. P. L., & Schifferstein, H. N. J. (2005). Design strategies to postpone consumers product replacement: The value of a strong person-product relationship. *The Design Journal, 8*(2), 38–48. https://doi.org/10.2752/146069205789331637

Mugge, R., Schoormans, J. P. L., & Schifferstein, H. N. J. (2009). Emotional bonding with personalised products. *Journal of Engineering Design, 20*(5), 467–476. https://doi.org/ 10.1080/09544820802698550

Mugge, R., Safari, I., & Balkenende, A. R. (2017). Is there a market for refurbished toothbrushes?: An exploratory study on consumers' acceptance of refurbishment for different product categories. Product Lifetimes and The Environment - PLATE 2017 Conference Proceedings - 8–10 November.

Nes, N. V., & Cramer, J. (2005). Product lifetime optimization. *A Challenging Strategy Towards More Sustainable Consumption Patterns, 14*, 1307–1318. https://doi.org/10.1016/j. jclepro.2005.04.006

Nußholz, J. L. K. (2018). A circular business model mapping tool for creating value from prolonged product lifetime and closed material loops. *Journal of Cleaner Production, 197*, 185–194. https://doi.org/10.1016/j.jclepro.2018.06.112

Philips. (2020). *Annual report.* Available at: https://www.annualreports.com/HostedData/Annual Reports/PDF/NYSE_PHG_2020.pdf Accessed October 25, 2021.

Prendeville, S., O'Connor, F., & Palmer, L. (2014). Material selection for eco-innovation: SPICE model. *Journal of Cleaner Production, 85*, 31–40. https://doi.org/10.1016/j. jclepro.2014.05.023

Ritchie, J., & Ormston, R. (2013). The Applications of Qualitative Methods to Social Research. In J. Ritchie, J. Lewis, C. McNaughton Nicholls, & R. Ormston (Eds.), Qualitative Research Practic: A Guide for Scoial Science Students and Researchers (pp. 27–44). SAGE Publications. London.

Rizos, V., Egenhofer, C., & Elkerbout, M. (2019). Circular economy for climate neutrality: Setting the priorities for the EU. CEPS Policy Brief, *1*(2019), 1–11.

Salari, M., & Bhuiyan, N. (2018). A new model of sustainable product development process for making trade-offs. *The International Journal of Advanced Manufacturing Technology, 94*(1), 1–11. https://doi.org/10.1007/s00170-016-9349-y

Sheldrick, L., & Rahimifard, S. 2013, April 17–19. *Evolution in ecodesign and sustainable design methodologies.* Proceedings of 20th CIRP international conference on life cycle engineering, Singapore.

Shevchenko, T., & Kronenberg, J. (2020). Management of material and product circularity potential as an approach to operationalise circular economy. *Progress in Industrial Ecology, 14*(1), 30–57. https://doi.org/10.1504/PIE.2020.105193

Signify. (2020). *Annual report.* Available at: https://www.signify.com/static/2020/signify-annual-report-2020.pdf. Accessed October 25, 2021.

Sumter, D., Bakker, C., & Balkenende, R. (2018). The role of product design in creating circular business models: A case study on the lease and refurbishment of baby strollers. *Sustainability (Switzerland), 10*(7). https://doi.org/10.3390/su10072415

Starritt, A. (2016). *Sweden is paying people to fix their belongings instead of throwing them away.* Weforum.Org. 2016. https://www.weforum.org/agenda/2016/10/sweden-is-tackling-its-throwaway-culture-with-tax-breaks-on-repairs-will-it-work/

Tseng, M. L., Chiu, A. S. F., Tan, R. R., & Siriban-Manalang, A. B. (2013). Sustainable consumption and production for Asia: Sustainability through green design and practice. *Journal of Cleaner Production, 40*(1–5). https://doi.org/10.1016/j.jclepro.2012.07.015

The Great Recovery. (2013). Investigating the role of design in the circular economy (pp. 1–46). Retrieved at: https://www.thersa.org/globalassets/images/projects/rsa-the-great-recovery-report_131028.pdf

Vallet, F., Eynard, B., Millet, D., Mahut, S. G., Tyl, B., & Bertoluci, G. (2013). Using eco-design tools: An overview of experts practices. *Design Studies, 34*(3), 345–377. https://doi.org/10.1016/j.destud.2012.10.001

Vanegas, P., Peeters, J. R., Cattrysse, D., Tecchio, P., Ardente, F., Mathieux, F., & Duflou, J. R. (2018). Ease of disassembly of products to support circular economy strategies. *Resources, Conservation and Recycling, 135*(July 2017), 323–334. https://doi.org/10.1016/j.resconrec.2017.06.022

Saskia van den Muijsenberg

9 Biomimicry and the Circular Economy

Abstract: This chapter serves as an introduction to how the discipline of biomimicry – innovation inspired by nature – can guide the Circular Economy (CE). It gives a broad overview of where and how biomimicry can be applied to inspire and help (business) to accelerate the transition towards a CE and ensure restorative and regenerative outcomes. It offers ideas on how biomimicry can be applied for both technological innovation (with a focus on materials and life-friendly chemistry) and organisational innovation (looking at ecological functioning).

Biomimicry makes it possible to rethink the CE. Unlike some visions of the CE, with its simple and closed loops, biomimicry allows us to see that the CE of nature works with continuous, open-ended and often complex loops. Finally, it is argued how we can and should take nature, not only as a model for innovation but also as a measure or benchmark, for our business performance.

Keywords: biomimicry, regenerative design, circular economy, life's principles, materials, life-friendly chemistry, ecological functioning, emergence

Introduction

The demand for alternative ways of production, with the use of recyclable and biodegradable materials, and managing the flow of materials and energy, is increasing because of the growing pressures on resources and the environment. It is in this context that the CE is currently one of the most discussed terms and a focus of the European Union Horizon 2020 strategy. It is argued that "in a world with growing pressures on resources and the environment, the EU has no choice but to go for the transition to a resource-efficient and ultimately regenerative circular economy" (European Commission, 2012, p. 1).

The CE claims to be restorative and regenerative by design because it mimics natural cycles, in which all resources flow continuously in loops. However, few enterprises have been able to reinvent themselves to take full advantage of the CE's potential. In the search to come to useful and practical ways to stimulate the transition to a CE, this chapter seeks to provide some perspectives on the discipline of biomimicry, as we understand and practice it, and link this to the subject of CE. The author chose to provide a broad overview of where and how biomimicry can be applied to inspire and help (business) to accelerate the transition towards a CE and ensure restorative and regenerative outcomes. The objective of this chapter is to establish a coherent basis for further academic discussion as well as to allow practitioners to better understand how they can use biomimicry to inform the CE, and act within it.

https://doi.org/10.1515/9783110723373-012

The chapter starts with an overview of what biomimicry is, its essential elements and how biomimicry can be connected to CE. It then presents the core framework of biomimicry: the 'Life's Principles' and how they can guide the CE. It starts by first zooming in on some principles, especially for resource efficiency (materials and energy) and the chemistry to address technological innovation (in CE), and finally zooming out to showcase how biomimicry could also be used to address organisational challenges, as they are occurring in the CE.

Circular Economy: schools of thought

One of the main advocates of the CE is The Ellen MacArthur Foundation (EMF), a non-profit organization that strives to develop the framework of the CE. On its website, the foundation states that the concept of CE has deep-rooted origins and cannot be traced back to one single date or author, and lists the 'schools of thought' and authors who have contributed to refining and developing the CE economy. Amongst these are the blue economy (Pauli, 2010), cradle-to-cradle (McDonough & Braungart, 2002), performance economy (Stahel & Reday, 1976; Stahel, 2010), natural capitalism (Hawken, A. Lovins and H. Lovins), regenerative design (Lyle, 1996), industrial ecology (Lifset & Graedel, reverse logistics (Agrawal et al., 2015), and biomimicry (Benyus, 1997).

What is biomimicry

Biomimicry can be understood as emulating or being inspired by nature's designs to develop environmentally sustainable innovations (Reap et al., 2005). The premise is that Homo sapiens are a young species and that other species, with the same needs and subject to the same set of operating conditions of planet Earth, have developed well-adapted strategies and mechanisms to meet those needs, over millions of years of evolution. It has been proposed that we can investigate and assess the many regenerative models that have already emerged within living systems over 3.8 billion years of evolution on Earth, for their applicability to design of products, processes and systems (van den Muijsenberg et al., 2013). When we understand nature's mechanisms and deep principles, we can also learn how to fit ourselves on this planet for the long haul. Benyus (1997) states that the goal of biomimicry is to create products, processes, policies and systems that function like natural components of the ecosystem, and, like nature, 'create conditions conducive to life.'

Within the field of biomimicry, it is important to address distinctions in terms, for clarity and systematic literature review (Hayes et al., 2020). As outlined in BS ISO18458:2015 Biomimetics – Terminology, concepts and methodology (BSI Standards, 2015), 'Biomimicry' refers to the *"philosophy and interdisciplinary design approaches, taking nature as a model, to meet the challenges of sustainable development (social, environmental, and economic)."* Within this framing, 'Biomimetics' (including 'biomimetic design') describes the *"interdisciplinary cooperation of biology and technology or other fields of innovation, with the goal of solving practical problems through the function analysis of biological systems, their abstraction into models, and the transfer into and application of these models to the solution"* (BSI Standards, 2015). 'Bionics' then refers to the *"technical discipline that seeks to replicate, increase, or replace biological functions by their electronic and/or mechanical equivalents"* (BSI Standards, 2015).

Other terms associated with biomimicry are bio-inspired design, nature-inspired design, biologically informed design, ecological design, ecomimicry, etc. Mead (2014) argues that while practitioners use the many terms interchangeably, there are effectively two schools of thought that divide the bio-inspired innovation space. The first is the use of biological models to guide design solutions because of the unique and outstanding physical properties of biological systems and, without explicit considerations, for the ethical implications of the outcomes; for example, just for the advancement of technology. The second is the use of biological models to guide design solutions to solve challenges of sustainability and human adaptability. Table 9.1 provides an overview of some of the terms and their intentions/goals.

Table 9.1: Historical timeline of the emergence of bio-inspired design.

Term (year)	Attributed to	Training of originator	Definition (where available)	Connection to sustainability
Bionics (1958)	Jack E. Steele	Medical doctor	"The science of systems, which have come function copied from nature, or which represent characteristics of natural systems or their analogues." [42]	No
Biomimetics (1969)	Otto Schmidt	Engineer/ Biophysicist	"The study of the formation, structure, or function of biologically produced substances and materials (as enzymes or silk) and biological mechanisms and processes (as protein synthesis or photosynthesis) especially for the purpose of synthesizing similar products by artificial mechanisms which mimic natural ones." [43]	No
Design with Nature (1969)	Ian McHarg	Landscape architect	"The ecological view requires that we look upon the world, listen and learn." [39]	Yes

Table 9.1 (continued)

Term (year)	Attributed to	Training of originator	Definition (where available)	Connection to sustainability
Ecological Design (1970s)	John Todd	Biologist	Defined by a set of 9 principles in the book "From Eco-Cities to Living Machines: Principles of Ecological Design" (see [37]).	Yes
Biomimicry (1997)	Janine Benyus	Biologist	"An innovation method that seeks sustainable solutions to human challenges by emulating nature's time-tested phenomena, patterns and principles. The goal is to create well-adapted products, processes, designs and policies by mimicking how living organisms have survived and thrived over the 3.8 billion yeas life has existed on Earth." [44]	Yes
Ecomimicry (2007)	Alan Marshall	Social scientist	"Ecomimicry is the practice of designing social responsive and environmental responsible technologies for a particular locale based upon the characteristics of animals, plants and ecosystems of that locale." [41]	Yes
Nature-Inspired Design Strategies (2010)	Pauw et al.	Industrial designer	"Nature-inspired design strategies are design strategies that base a significant proportion of their theory on 'learning from nature' and regard nature as the paradigm of sustainability." E.g. biomimicry, cradle-to-cradle, and natural capitalism [3]	Yes
Biologically Informed Discipline (2014)	Alena Iouguina	Industrial designer	"The informed interpretation of biological research in order to address human challenges for the purpose of innovation that may or may not result in sustainable solutions." [37]	Yes and No

Source: Mead (2014).

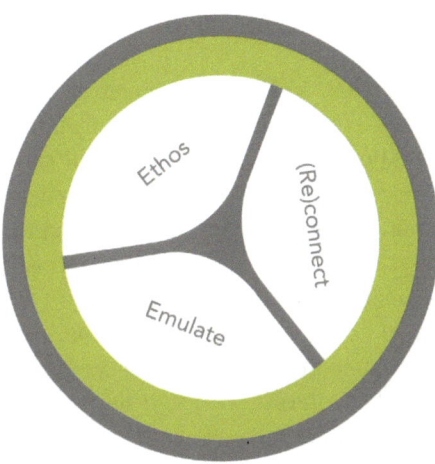

Figure 9.1: Summary of core principles for biomimicry.
Source: Biomimicry3.8, US-based biomimicry consulting group.

Three levels of biomimicry

Biomimicry does not necessarily yield environmentally sustainable solutions (Kennedy & Marting, 2016; Mead & Jeanrenaud, 2017). It depends on the level of application.

The practice of biomimicry embodies three interconnected, but unique ingredients; the three Essential Elements of Biomimicry (Figure 9.1) represent the foundation of the biomimicry meme. By combining the essential elements, bio-inspired design becomes biomimicry.

- The ethos element forms the essence of our ethics, our intentions, and our underlying philosophy for why we practice biomimicry. Ethos represents our respect for; responsibility to, and gratitude for our fellow species and our home.
- The (re)connect element reinforces the understanding that, while seemingly "separate", people and nature are actually deeply intertwined. We are nature. (Re)connecting is a practice and a mindset that explores and deepens this relationship between humans and the rest of nature.
- The emulate element brings the principles, patterns, strategies, and functions found in nature to inform design. Emulation is about being proactive in achieving the vision of humans fittin in sustainably on earth.

Biomimicry can be applied at different levels. The first level is the mimicking of the *natural form*. A good example for this is the innovation of Velcro. George the Mestral was inspired by the burrs he found on himself and on his dog after a walk. He mimicked the small hooks of the burr and loops of the fur/fabric that allowed the burr to adhere exceedingly well. Copying burr design is just the beginning because it may or may not yield something sustainable.

For a deeper biomimicry, a second level is added, which is the mimicking of the *natural process,* or how it is made. The burr self-assembles at ambient temperature without toxins or high pressures, by way of nature's chemistry. The field of green chemistry attempts to mimic these benign recipes. An example is a material made by the Wyss institute, called Shrilk, which is made from the exoskeletons of shrimps and is a replacement for traditional plastics. It is as strong as aluminium and only half the weight. It is biodegradable and biocompatible, and micro mouldable.

The third level is the mimicking of *natural ecosystems* – The burr is part of a plant, which is part of a forest that is part of a biome that is part of a sustaining biosphere. In the same way, Velcro should be part of a larger economy that works to restore and regenerate, rather than deplete. Even if Velcro was made using green chemistry, but you have underpaid workers making it in a sweatshop, and you ship it long-distances using heavy polluting trucks, you have missed the point. To mimic a natural system, you must ask how each product fits in – is it necessary, is it part of a nourishing food web of industries, and can it be transported, sold and reabsorbed in ways that foster a forest-like economy? (Van den Muijsenberg et al., 2014).

Numerous authors highlight the importance and opportunity to shift towards the ecosystem level (Benyus, 2011; Pedersen Zari & Storey, 2007), as an application at the 'form' and 'process' level do not necessarily yield sustainable solutions, especially, as there is a tendency to emulate a few select features of a particular organism. A more holistic interpretation and practice of biomimicry is promoted – one that considers the application of biomimetic solutions across multiple spatial, temporal and organisational scales (de Pauw et al., 2014; Hayes et al., 2020; Pedersen Zari & Storey, 2007).

Connection to CE

Dicks states that "Biomimicry is often reduced to a narrow vision that seeks to emulate the 'innovations' of different species" (2018). Others see biomimicry as a narrow subset of CE (Geisendorf & Pietrulla, 2018); as a specific tool to reduce the environmental impact of the industry by substituting with renewable resources and natural processes to create significantly more environmentally benign industrial processes (Bocken et al., 2014).

Most biomimicry practitioners have a broader view and argue that there is more to be gained by looking at the functioning of ecology. The latter is based on two major principles: the use of renewable energies and recycling. They state that the circular economy is a biomimetic concept at the systems level, inspired by the flow of resources and energy within ecosystems (The Biomimicry Institute, Biomimicry Norway; Wautelet, 2018). In their view, a circular economy already exists in nature.

There is no waste in nature. The waste of one life form is food for another. All resources flow continuously through natural cycles, with the involvement of animals, plants, fungi, bacteria and protists (Margulis, 2009). All other life forms engage in a circular way of operation; the human species is the only one that doesn't fit this circular system. So conceptually, the CE can be seen as an example of biomimicry at the systems level. And, biomimicry makes it possible to rethink the circular economy. Unlike some visions of the circular economy with its simple and closed loops, biomimicry allows us to see that the circular economy of nature works open-ended and often in complex loops. Biomimicry calls for rethinking the current model of the circular economy, including advocating more open-ended, complex and flexible loops (Dicks, 2018).

Next, biomimicry can also be seen as a design discipline, a branch of science, a problem-solving method, a sustainability ethos, a movement, a stance towards nature, a new way of viewing and valuing biodiversity (Benyus, 2011).

Biomimicry as a departure point for CE innovation

Circular business is currently mainly limited to recycling, energy efficiency and raw materials reduction (Jonker et al., 2017). Recycling can be seen as a first step towards sustainability, but CE transitions based on higher circularity strategies call for more radical socio-institutional change (Constanza et al., 2015). Without coupling business practice to the Earth's planetary boundaries (Rockstrom, 2009), CE remains a metaphor.

Biomimicry is a source of design inspiration for a new industrial and economic paradigm that seeks to work with the laws of nature to identify solutions for sustainability challenges (Mead, 2017, p. 113). We suggest biomimicry as a departure point for CE innovation, product design and new business models and strategies, as biomimicry views nature both as a model for innovation as well as a measure/standard, which contains a normative or ethical principle for sustainability (Benyus, 1997; Blok et al., 2016; Dicks, 2017). In the end, biomimicry has the potential to change our worldview as well as our world.

Biomimicry innovation for the circular economy: materials and energy

> Biology, like technology, is reliant on materials for the structures it makes. These structures also have to be cheap and reliable. Evolutionary fitness (and therefore survival) is, in part, value based – the survival of the cheapest. Success requires the ability to compete for, and survive upon, resources that may be scarce. Julian Vincent (2010)

It is easy to see how this connects to the CE, especially replacing 'biology' with 'economy' and 'structures' with 'products (Schumer, 2011).' This sentence doesn't only address competition and scarce resources; it also mentions value and long-term continuity, with the concept of evolutionary fitness. These terms are at the core of CE as well. However, a CE is not only about the types of materials and energy. It is also about the *organisation of the flow* of these materials and resources at the systems level.

Life's principles

To help create well-adapted products, processes and systems (such as an economy), biomimicry uses a framework called 'life's principles' (Biomimicry Resource Handbook, 2013). This framework can be used as a design tool to ideate novel and sustainable ideas (based on the deep principles and strategies found throughout organisms and living systems) and as a benchmark to assess our designs against these sustainable benchmarks.

Life's principles are the deep principles of nature that fuel and inspire deep sustainability, or whatever is beyond that concept. These principles (Figure 9.2) are present in almost all organisms at multiple scales and levels. They are the deep criteria for thriving and surviving on Earth, while creating conditions conducive to life. It is through these principles that work is done to prevent superficial biomimicry, because each principle challenges humans to think systemically within a broader context, than a single organism. Every principle (with icons) is met by organisms through various strategies (inner section). By taking these strategies and principles as our aspirational goal, we too can begin to create 'conditions conducive to life.' The phrase 'life creates conditions conducive to life' is at the centre of the Life's Principles circle. This concept is the goal and the reason for using Life's Principles as a tool.

Our economy – whether it's linear or circular – is a subsystem of the global society, which in turn is embedded in nature. This embedded view brings about the notion that business, societal and biospheric systems are not only interrelated, but are also most realistically viewed as nested systems (Marcus et al., 2010). From this embedded systems perspective, CE cannot and should not be seen in isolation from its context. Life's strategies have evolved to function well in the same context that human designs, and the CE must – the context of our planet. Thus, the CE needs to fit within and meet the Earth's operating conditions (sunlight, water and gravity; dynamic non-equilibrium; limits and boundaries and cyclic processes) in a way that does not jeopardise future generations.

Life's principles provide strategies for how to deal with the operating conditions of system Earth. They are the success factors of evolution.

Figure 9.2 shows the six life's principles along with their underlying strategies, and Figure 9.3 provides some more detail to the strategies. It should be noted that just as no principle stands alone, all principles are interconnected. The case can be made for a rearrangement of the sub-principles and master principles. Practitioners can use the tool and the diagram as it suits them, yet should keep in mind that it is the integration and optimisation of the collective and collaborative suite of principles that yields life its successes.

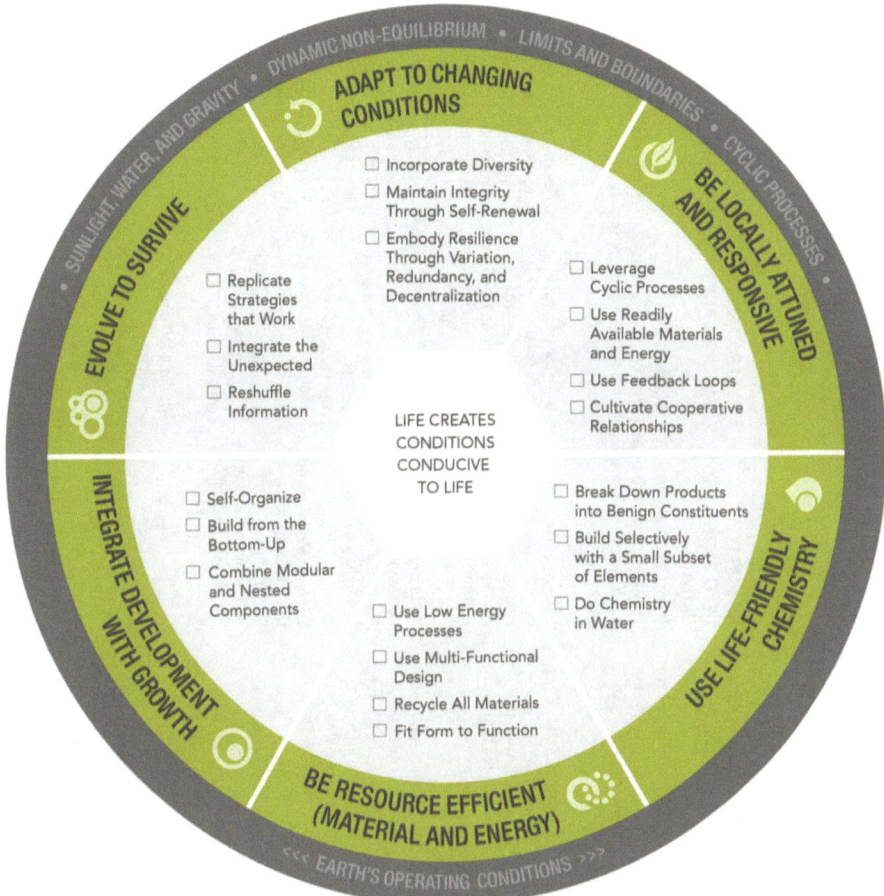

Figure 9.2: Life's Principles g6.
Source: Biomimicry 3.8 (2013).

As pointed out before, nature is reliant on materials for the structures (products) it makes. So are humans (since we are nature too) and so is an (circular) economy. Processing materials plays a very important role in nature, and all six life's principles (LPs) are in one way or other related to materials and materials processing.

Life's principle 5, be resource efficient (material and energy) and 6, use life friendly chemistry, both address materials and materials processing and give clear directions about what strategies could be applied. 'Recycle all materials' and 'use low-energy processes' may seem obvious, as well as 'fit form to function' and 'use multifunctional design.' However, when looking at the products in our own household, these strategies often have not been considered in the design process. Also, the ways these strategies are executed in biology are different from what is common

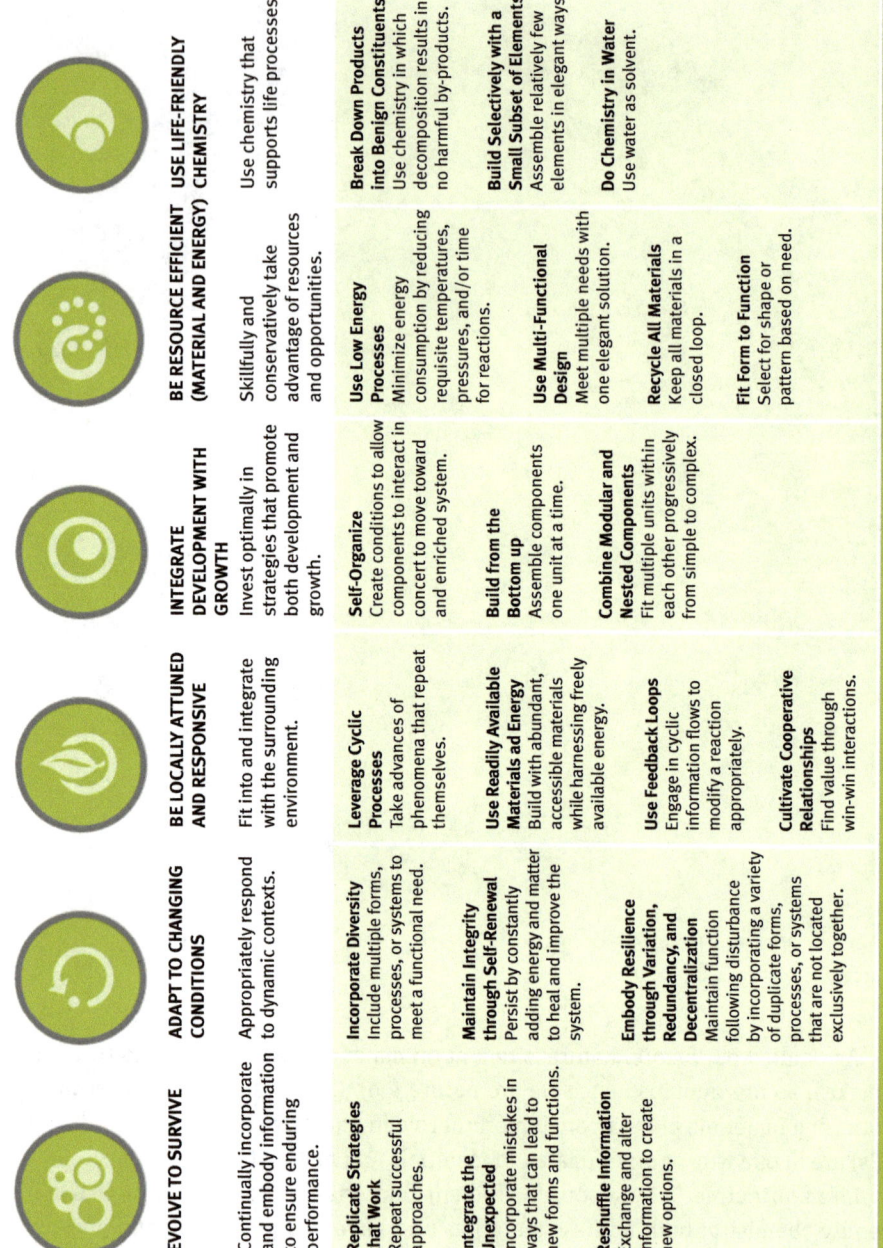

Figure 9.3: Life's principles.
Source: Biomimicry 3.8 (2013).

in our technology. This is covered in the next section. Making use of water-based chemistry is challenging for us. Using water as a solvent is difficult, and more research is desirable in this area. All life is based on water-based chemistry. Gaining a better understanding of how this works and could be mimicked would offer advantages for biological and chemical nutrient cycles.

(Human) Technology vs. nature

We view our human technology as very advanced. However, it doesn't always do a lot of favours for the long-term continuity of our species. Our industrial-age technologies are often ecosystem-destructing (Blok & Gremmen, 2016). We are currently depleting our resources at a speed that is far from sustainable. So what then are the differences between materials processing in (human) technology and nature?

Comparing materials used in nature and technology

Let's consider the elements from the periodic table that are used in nature and in human technology. All animals and plants are constructed from materials that are available in abundance, such as carbon, hydrogen, nitrogen and oxygen. Some of them combine them with the elements found on the second line of Figure 9.4. Even fewer organisms contain the elements on the third row, and the elements on the bottom line are found in very small amounts.

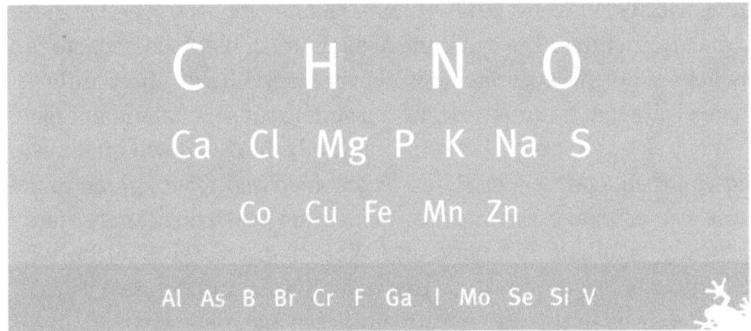

Figure 9.4: Elements found in organisms.
Source: Biomimicry 3.8.

In engineering, many heavy elements of the periodic table are used, which are the key for our technology and are rare or increasingly becoming rare (see Figure 9.5).

THE PERIODIC TABLE'S ENDANGERED ELEMENTS

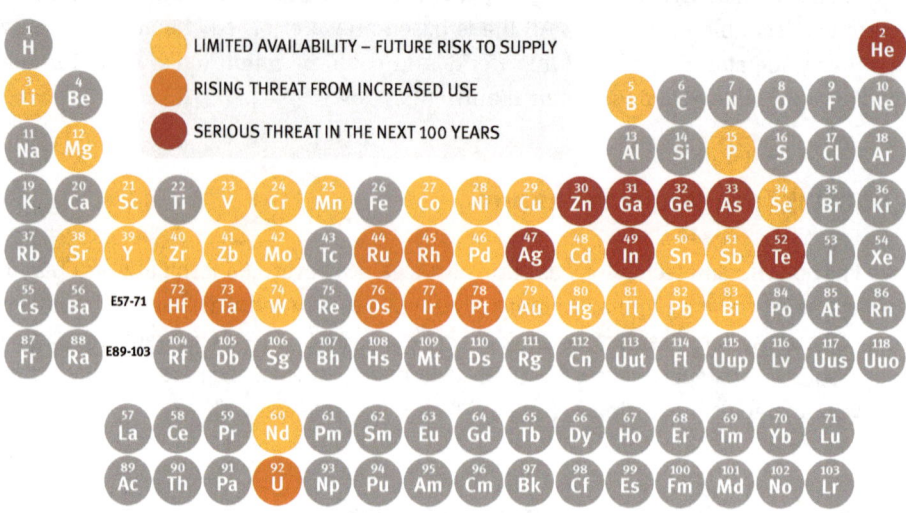

Figure 9.5: We are running out of key elements.
Source: ACS, Green Chemistry Institute.

Another difference is that nature uses just two basic building blocks (polymers; protein and polysaccharide) to create biological structures. These are so variable because of the information they contain, and they can achieve more than any man-made polymer. This information allows them to self-assemble into structures, such as the cell wall of a plant or the cuticle of an insect or lobster, which provide yet more functional versatility (Vincent, 2012). Human technology uses over 350 different polymers to make structures. The declining availability of these elements to produce these polymers is posing a problem. Moreover, recycling all these different polymers is a huge challenge. Nature works with a simple subset of abundant materials (because it requires less energy to obtain) that can be broken down into benign constituents and can then again endlessly be recombined and reconfigured to create something else. Our economy and business models are based on scarcity, nature works what is locally available in abundance. The simple (Lego-like) building blocks enable a large set of organisms to be part of food webs and natural cycles.

We also see differences in the way nature processes materials when compared to human technology. In nature, materials (e.g. spider silk and abalone shells) are processed at ambient temperatures and pressure with water as a solvent, whereas human-made materials require a lot of heat, beat and treat (Benyus, 1997). Another significant difference between biological materials and engineered materials is the fact that biological materials contain water, whereas that is usually absent in human-made materials. Water is often perceived as an agent for degradation. It would give us a lot of benefits to develop a system of materials synthesis and

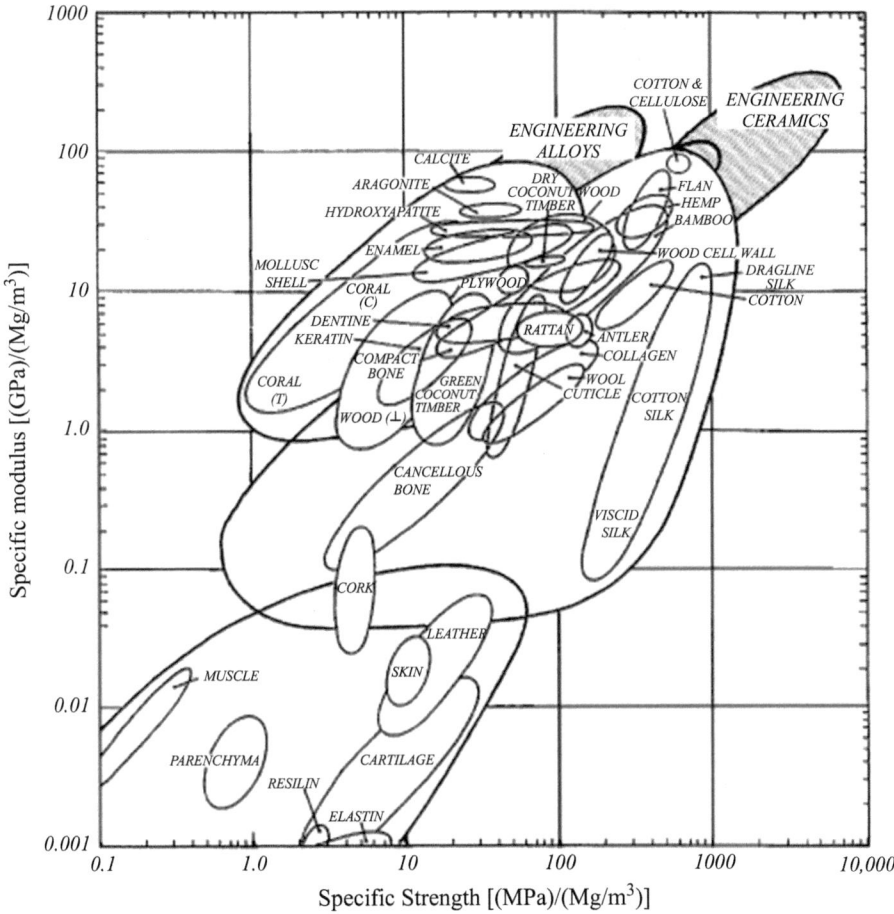

Figure 9.6: Comparison of specific properties of biological and engineering materials.
Source: Vincent (2008); Constructed from data by Wegst and Ashby.
Note: Both groups of materials are plotted, showing that they cover nearly the same ranges except for high-performance ceramics and alloys.

materials processing based on water. Water would then no longer be seen as the disruptive agent that it currently is. It would also rule out processes that currently require high temperatures, high pressures and toxic chemical solvents or reactants, such as in metal and ceramic production (J. Vincent, 2010).

There are also some similarities between natural and human-made materials. Figure 9.6 shows that natural and human-made materials, in terms of strength and stiffness, cover much the same area, indicating that they deliver the same mechanical properties (when density is considered). This indicates that we may be able to substitute quite a few technical loop materials for biological ones. This would probably require altering our materials processing processes.

Technology vs. biology in engineering and materials processing

Not only are materials used by nature and technology different, the way they are created/ processed is also different. How different are they really, what aspects are different and how could these insights contribute to CE?

Vincent et al. (2006) compared solutions to problems addressed through biology and (human) technology. By looking at things (substance, structure), doing things (energy, information), somewhere (time and space), many examples of problems in biology (2,500) and technology (5,000) were categorized according to the parameter, the source of the main control variable that solved the problem. The examples covered sizes ranging from molecular (nm) to civil engineering or ecosystem (km) scale. They found that biology solves the same types of problem that we solve with engineering, but the method of doing so is similar only 12% of the time. Vincent and colleagues had a more important result (see Figures 9.7 and 9.8).

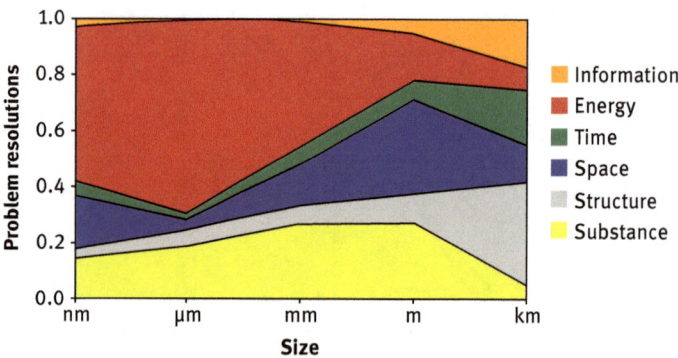

Figure 9.7: Problem resolution in technology.
Source: Vincent et al. (2006).

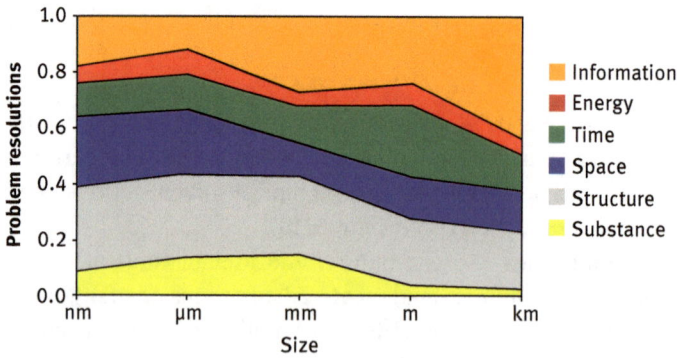

Figure 9.8: Problem resolution in biology.
Source: Vincent et al. (2006).

The two figures show that human technology relies heavily on the manipulation (not necessarily the amount) of energy to solve problems, especially at the μm level where energy is used as the main variable to solve up to 75% of the problems. In contrast, energy is the least-used controlling variable in biology (only 5% at any size). Energy is being replaced at the materials-processing level by information (15%) and structure (30%). This is an important finding; although we've known for many years that our technology is too reliant on energy, there has not been a proper indication of what might replace it (Vincent, 2012). To give a concrete example: a bag of crisps contains many different layers of materials to fulfil various functions, such as impact-resistance, water-tightness, airtightness, provide colour, etc. The exoskeleton of a beetle achieves these functions as well with just one material, chitin. It is because of the structural manipulation of the material that the functions are achieved. Another example: to colour something, we add pigments/heavy metals that end up in the living environment. Many birds and butterflies achieve colour through (nano)structure. The shape of the surface reflects light so that we perceive different colours. With biomimicry, interventions can be done early in the design process, reducing the need to recycle. Nature's golden rule: shape is cheap, material is expensive.

By learning from nature in the realm of materials and materials processing, we can do more with less resources, and make recycling less complicated – at least in terms of different materials that need to be recycled. When we make use of materials that are readily available, we don't need as much energy. We have seen in Figure 9.6 that the properties of biological materials are in the same range as those of human-made materials. However, biological materials are processed differently. Doing more research in this area could be a big contribution to CE.

Biomimicry and rethinking the current model of the circular economy

As pointed out earlier, biomimicry may also be of use when it comes to the 'organisational challenges' that come with a CE, and calls for rethinking the current model of the circular economy.

Biomimicry shows that the current thinking on the CE is too limited. It is too focused on material cycles and old economic principles. The role of natural ecosystems is missing herein (Jonker & van den Muijsenberg, 2016).

Cycles and ecosystems

Virtually all ecosystems in nature, no matter how different, have the same organisation. Producers (e.g. trees) convert CO_2 and water with the aid of sunlight into sugars. Primary consumers (e.g. butterflies) eat plants and take advantage of the sugars made by the plants. Secondary consumers (carnivores, such as the cuckoo), eat the primary consumers and take advantage of the energy they have stored in their bodies. And decomposers (like the dung beetle) are garbage eaters, processing organic waste into minerals for plants. So these are 'prosumers' at its best, because everyone is a producer and a consumer at the same time! But, how to translate these ideas into the organisation of daily life? The following three aspects of biomimicry place the CE in a different perspective.

In the CE, we will produce less

Some see only advantages of CE. It would create new jobs and new economic growth. Others say that CE is more about a shrinking economy. It is, after all, about the efficient use of raw materials, resulting in extraction of fewer raw materials. This thinking in the CE is still based on the current one-dimensional thinking about economic growth. In our economy, there are no decomposers. The development of that function offers enormous potential. If that function is integrated well, we will not produce less, but produce different. Production will get the real meaning of 'adding value.' Because there will be 'produced' at any point in the cycle. What is shrinking is the mining of raw materials and waste. Applying that to the CE can lead to economic development – translated into other types of work. The organisation of decomposition demands more work in the area of breaking down, processing and upcycling materials and raw materials to make them of constant value. The challenge is to do so, as in nature, by reducing material flows to a few 'Lego bricks.' By interacting these 'simple' components with each other, both the ecological and economic system will be richer. That is an investment in innovative development and stable and sustained growth.

Extend the lifetime of products

A common view is that the CE aims to make the lifetime of products as long as possible to keep precious resources. Who looks at nature's cycles and wonders: why should the lifetime be as long as possible? We're missing a golden rule of life here: 'life meets function in context.' Take a water bottle that you buy at the station because you're thirsty. That bottle is empty within 15 min, but it takes forever before it is decomposed. Within the CE, we do think in cycles over time, but our materials are not suited for that yet.

Not all acorns become oaks

Products in the CE that have reached their end of lifetime contain raw materials for new products. Collection, disassembly and recycling ensure a continuous flow.

Reuse is the cornerstone of the CE. But our thinking about this is limited. In our economy, plastic turns into plastic, and glass becomes glass. The acorns from an oak tree will almost never become an oak tree again. Because nature works with only a few basic building blocks, it is much easier to break materials and resources down and reuse them again. By combining and reconfiguring them differently, endless opportunities are created. This makes scaling-up and recycling much easier.

Finally, on ecosystems

Ecosystems contain emergent properties; they 'arise' from the interactions between actors in the system. No player oversees the ecosystem nor can find solutions on its own, solutions that the ecosystem as a whole can indeed create. That is an idea that calls for emulation. The organisational quest of the CE would have to focus on how we can make it emerge, rather than design and impose it top-down. The Life's Principles are a helpful tool for this thought. When all our products, processes and systems integrate, we are creating conditions conducive to life through the six principles.

Next to looking at ecosystems as models for organising the CE, we can also create place-based organisations, such as factories and cities, to *function* as biophysical ecosystems. Biomimicry 3.8, the leading biomimicry consulting group, recently launched 'Project Positive.' The work of Project Positive builds on the vision of Janine Benyus: "when a city (or manufacturing site) is functionally indistinguishable from the forest or wildlands next door, we've reached true sustainability." To be 'functionally indistinguishable' requires that we design and create spaces that perform as well as an existing local healthy ecosystem. Healthy ecosystems produce ecosystem services – these are the benefits humans get for free and that are essential to life – that is, clean air, filtered water, healthy soils, pollination, etc. Buildings and cities can also contribute to ecosystem services, shifting human impact from negative to positive. The initial pilot was a project in partnership with Interface, called Factory as a Forest. The multi-year pilot produced positive performance outcomes for Interface *and* the ecosystem, along with a framework and methodology to scale the application and vision.

Conclusions, food for thought and recommendations for the future

Earth operating guidelines (Life's principles) give directions for CE to be able to achieve a restorative and regenerative industry. They guide which system conditions should be met by CE to function within system Earth and to achieve its restorative and regenerative goal. One way to achieve this might be to introduce Life's principles into strategic planning processes.

1. Comparing biological materials to human-made materials show that they cover the same area in terms of strengths vs. stiffness. This indicates that, theoretically, it is possible to reduce materials in the technological cycle and replace them with biological substitutes. The advantages are multifold:
- Recycling fewer different types of materials.
- Functional molecule mining and, eventually, more water-based chemistry.
- Materials that are currently used in the technical loop are getting increasingly scarce; imagine replacing these with materials that are available in abundance.
    - Recommendation for the future: It is obviously recommended to carry out more research on materials processing to create biological substitutes. This should be connected to business model innovation, as working with few but abundant elements is a totally different way of delivering value from our current business practice.

2. The way of thinking that got us in trouble in the first place will probably not get us out of it again. Currently, much of the literature is aimed at improving material efficiency. It does not radically change the way we do things though; we just focus on more efficiency. It is to be seen if this approach will eventually lead to a CE that works. Biomimicry brings a different approach. It can bring about interventions much earlier in the design process and provides answers to our most pressing material and material-processing-related challenges.
- Recommendation for the future: New ways of thinking are best integrated early on in our school system. It is recommended to integrate biomimicry in formal education. This can start at the primary education level. Research is needed to understand how biomimicry is best taught at the different school levels, and how students best learn and improve their biomimicry skills.

3. Energy, in nature, is not so much an issue as it is in human technology. Only what is locally available is used and in no more than 5%. It is the main variable for problem resolution. This might imply that if the 'biology way' is adapted in CE, we will be better able to work with the amount of energy that is locally available, and we might need considerably less energy in the future.
- Recommendation for the future: Research scalability of energy solutions in nature for our human needs.

4. It is often said about recycling in the CE consumption – it is not a problem if we keep the materials in a loop. This line of thinking may get us in trouble again. This assumption implies 100% efficiency in the loops and overlooks the fact that materials can be 'stuck' in a loop for a long time, which prevent them from being available for production on time. This puts a strain on the availability of recycled materials for the production process. This applies especially to the technological cycle. Ashby's materials properties map and Vincent's graphs suggest that it might be very well possible to shift from the technological cycle to the biological one. In terms of

CE, it might be important not to have a closed technical and biological loop as a goal, but to minimise the technological loop, and focus on the biological loop. This could also have a major impact on energy use, not only in terms of materials and materials processing, but also on the amount of energy necessary for recycling and the related transportation. Material ownership issues will change too, since biological materials, in general, are available in abundance.

– Recommendation for the future: A combination of points 1 and 3.

5. Biomimicry links directly with the R-strategies of CE. With the help of 3.8 billion years of R&D, nature has become proficient in reduce, reuse and repair. Our species is equipped with an exceptional brain that we can use for reflection and understanding (refuse) and for finding alternative solutions (rethink & redesign). The three essential elements of biomimicry are 'ethos,' 'reconnect' and 'emulate.' Ethos concerns why we do biomimicry. Reconnect is the realization that we are also part of nature and are inextricably linked to it. Emulate is an explicit set of activities. Together, the three elements focus on reaching the highest steps.

– Recommendation for the future: MacKinnon et al. (2020) show that the outcome of a biomimetic design changes significantly, depending on the imaginaries and discourses of biomimicry, as used by practitioners. Some view nature as a technological entity for exploitation; deem it perfect; acknowledge it as a complex, deficient system; or consider it as a part of and not separate from culture. Design, at its root, is the expression of intentionality through interaction. It is recommended for a biomimicry or CE practitioner or an aspiring one, to reflect on how different intentions within the field could result in largely contrasting outcomes, or even ones that contradict the initial promise. It is equally important for scholars to take into account their epistemological assumptions when doing research.

6. By taking nature as model, mentor and a measure (as in Project Positive, in which ecological performance standards become the benchmark for business operations), we can ensure CE operates within the planetary boundaries and does not remain a metaphor.

– Recommendation for the future: Perform research into how CE activities can be connected (more) explicitly to biological ecosystems and how that impacts both business as well as ecological performance.

References

ACS, Green Chemistry Institute. https://www.acs.org/content/acs/en/greenchemistry/research-innovation/endangered-elements.html (accessed September 16, 2022).

Agrawal, S., Singh, R. K., & Murtaza, Q. (2015). A literature review and perspectives in reverse logistics. *Resources, Conservation and Recycling, 97*, 76–92. https://doi.org/10.1016/j.resconrec.2015.02.009

Benyus, J. M. (1997). *Biomimicry: innovation inspired by nature*. New York: Perennial.

Benyus, J. M., 2011. A biomimicry primer. Biomimicry 3.8 Resource Handbook. pp. 1–10.

Biomimicry Institute and Biomimicry Guild. (2011). *Biomimicry resource handbook: A seed bank of knowledge and best practices*.

Blok, V., & Gremmen, B. (2016). *Ecological Innovation: Biomimicry as a New Way of Thinking and Acting Ecologically*. Journal of Agricultural and Environmental Ethics, 29(2), 203–217. https://doi.org/10.1007/s10806-015-9596-1

Bocken, N. M. P., Short, S. W., Rana, P., & Evans, S. (2014a). A literature and practice review to develop sustainable business model archetypes. *Journal of Cleaner Production, 65*, 42–56. https://doi.org/10.1016/j.jclepro.2013.11.039

Constanza, R., Cumberland, J. H., Daly, H., Goodland, R., Norgaard, R.B., Kubiszewski, I., Franco, C. Second edition, 2015. *An Introduction to Ecological Economics*. CRC Press. Taylor & Francis Group. Boca Raton.

De Pauw, I. C., Karana, E., Kandachar, P., Poppelaars, F., 2014. *Comparing biomimicry and cradle to cradle with ecodesign: a case study of student design projects*. J. Cleaner Prod. 78, 174–183. https://doi.org/10.1016/j.jclepro.2014.04.077.

Dicks, H. (2018). *Nature as Mentor: Foundations of Biomimetic Epistemology*.

Dicks, H. (2017). Environmental ethics and biomimetic ethics: Nature as object of ethics and nature as source of ethics. *Journal of Agricultural & Environmental Ethics, 30*, 255–274. https://doi.org/10.1007/s10806-017-9667-6

European Commission. (2012). *Manifesto for a resource-efficient Europe*. http://europa.eu/rapid/pressrelease_MEMO-12-989_en.htm (last accessed July 9, 2021).

Geisendorf, S. & Pietrulla, F., 2018. *The circular economy and circular economic concepts – a literature analysis and redefinition*. Volume 60, Issue 5. Special Issue: Companies in the Circular Economy. September/October 2018. Pages 771–782. Wiley Periodicals.

Hayes, S., Desha, C., & Baumeister, D. (2020). Learning from nature – Biomimicry innovation to support infrastructure sustainability and resilience. *Technological Forecasting and Social Change, 161*, 120287.

http://zeri.org/vision.html

https://asknature.org/resource/re-thinking-progress-the-circular-economy/

https://bouncingideas.wordpress.com/category/scenarios-of-sustainability/

https://www.biomimicrynorway.com/circular-economy

[Insights] Biomimicry and circular economy. by Rédaction ECLAIRA. 2018-08-2815:09:01. Eclaira. org. https://www.eclaira.org/articles/h/insights-biomimicry-and-circular-economy.html (accessed July 9, 2021).

Jonker, J., & van den Muijsenberg, S. C. B. (2016). *Organiseren in de natuur*. http://www.managementimpact.nl/organisatie/artikel/2016/7/organiseren-in-de-natuur-1016232

Jonker, J., Faber, N., Stegeman, H., 2017. *The Circular Economy – Developments, concepts, and research in search for corresponding business models*. Working paper.

Kennedy, E., Fecheyr-Lippens, D., Hsiung, B.-K., Niewiarowski, P. H., & Kolodziej, M. Biomimicry: A path to.

Lyle, J. T. (1996). *Regenerative design for sustainable development*. New York: John Wiley.

MacKinnon, R. B., Oomen, J., & Zari, M. P. (2020). Promises and presuppositions of biomimicry. *Biomimetics, 5*(3), 33. https://doi.org/10.3390/biomimetics5030033

Marcus, J., Kurucz, E., & Colbert, B. (2010). Conceptions of the business-society-nature interface: Implications for management scholarship. *Business and Society, 49*, 402–438. 10.1080/0969160X.2011.556422

Margulis, L., & Chapman, M. J. (2009). *Kingdoms and domains: An illustrated guide to the phyla of life on earth.* Academic Press.

McDonough, W. (2002). *Cradle to cradle: Remaking the way we make things.* New York: North Point Press.

Mead, T. L. (2014). Biologically-inspired innovation in large companies: A path for corporate participation in biophysical systems? *International Journal of Design & Nature and Ecodynamics, 9*(3), 216–229. WIT Press, http://www.witpress.com ISSN: 1755-7437 (paper format), ISSN: 1755-7445 (online) http://journals.witpress.com 10.2495/DNE-V9-N3-216-229.

Mead, T. & Jeanrenaud, S. (2017). The Elephant in the Room: Biomimetics and Sustainability?. Bioinspired, Biomimetic and Nanobiomaterials. 6. 1–36. 10.1680/jbibn.16.00012.

Murray, A., Skene, K., & Haynes, K. (2017). The circular economy: An interdisciplinary exploration of the concept and application in a global context. *Journal of Business Ethics: JBE, 140*, 369–380.

Pauli, G. A. (2010). *The blue economy: 10 years, 100 innovations, 100 million jobs.* Taos, NM: Paradigm Publications.

Pedersen Zari, M., Storey, J. B., 2007. *An ecosystem based biomimetic theory for a regenerative built environment.* In: International Conference on Sustainable Construction, Materials and Practices: Challenge of the Industry for the New Millennium, Portugal SB 2007. Minho. IOS Press. pp. 620–627.

Reap, J. J., Baumeister, D., & Bras, B. (2005). Holism, Biomimicry and Sustainable Engineering.

Rockström, J., Steffen, W., Noone, K., Persson, Å., Chapin III, F. S., Lambin, E., Lenton, T. M., Scheffer, M., Folke, C., Schellnhuber, H., Nykvist, B., De Wit, C. A., Hughes, T., Van der Leeuw, S., Rodhe, H., Sörlin, S., Snyder, P. K., Costanza, R., Svedin, U., Falkenmark, M., Karlberg, L., Corell, R. W., Fabry, V. J., Hansen, J., Walker, B., Liverman, D., Richardson, K., Crutzen, P., Foley, J. A safe operating space for humanity. Nature, Vol 461/24 (2009), pp. 472–475 (September 2009).

Schumer, A. (2011). Biomimicry, materials, energy and Circular Economy, assessment module. *2*.

Stahel, W.R. and Reday, G. 1976 the potential for substituting manpower for energy, report to the Commission of the European Communities, Brüssel.

Stahel, W. R. (2010). *The performance economy.* Basingstoke, England: Palgrave Macmillan.

Standards, BSI, 2015. BS ISO18458:2015 Biomimetics – Terminology, Concepts and Methodology. BSI Standards

Van den Muijsenberg, S., Appelman, J., Baumeister, D. Chapter 10|10 pages Biomimicry: Design and innovation that help reach eco-effective solutions. In: Appelman, J. H., Osseyran, A., & Warnier, M. (Eds.). (2013). Green ICT & Energy: From Smart to Wise Strategies (1st ed.). CRC Press. https://doi.org/10.1201/b16361

Vincent, J. F. V., Bogatyreva, O. A., Bogatyrev, N. R., Bowyer, A., & Pahl, A.-K. (2006). Biomimetics – Its practice and theory. *Journal of the Royal Society Interface, 3*, 471–482.

Vincent, J. (2008). Biomimetic materials. *Journal of Materials Research, 23*, 3140–3147. 10.1557/JMR.2008.0380

Vincent, J. F. V. (2010). Chapter: New materials and natural design in: Bulletproof feathers : how science uses nature's secrets to design cutting-edge technology. In R. Allen (Ed.), *New materials and natural design, bullet proof feathers.* The university of Chicago Press, Chicago.

Vincent, J. (2012). *Structural biomaterials*. Princeton: Princeton University Press. https://doi.org/
10.1515/9781400842780

Wahl, D. C. Bionics vs. biomimicry: From control of nature to sustainable participation in nature.
*WIT Transactions on Ecology and the Environment, Vol. 87, Design and Nature III: Comparing
Design in Nature with Science and Engineering*, p. 289–298. doi:10.2495/DN060281

Wautelet, T., 2018. *The Concept of Circular Economy: its Origins and its Evolution*. Working Paper,
January 2018. DOI: 10.13140/RG.2.2.17021.87523

Daniel Guzzo, Janaina Mascarenhas, Allen Alexander

10 The transformational power of Circular Innovation

Abstract: The strategic economic and societal impact of innovation across our global economies is indisputable. In considering Circular Economy (CE) transitions, the role that innovation can play should not be underestimated; the critical question is how, if today's economy is still linear? High numbers of new business opportunities, created to meet user, market and societies ever-novel demands, have increased the emphasis placed on responsible and sustainable products and services, but will extended responsibility and sustainable goals create a timely enough transition? This chapter considers innovation from inside and outside the organisation's boundaries and explores circular innovation – where refocused, regenerative and balanced outcomes are created by realising new opportunities, generating novel ideas and developing radical solutions. We discuss the use of state-of-the-art approaches for strategic and innovation management, guided by the CE principles, to uncover how companies can navigate the innovation ecosystem to couple firm-level CE transformation with CE transitions. Our rich examples of different types of circular products, services and processes aim to demystify the complexity of circular innovation and show that it is possible to achieve significant results, even when starting in a linear regime. Our reflections on the power of creating a circular innovation journey and a culture that is genuinely committed to experimentation pave the way for the circular transformation of the company.

Keywords: strategic management, innovation portfolio, socio-technical transitions, innovation roadmap

Introduction

Companies' roles and expectations towards them have evolved. An entirely shareholder-centric view is being replaced by incorporating stakeholders' perspective, triple-bottom-line (TBL) impact and now environmental, social and corporate governance (ESG) requirements for investment. Complementarily, regulations and user demand constantly integrate the need for more sustainable forms of consumption and production. It is increasingly clear that extending a company's purpose can yield increased competitive advantages (Barney, 2018; Fink, 2021; Gibson et al., 2021). Also, we see emerging arguments linking a firm's sustainability efforts to its economic success (Schaltegger et al., 2012). The rise in the expectations of a proactive attitude from modern firms towards sustainable impacts becomes a critical

https://doi.org/10.1515/9783110723373-013

factor for the long-term survival of companies and even sectors. The global pandemic amplified the rising expectations, as it made society further aware of our vulnerabilities.

The Circular Economy (CE) emerges as a new economic model that sets clear regenerative aspirations, with the potential to largely contribute to achieving sustainable development (Geissdoerfer et al., 2017; Kirchherr et al., 2017). Product, service, technology and business-model innovations are central enablers towards circularity (Bocken et al., 2016; Lüdeke-freund et al., 2018). CE provides opportunities for fast-expanding markets with potential for positive economic and social externalities that lie undetected under the linear economy paradigm (Esposito et al., 2015). Many companies have realised the potential of circular innovation to sustain their long-term prosperity. They have been engaging in multiple forms: in knowledge development as strategic partners of think tanks, by experimenting with new business models (such as IKEA's buy-back scheme) and by committing to bold circular transformations (for example, Adidas' clime-neutrality goal through the three-loop strategy). Meanwhile, companies' CE initiatives are driven by profits and growth, leading to accelerated environmental degradation through rebound effects (Whalen & Whalen, 2020). This tension makes us wonder: will the current pace towards circularity create a timely enough transition?

We assert that companies need to fully transform their linear portfolio, as providing a few circular band-aids might not be enough to deal with the potential resource scarcity and the effects of the climate crisis. A balanced portfolio includes investing in long-term and short-term projects, and harmonising significant innovations and minor modifications (Cooper et al., 2001). A complete CE transition requires envisioning and building a new regime that can inject regenerative aspiration and practices into the dominant systems. It is equally important to fuel radical innovation efforts, perhaps operating in a niche, whilst encouraging more incremental innovations from the incumbent regime players towards a future circular reality. However, the challenge of transforming companies' portfolios is a significant source of uncertainty (Hopkinson et al., 2018; Linder et al., 2017). Although they see potential in this new paradigm, it is natural that they seek to cope with the risks of changing, set into the risks of sticking to the linear mindset and practices. Therefore, managing a circular innovation portfolio "means making repeated, coherent strategic investments in markets, products, and technologies" (Hauser et al., 2006, 698) capable of de-risking the CE transformation.

This chapter makes a case for the proactive companies that realise they need to embark on circular transformation. Thus, the goal is to discuss innovation management practices that guide committed companies in circular business transformation. State-of-the-art approaches for strategic and innovation management, guided by the CE principles, uncover how companies can navigate the innovation ecosystem to couple firm-level CE transformation with system-level CE transitions. As a result, companies will proactively assess their innovation portfolio, identify circular opportunities, and strategically select and implement them alongside a broad set of stakeholders. Discussing circular innovation at a portfolio level and alongside the business ecosystem is critical, as the CE transition cannot happen only through a collection of niche initiatives or by a company alone. Instead, efforts must aim and coordinate the rise of a circular regime where industrial practices, applied knowledge and changes in consumption habits will make it possible to achieve full CE's regenerative aspiration.

Adopting a systems approach for circular innovation

A proactive business transformation requires a systems approach for circular innovation that responds to the pressures and opportunities for shifting away from a linear economy. Efforts should aim at appreciating the socio-technical aspects of transitions and cope with the complexity and resistance to change. The Multi-level perspective (MLP) articulates transitions as the interplay of processes happening on the niche, the regime and the landscape levels (Geels, 2011). A transition results from the combination of more radical innovation in the niche, changes in the landscape pressuring the regime, followed by the destabilisation of the current regime, creating space to shift to a new configuration. The initial steps for committed companies includes understanding the wider context of the linear economy pressures in the landscape, which influences both niche and regime dynamics. Additionally, it is critical to taking a life cycle thinking perspective to deeply understand the processes leading to systemic sources of waste in their sector and the reasons for their existence.

Linear economy pressures in the landscape

Circular innovation requires sensing and internalising global trends and drivers into business dynamics to enable timely and proactive responses (Antikainen & Valkokari, 2016; Gorissen et al., 2016). Developments at a landscape level create shocks, pressures and windows of opportunity for sectors and companies alike. Continuously mapping external factors helps a firm deal with the dynamic complexities that emerge from the constant change and hypothesise how these changes influence the regime stability and the possibilities for innovation. The PESTEL analysis framework is already extensively applied in practice and can help map a sector or a company landscape (Rastogi & Trivedi, 2016), elucidating reasons to engage in a CE transition. To illustrate this, Table 10.1 shows specific **political, economic, social, technological, environmental** and **legal (PESTEL)** issues that influence the operating context of businesses, based on the existing analyses of drivers and barriers to CE implementation (Circle Economy, 2021; Esposito et al., 2015; European

Investment Bank, 2019; Ramkumar et al., 2018; Sariatli, 2017; Stumpf et al., 2021; Veleva & Bodkin, 2018). While all the identified factors potentially disrupt the competitive forces within the sector, some of them are apparent risks, while others present opportunities for businesses.

Table 10.1: A PESTEL analysis of the external factors affecting businesses under a CE transition lens.

PESTEL dimension	Guiding question	CE-related identified influencing factors
Political	What changes in the topics to which governments and groups of individuals pay serious attention are likely to affect my organisation or sector?	– Restricted access to markets, if not sustainability compliant – e.g. EU/ Mercosur trade agreement jeopardised by the Brazil deforestation – The sustainability agenda is central to an increasing number of nations – Conflicts over natural resources
Economic	What changes in economic trends and indicators throughout the whole life cycle of products and services are likely to affect my organisation or sector?	– Scarcity in input materials – Volatility of prices in energy and materials – The value-mass-carbon nexus, i.e., the dependence of economic success on extraction and emissions, challenges to the current economic model – Proliferation of ESG funds – Series of economic crisis – 2008 financial crisis, COVID-19 recession – Low hanging fruit opportunities – e.g. remanufacturing and recurring revenue models
Social	What changes in demographics and people's behaviour are likely to affect my organisation or sector?	– World population is still increasing – projections show stabilisation at 11 billion people by 2100 – Increasing worldwide urbanisation – Emerging middle class – especially in Asia – Increasing demand for sustainable solutions – Increasing customers' expectation towards companies' role

Table 10.1 (continued)

PESTEL dimension	Guiding question	CE-related identified influencing factors
Technological	What changes in the techniques, methods, and processes used in producing and consuming products and services are likely to affect my organisation or sector?	– Digital transformation enables real-time information of resources – New materials – e.g. biobased plastics – New manufacturing processes – e.g. 3D printing – Sustainability concepts gaining critical mass – CE, life cycle thinking, sustainability accounting, etc.
Environmental	What changes in the Earth systems, processes and stocks are likely to affect my organisation or sector?	– Trespassing of planetary boundaries – Resource scarcity – e.g. land, water, minerals and other elements – The consequences of climate change led us into a climate emergency – floods, droughts and extreme heat events – High and accelerating rates of species extinction
Legal	What changes in the system of institutional rules and legislation are likely to affect my organisation or sector?	– Increasing environmental regulation – e.g. EPR schemes – Increasing regional and national development policies linked to sustainability – e.g. the NextGenerationEU and the American Jobs Plan integrate sustainability aims
General	How are these changes connected to the prevailing linear economy mindset?	– How can such changes enable a CE transition?

For example, the **social** dimension reveals an ever-growing worldwide population, increasingly urban, and reaching middle-class standards, which leads to pressures in demand for products and services. Such social changes reinforce critical **economic** aspects, such as the volatility of prices in energy and materials due to scarcity of input materials or the occurrence of extreme weather events. Such threats are increasingly on the companies' radars. Meanwhile, people are progressively demanding sustainable solutions and expecting companies to lead change. Such demands can enact new regional and national policies, making specific practices mandatory. Indeed, the **political** and **legal** dimensions show the sustainability agenda becoming central to an increasing number of nations. Also, from the

economic perspective, the proliferation of ESG funds brings additional incentives to spearhead changes towards sustainability.

It should be clear to businesses that a CE transition cannot occur in a vacuum. The dynamic business conditions, often "external" to the organisation and the sector, should be internalised to respond to the trade-off in short-term gains (and uncertainties) of exploiting the linear economy paradigm and the long-term prosperity opportunities by following a regenerative approach. However, critical factors are often mapped with a narrow lens, and companies ignore other relevant trends. For instance, what factors of the current digital transformation might hinder or accelerate a CE transition? Committed companies continuously assess the context to internalise the pressures for change and proactively respond to changes.

Taking a life cycle perspective to identify hotspots

Life cycle thinking is critical to developing a deep understanding of the flows of resources, co-product generation and emissions throughout sourcing, manufacturing, distribution, consumption and post-life destination. In addition, it is central to identifying systemic sources of waste across the company portfolio and along sectors bordering the company (Heiskanen, 2002). Systemic sources of waste are processes that generate negative impacts on the environment happening at the Beginning of Life (BOL), Middle-of-Life (MOL) and End-of-Life (EOL) and represent a potential value capturing opportunity (Yang et al. 2017). Therefore, adopting life cycle thinking and mapping can reveal potential systemic sources of waste, for example:

- In the food industry, fossil-fuel-based fertilisers constitute a significant source of systemic waste in the BOL as it uses scarce, toxic or non-renewable inputs instead of regenerative substitutes.
- In the construction industry, office and residential buildings present low utilisation rates, while suitable for use as clear systemic waste in the MOL.
- In the electronics industry, post-consumption electronics are still illegally exported to developing economies in the EOL, receiving low-quality treatment, and causing harm to human health and the environment.

Life cycle thinking should transcend the company and nations' borders as the consumption and production systems exist to address core societal needs such as nutrition, mobility, clothing and housing (Circle Economy, 2021). Value chains that enable such systems as well as the environmental impacts caused by them transcend those borders, too. The responsibility for the as-is structure of resource flows should be shared among the network of actors operating and relying on those activities. Therefore, the aim is to seek waste in internal operations and take a stewardship role by influencing processes under the responsibility of suppliers, manufacturers, service providers, distributors and consumers (Hart & Milstein, 2003).

A life cycle perspective in the value network will identify critical hotspots, i.e., the most problematic points regarding using resources and generating emissions that the company can contribute to through circular innovation. Meanwhile, cultural barriers, lack of consumer interest, and company hesitation are critical barriers in transitioning to the CE (Kirchherr et al., 2018), meaning that it is not only a matter of mapping the systemic sources of waste. A clinical view is required to investigate why the current regime is linear, and what in the current industrial setting, consumption habits and policy frameworks sustain the 'take-make-dispose' paradigm. Investigating the stakes of critical organisations and individuals involved in the processes leading to systemic sources of waste can also enlighten the reasons for the linear regime behaviour.

The CE principles and strategies leading the way

The CE offers the rationale for transforming the linear economy pressures in the landscape and systemic sources of waste into systemic opportunities. It requires facing the production and consumption systems as living systems and being deeply mindful of their inputs, transformations and outputs (Ellen MacArthur Foundation, 2015). Three principles permit rethinking the production and consumption systems to achieve regenerative potential.

The **first principle** requires a careful selection of materials and the energy to be used as inputs. Except for sunlight, biomass, rain and wind, the resources available on Earth are finite and non-renewable. Therefore, renewable inputs are the priority and should replace non-renewable sources in the production and consumption systems.

The **second principle** requires us to distinguish the biological and technological transformations within the production and consumption systems and keep them in use as effectively for as long as possible. In the biological cycles are the nutrients that nature produces and that nature alone can metabolise. We can help by not wasting nutrients and circulating any by-products back to the biosphere as food. In the technical cycles are nutrients transformed through anthropogenic processes and those that nature cannot metabolise. We must use our manufacturing and operation capabilities to keep technical nutrients in use as effectively as possible.

The **third principle** leads to the desired outcomes of a circular system that must ensure that food, housing, mobility, clothing and other necessary systems meet human needs while nature is regenerated. Changing the mindset regarding waste and negative externalities is critical, as any negative impact is an imperfection of the consumption and production system that must be addressed.

Table 10.2: Aims and strategies of known forms of circularity based on Guzzo et al. (2019).

Forms of circularity	Aim	Strategies
Inner Circles	Enhance the level of utilisation of products	– Encourage reduced consumption – Sharing of products – Product-as-a-service
Circling Longer	Enhance the lifetime of products, parts and components	– Maintenance services – Product upgrading – Remanufacturing – Optimal lifespan design
Cascade Use	Recover discarded materials and energy	– Industrial and urban symbiosis – Recycling – Biodegradation and composting – Energy recovery
Pure Inputs	Enhance the application of materials and energy	– Use of renewable energy – Bio-based materials – Fully recyclable materials – Dematerialisation

Several strategies can be applied to implement the three principles and sustain the CE rationale (Guzzo et al., 2019). See Table 10.2. These strategies operationalise the circular flows by focusing on the different life cycle stages. In general, the embodied energy preserved when providing additional lifetimes to resources (emergy) sets the rule of thumb to prioritise them. Also, from the value perspective, utilisation-oriented strategies lead to pursuing value maximisation throughout the whole live cycle of resources (Stahel, 2016). Therefore, trade-offs and synergies emerge when applying circular strategies, which ought to (and, in many cases, should) be used in combination.

Innovators should internalise the CE principles and strategies to identify circular opportunities and paths for a systematic business transformation.

Setting the options for circular innovation

The circular Innovation portfolio of an organisation relies on a continuous search and selection of initiatives. The search consists of mapping the organisation's innovation space, considering the CE principles and strategies, the areas and the degrees for innovation. Divergent thinking prevails in searching, as practitioners seek to create options. The selection, in turn, consists of continuously assessing and prioritising circular innovation based on their impact potential, ecosystem readiness and market potential. At this point, convergent thinking prevails, as the aim is to narrow down options to the set of innovations that will likely enter the portfolio. Still, as new knowledge arises from the selection process, room is created to alter the innovations to better impact the odds of success.

Mapping the circular innovation space

The search for circular Innovation occurs through a continuous mapping of the organisation's innovation space, i.e., the operating space of an organisation where it can create something new, either as incremental or radical innovation (Tidd & Bessant, 2009, Section 1.5). Companies engaged in CE transformation should identify opportunities to create more circular solutions and benefit from introducing that innovation into the market. Circular innovation occurs in association with technological or behavioural changes and may be positioned in the interfaces with the customer or in the processes of value creation and delivery (de Jesus & Mendonça 2018).

The area defined by the semi-circle in Figure 10.1 shows the circular innovation space of an organisation. Three areas for circular innovation exist[1]:
- Product-service innovation: Changes in what the company offers to satisfy its customers' needs
- Process innovation: Changes in the way products and services are created and delivered, including processes and technologies
- Paradigm innovation: Changes in the mental models and beliefs that support existing behaviour

New versions and optimisations hold potential for climbing the hill of performance, and radical innovation might unfold a performance leap (Norman & Verganti, 2014). A continuum among incremental and radical innovation applies to the three areas. At one end, **incremental circular innovation** enhances the use of resources –improving existing product-services, continuous improvement of processes or minor adjustments in behaviour. At the other end, **radical circular innovation** constitutes changes leading to high circular performance potential, based on discontinuities in products, processes, and paradigms that are new for the sector and to the world.

A thorough mapping of the circular innovation space includes innovations under development by the company, the competitors, and the companies playing in the boundary sectors that might become a competitor in the future or provide insights towards a CE transformation of our portfolio.

---

1 Business-model innovation connects to the three types of innovation as product-service, process and paradigm innovation will lead to changes in the ways companies create, deliver and capture value from the innovation endeavour.

For instance, when investigating the possibilities for circular innovation in the electronics sector, one may find such cases as Fairphone, Apple's Daisy, Back Market and iFixit. The analysis of these cases in terms of the areas and degrees for innovation (Table 10.3) can provide practitioners with insights about the existing possibilities in the sector and sustain brainstorming of additional initiatives. For instance, circular innovation in the electronics sector will likely require a change in mindset, as customers must be willing to compromise functions to adopt more sustainable behaviour – Fairphone 3, start taking responsibility for repairing activities – iFixit, or buying reconditioned electronics – Back Market. On the flip side, the most radical process innovation – Apple Daisy – leads to less-intensive paradigm change as it is still connected to a recycling mindset, and there is no indication of using disassembled products for remanufacturing or refurbishing. Finally, the analysis shows that the three areas intersect as circular innovations contain concomitant changes in the product-service, process and paradigm with varying degrees of innovation.

Assessing circular innovations for impact

The many possibilities in the circular innovation space lead to varying levels of uncertainty in carrying out each project. The option to explore and exploit all the innovation space is a matter of strategy (Tidd & Bessant, 2009, Section 5.15), meaning that assessing and prioritising innovation projects is critical towards a balanced innovation portfolio. The replacement of the existing structure caused by adopting innovation will lead to a new set of benefits – most likely leading to winners and losers from change and the eventual rebound effects. In this matter, the Impacts-Readiness-Market potential matrix (Figure 10.2) is composed of three criteria to assess, scrutinise and prioritise circular innovation:
- Circular impacts potential or the regenerative potential of the solution, in comparison to the business-as-usual behaviour (y axis)
- Innovation readiness or the fitness of the proposed change of behaviour, in comparison to the status quo (x axis)
- Market potential or the revenue opportunity from the innovation (size of bubbles)

As innovation implies change, each innovation should be analysed in the face of the as-is behaviour and the impacts against could-be setups in which the solution at hand becomes part of the regime. Could-be setups are exercises of future scenarios, and they rely on the grasp of contextual temporality, i.e., changes over time that will make innovations more (or less) likely to succeed (Turnheim et al., 2015). For example, a solution that redesigns electronics for modularity, offered on a subscription contract, accompanied by an infrastructure to easily disassemble and upgrade them for the next user could be rather promising in terms of impacts and market potential. Meanwhile, it requires fundamental paradigm changes in consumers and providers,

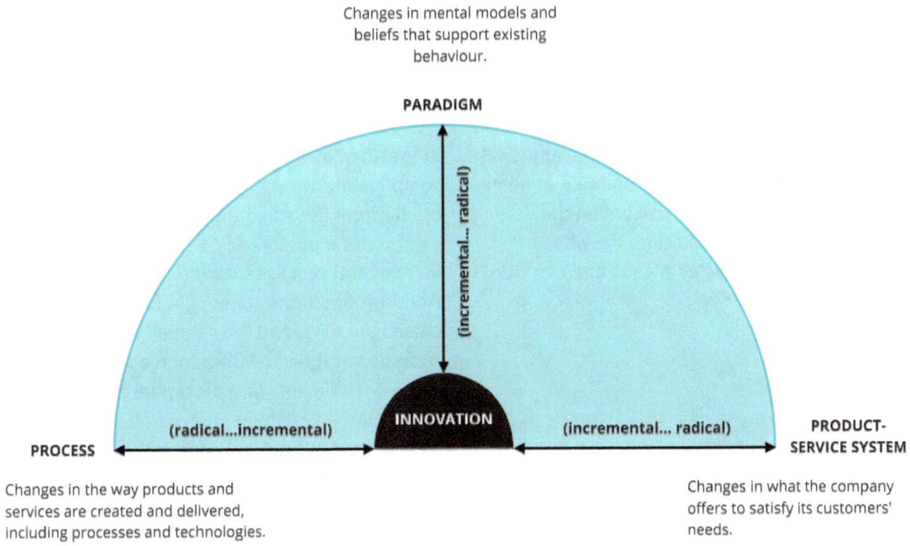

Changes in mental models and
beliefs that support existing
behaviour.

PARADIGM

$(incremental... radical)$

INNOVATION

PROCESS · (radical...incremental) · (incremental... radical) · **PRODUCT-SERVICE SYSTEM**

Changes in the way products and
services are created and delivered,
including processes and technologies.

Changes in what the company
offers to satisfy its customers'
needs.

Figure 10.1: Circular Innovation space considering areas and degrees for innovation.
Source: Adapted from (Tidd & Bessant, 2011).

Table 10.3: Analysis of electronics equipment innovation cases, considering the areas and degrees for innovation.

Innovation case	Areas and degrees of innovation
Apple Daisy: robot that uses a four-step process to disassemble iPhones and separate parts for recycling.	Although the robot Daisy has been recurrently linked to the company's take-back program, Daisy by itself holds no interaction with customers, so **no product-service innovation** was identified. There are elements of **radical process innovation,** as disassembling complex products in seconds involves critical technical and operational challenges. There are elements of **incremental paradigm innovation,** as the possibility of quickly disassembling smartphones might lead to interesting possibilities for the electronics sector. Meanwhile, the robot is still linked to recycling, instead of remanufacturing or product upgrading.

Table 10.3 (continued)

Innovation case	Areas and degrees of innovation
Fairphone 3: modular smartphone that enables customers to upgrade functionalities and swap broken parts. The phone is repairable at home, and replacing parts are available on the company website. In addition, the phone is made from 40% recycled plastic and conflict-free minerals.	There are elements of **radical product-service innovation,** as the modular design and complimentary repairing and replacing services are far from the standard practice in the electronics industry. **Incremental process innovation** occurs, as good working conditions are sought for the supply chain members, and the company creates the conditions for responsibly sourced materials. There are elements of **radical paradigm innovation,** as customers should be willing to compromise a few functionalities, to make an ethical choice for electronics.
Back Market: marketplace for reconditioned electronics. Guarantees checking and restoration by experts and provides a grading system for the appearance and technical conditions of the devices.	While the marketplace is rather like other marketplaces, there are elements of a **somewhat radical product-service innovation** due to the grading system that enables further trust in the reconditioned products. There are **somewhat radical process innovation** elements, as the reconditioning factory and processes are critical to enabling the innovation and are not common practice in the electronics industry. There are **somewhat radical paradigm innovation** elements, as the adoption by customers and manufacturers to the reconditioning model requires a new mindset.
iFixit: a web platform that provides free repair guides for consumer electronics and sells repair parts and tools to support repairing activities.	There are **somewhat radical product-service innovation** elements, as community platforms are no new concept; meanwhile, the composition of guides, tools, a community, and even a repairability score for products provides a unique value proposition to empower owners to repair their products. There are **incremental process innovation** elements, as the repairing processes used are like the ones used by professionals. There are elements of **radical paradigm innovation,** as the solution requires and instigates a substantial shift in mental models for both consumers and providers, by advocating the right to repair.

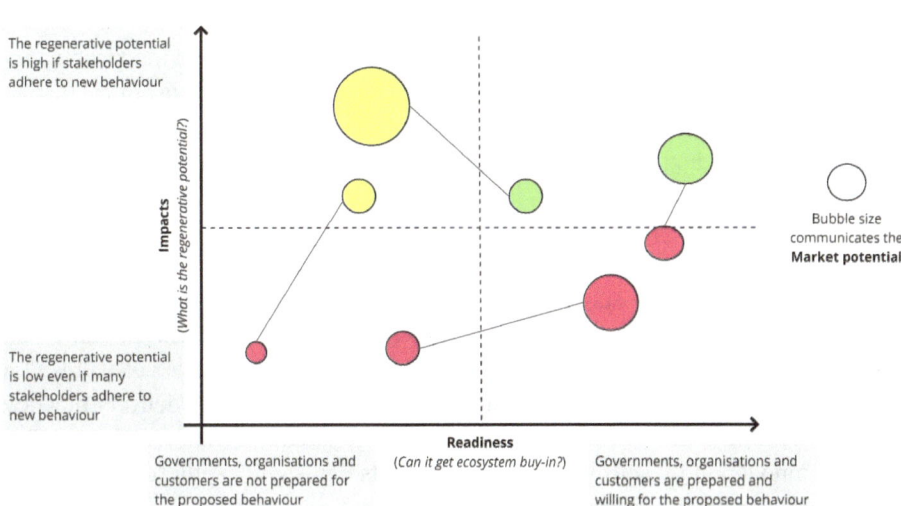

Figure 10.2: The Impacts-Readiness-Market potential matrix for assessing, refining and prioritising Circular Innovations.
Source: Authors.

which could lead to low readiness. On the other hand, a trade-in scheme for collecting end-of-life smartphones, aligned with disassembling for recycling, might lead to lower positive impacts, lower market potential, but higher readiness.

Rough estimates for impacts can support decision-making. The number of instances of change caused by the innovation multiplied by the regenerative potential from each change can lead to those estimates. For instance, the number of clients expected to adopt the trade-in scheme multiplied by the amount of material saved per recycled device, and considering the expected lifetime for smartphones, provides a rough estimate of material savings for that innovation per year. Meanwhile, the market potential leads to the revenue obtained from each of those same instances of change. Thus, the amount paid to customers, the costs for operating the system and the value of the materials matter. Finally, an active investigation for potential rebound effects (Makov & Vivanco, 2018; Whalen & Whalen, 2020; Zink & Geyer, 2017) helps achieve a reliable regenerative potential analysis. For example, could the trade-in scheme accelerate the rate of buying new smartphones? How does that impact the initial estimative?

The map obtained from positioning innovation in the Impacts-Readiness-Market potential matrix helps discuss a company-level strategy towards circular transformation. Discussions emerging from the positioning process can eventually lead to new ideas and feedback as opportunities to iterate and adapt them. In combination, the three criteria enable a proactive approach for systemic benefit that makes explicit the trade-offs among initiatives – leading to the critical question: which of those innovations should be prioritised?

Companies need to set an innovation ambition and balance the incremental and radical innovation (Nagji & Tuff, 2012). Ambidexterity capabilities[2] are critical to managing radical innovation needed for CE transitions while finding incremental quick-wins in their existing portfolio to enhance engagement (Esposito et al., 2018; Ritzén & Sandström, 2017). For instance, incremental innovation might help to evolve the industrial practices and consumption habits in the regime. Moreover, continuous incremental innovation can culminate in radical innovation through a gradual process of evidencing and legitimising the opportunity for substantial change within business proponents, partners and customers (Bocken & Snihur, 2020). Meanwhile, CE constitutes a new mode of consumption and production; much of the changes will lead to entirely new value propositions to customers or value creation and delivery practices within the value chain that require more significant paradigm changes. Thus, radical circular innovation by start-ups and new entrants can significantly threaten incumbents' adherence to linear economy practices.

A few guidelines may guide the company choices after assessing innovations separately:

- **Regenerative and ready innovations (top-right quartile):** Companies should strive for short-term adoption of that innovation, looking for ways to accelerate scaling through the regime.
- **Regenerative and unready innovations (top-left quartile):** Companies should allocate them as niche innovations and aim for launching them on a small scale to enable learning more effective paths for full adoption. Experimentation should answer the questions "how can we influence regime readiness?" or "how to make this innovation profitable?".
- **Low-impact and ready innovations (bottom-right quartile):** Companies should actively seek ways to enhance the regenerative potential of change.
- **Low-impact and unready innovations (bottom-left quartile):** Keep them as ideas. There might be ways to move them up and right in the matrix.

All things considered, how much each company will stand out for nurturing less-ready innovations with high regenerative impacts or aim for more incremental innovations leading to smaller gains is a matter of strategy. The Impacts-Readiness-Market potential matrix should help companies set a balanced circular innovation portfolio that proactively responds to the urge for a company-wide CE transformation.

---

2 Ambidextrous organisations are able to exploit shorter-term opportunities through incremental innovation in the portfolio and explore new possibilities towards more radical innovation (Andriopoulos and Lewis 2009).

Building an open circular innovation ecosystem

It is rather unlikely that one company will master the required knowledge, invest the resources, and generate legitimacy and the demand to the change required in a circular transition by itself and on time. Some producers might not have the ability to pressure more dominant suppliers. Start-ups might possess the right idea – impactful, ready, economically viable – but lack the resources to implement it. Thus, in addition to systematically choosing the appropriate set of circular innovations, a circular transition requires cooperation among complex networks of stakeholders – the circular innovation ecosystems. Cooperation in these ecosystems happens by identifying the potential roles, benefits and the trade-offs at stake for each participant, responding to the reasons to engage in collective learning and novelty development. Strategic understanding of the innovation ecosystem permits building trust and commitment within that network of stakeholders and determining the efforts that each one – organisations and individuals – should strive for in circular innovation.

Mapping the components and stakes in the circular innovation ecosystem

An innovation ecosystem relies on a group of stakeholders with different goals that interact – quite often unintentionally – to develop, introduce and diffuse innovations (Bergek et al., 2008). In Circular innovation ecosystems, such actors, networks and institutions might share the regenerative aspirations but still hold quite different views and means towards circularity. Thus, mapping and engaging with the ecosystem is central. The sectors the innovation will influence and the regions where those innovations will be developed and introduced help define boundaries for the innovation ecosystems. A set of regime and niche stakeholders might act and influence in the innovation ecosystem: suppliers, manufacturers, distributors, service providers, start-ups, users or consumers, research institutes, financiers, public authorities, civil society organisations and industry associations (Geels, 2002, figure 2; Bergek et al., 2008).

The networked compositions of innovation ecosystems imply formal and informal relationships among those stakeholders (see Figure 10.3). The relationships comprise exchanges of material and information, as well as influences between the network components. A value chain perspective helps determine the structure set to operate the production and the consumptions system, i.e. how suppliers, manufacturers, distributors and consumers are connected to continuously provide products and services (Kelly & Marchese, 2015). Changes in the flow of resources will lead to new arrangements in the value chain. For example, activating reverse flows as the collection of smartphones for recycling might require collectors' involvement from design to operation. Also, customers will gain additional responsibilities. The introduction of pure

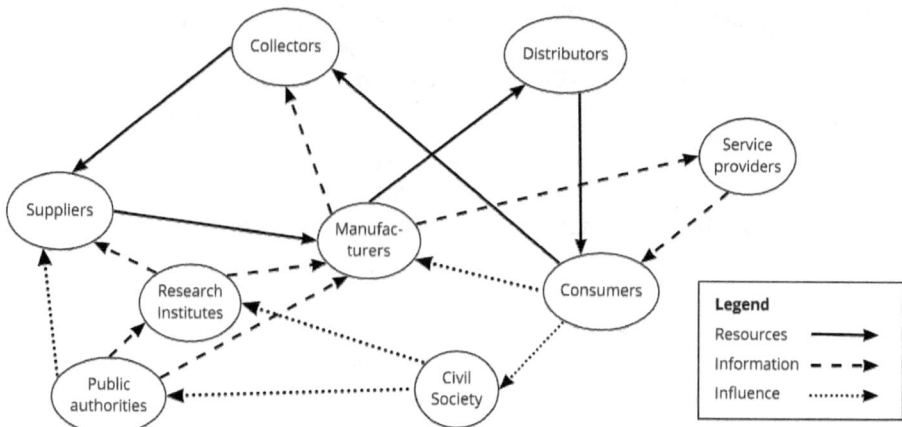

Figure 10.3: The components and structure of a Circular Innovation ecosystem. Source: Authors.

inputs as bio-based materials might require new practices and technological capabilities from suppliers. Such changes will lead to new duties, benefits and trade-offs, often connecting previously disconnected stakeholders.

In addition to changes in the value chain structure, socio-technical alignment is needed for the diffusion of innovations (Bergek et al., 2008) as circular innovation is about technical changes and individual and collective behaviour change. In other words, businesses can only be sustainable if the social system is structurally and culturally prepared for the required changes (Laukkanen & Patala, 2014). Barriers to consumer acceptance, financial risks to organisations and a lack of regulatory frameworks or economic incentives might jeopardise sustainable innovation adoption (Laukkanen & Patala, 2014). It is, thus, critical to understand the dynamics that might make space for the assimilation of norms, habits and practices other than the ones in the dominant linear arrangement. Acceptable practices might become outdated or even illegal. Also, legitimacy needs to be conquered for new habits to take place. Ecosystem stakeholders have to explore and nurture the flow of influences towards the necessary behaviour change.

Stakeholders will engage on the proposed change, based on the potential benefit they will capture from the innovation compared to other options at hand; thus, the ecosystem should adequately mind the benefits and trade-offs. Table 10.4 shows that different perspectives coexist in such networks, requiring tangible and intangible forms of value to be captured. While individuals benefit from the use-value, e.g. functional and emotional benefits from using a specific product or engaging in a specific activity (Almquist et al., 2016), organisations may benefit from the transaction value, including financial gains and intangible benefits from sustaining that mode of consumption and production (Schaltegger et al., 2012). Also, it is critical to consider the environment and society as active stakeholders in such networks, thoroughly considering the positive and negative impacts incurred to such entities (Evans et al., 2017; Pedersen et al., 2021). Explicitly making the society and the environment as stakeholders can help identify organisations and people who have a shared interest in internalising the benefits of making positive impacts and leverage synergies.

Table 10.4: Potential benefits for engaging in Circular Innovation.

Individuals	Organisations	Society	Nature
– Simplifies, reduces effort – Entertains, attracts – Belongs, provides hope – Self-transcendence	– Cost efficiency – Profit margins – Risk reduction – Reputation and brand – Attractiveness as employer – Innovative capabilities	– Employment development – Community development – Accessibility – Well-being	– Resource use – Greenhouse gases emissions – Land and water use – Waste generation

How ecosystems can leverage circular innovation

Engaging in circular innovation requires a deep understanding of the innovation ecosystem to orchestrate or play along with the setup at hand. In addition to identifying the composition and structure of innovation ecosystems, it is crucial to understand and influence the dynamics emerging within, to favour the development, introduction and diffusion of circular innovation. Innovation ecosystems perform critical functions that reinforce and provide feedback to each other, which occur as follows (Bergek et al., 2008; Hekkert et al., 2007; Laukkanen & Patala, 2014):

- **Knowledge development and diffusion:** learning from R&D activities and disseminating into the network
- **Influence on the direction of search:** prioritisation for knowledge development based on expectations, evidence and formation of critical mass operationalised via targets, incentives and legislation
- **Entrepreneurial activities:** actions that unfold the knowledge and networks created into business opportunities
- **Market formation:** creating conditions for the innovation to compete with scaled solutions, from building protected spaces to achieving mass markets
- **Legitimation:** appropriateness to habits and social norms, and acceptance by relevant actors to mobilise resources, form demand or act
- **Resource mobilisations:** the extent to which human capital, financial capital and infrastructure is mobilised to knowledge production and innovation adoption

Comparing the current socio-technical conditions of the ecosystem against the functions to be performed is a crucial step to achieve the intended change of behaviour. For instance, technology development has been widely discussed as a CE enabler. Artificial Intelligence and blockchain can facilitate managing the life cycle of products and, in combination with 3D printing greener materials, creating waste-free value chains (Howllet, 2019). R&D investment is, thus, essential to generate knowledge that will create the possibilities for innovations that will reach the market. Start-ups play a critical role in making available technology into actionable products and services. For example, Circularise uses blockchain to help companies trace materials and emissions throughout product life cycles. On the materials technology side, StoneCycling® transforms waste into bricks for the construction sector. Both companies were founded in close contact with universities. In both cases, reaching technological maturity and validating market needs should lead the ecosystem to orchestrate change in the dynamics of legitimisation, market formation and mobilising resources to scale.

In addition to making technology more actionable, start-ups founded under the CE principles often influence behaviour change towards CE by promoting education and events, and setting exemplary practices for customers and competitors (Rovanto & Bask, 2021). Driven by the business opportunity and the strong personal engagement of the founders, they showcase successful CE operations, demonstrating the possibility of pursuing more sustainable options. They strive to provide legitimacy on different fronts: to consumers, competitors and the market in general. Big (and somewhat unexpected) players strive to bring legitimacy to the urgent transition to a more effective consumption and production systems, too. For example, the Organising Committee of the Olympic and Paralympic Games, in partnership with a market-leader mobile phone operator in Japan, set up the Tokyo 2020 Medal Project, in which all medals in the Olympic games were made from electronic waste. This initiative might provide limited impact per se, but it is an incredible opportunity to educate about the e-waste issues to billions of people worldwide.

Innovation also happens on the blurring borders between companies. Open innovation strategies can leverage the existing knowledge and access to the market to accelerate the development and introduction of circular innovation. Organisations engaging in open innovation rely on the ecosystem of actors as sources of knowledge, and for increased market reach that is still far from the company boundaries (Chesbrough & Bogers, 2014, figure 5). For instance, ABInbev has partnered with Coca-Cola, Colgate-Palmolive and Unilever to set up the 100 + Accelerator, an acceleration program built upon the CE principles to identify and nurture start-ups that will help the four corporations achieve their sustainability targets, making space for cross-sector collaboration. In addition to fostering business opportunities from knowledge development, cross-sector collaboration empowers underrepresented stakeholders and mobilises resources that might enable scaling up circular innovations (Pedersen et al., 2021), acting as an essential motor for circular transitions.

However, a critical challenge is to select and scale CE innovations that lead to relevant ecological contributions (see Whalen & Whalen, 2020), if the intention to achieve the sustainability targets is real.

Finally, governments play a critical role in influencing innovation ecosystems. There is clear evidence that policy has strongly influenced recycling and after-use innovation in Europe (Cainelli et al., 2020). The CE action plan developed by the European Commission is strongly innovation-oriented, stimulating a more holistic application of CE strategies (Cainelli et al., 2020; Milios, 2021). Also, the European Commission helps entrepreneurs developing circular innovation, navigate heavily regulated fields to address legal barriers and reduce uncertainty (Bogers et al., 2018). Therefore, combining industry-led bottom-up initiatives and government-directed top-down innovation is vital to generate knowledge and encourage entrepreneurship towards market formation to enable transitions (Milios, 2021). While the dynamics for change might follow different paths in the functions of an innovation system, coordination among the different stakeholders – public institutions, corporations and start-ups – is critical to achieving CE transitions.

Setting the circular innovation journey

Developing a circular business vision alongside pathways towards such vision is the natural unfolding to materialise the organisation's efforts to pursue a circular portfolio in congruence to the innovation ecosystem. A detailed roadmap can help identify gaps in the organisations' efforts, and anticipate obstacles. Furthermore, it enables aligning evolving societal needs, innovation adoption and experimentation.

Dealing with time and timing

Timing is critical for a proactive path for the company's circular transformation as they must figure out when to design, test, launch and respond to societal needs. The time dimension brings into question the types of business choices a company can make to compete by aligning strategic planning and tactical responses to change (Casadesus-masanell & Ricart, 2010). Adequate appraisal of strategic and tactical responses helps align the choices and expected outcomes in the short-, medium- and long-term.

Setting the foresight lens when planning innovation is a fascinating challenge: the further you dive into the future, less data and evidence is available, leading to uncertainty about the assumptions made; on the flip side, the less you try to capture the potential future scenarios, further focus is given to respond to the constant change through short-term solutions (Webb, 2019). So, too much attention provided to faraway events might lead to putting all efforts in situations that might not happen, while full attention in actions having an effect for a short period will lead to continuous incremental change and firefighting. Thus, a time cone balances the foresight timeframe of actions to the uncertainty of events (Webb, 2019), offering specific types of choices:

- **Tactics (1–3 years):** short-term choices aimed at likely events, with existing data and evidence for their occurrence, including product-service redesign and targeting a new segment
- **Strategy (3–10 years):** medium-term choices that direct the organisation towards potential events, including defining priorities and allocating resources
- **Vision (10 + years):** long-term choices that drive the organisation towards a desirable state, preparing it for potential disruptions in the industry

The time horizon for the analysis and the meaning of short-, medium- and long-term is contextual to the sector, organisation and the team's understanding of the expected transition speed. Again, it is a matter of contextual temporality (see Turnheim et al., 2015). Discontinuities caused by the build-up of behaviour change (e.g. increasing preference for healthy and organic food) and technological convergences (e.g. maturing of Internet of Things and blockchain enabling material traceability) reinforce the value of long-term foresight, including the generation of visions to ride and drive change.

The power of a genuine circular innovation roadmap

A roadmap organises the prioritised set of innovations and the strategic efforts alongside the ecosystem, considering adequate time and timing. It enables setting a path towards a vision through strategically positioning the innovation efforts, enabling a logical plan among initiatives, while considering the trade-offs, constraints and delays inherent in the innovation process (Phaal, 2004). In circular innovation, road mapping involves appreciating and anticipating the dynamics of a circular transition and proactively responding to societal needs by enabling innovation adoption through continuous development of evidence and capabilities.

A Circular innovation roadmap (Figure 10.4) is composed of four complementary perspectives into a single diagram, addressing critical questions:

- **Timeframe:** When should we act? What are the foresight periods to guide actions and cope with the uncertainty of events?
- **Societal needs:** Why should we act? What are the opportunities and uncertainties emerging from changes in the socioeconomic dynamics that affect our sector and business?
- **Innovation adoption:** What should we do? What are the evolutions of the product-service, paradigm and process innovations to be launched and adopted?
- **Experimentation:** How can we enable that? What is the validated learning and capabilities needed to sustain the required innovation?

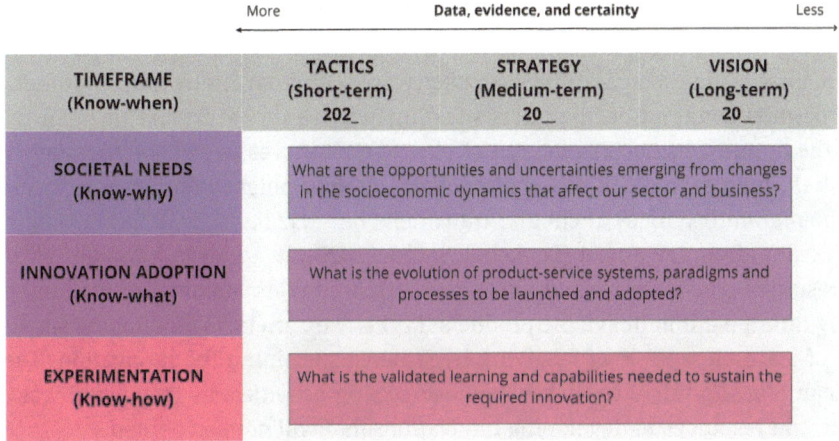

Figure 10.4: A Circular Innovation roadmap.
Source: Authors.

The circular innovation roadmap builds upon the traditional technology road-map – TRM (Phaal, 2004), by reframing the Market, Product and Technology layers. First, it expands the market to a societal needs dimension, recognising that "societal needs, not just conventional economic needs, define markets" (Porter & Kramer, 2011, 5). Such expansion permits dealing with hard drivers and barriers as technical and economic factors, and soft drivers and barriers as institutional and social factors (de Jesus and Mendonça 2018). It also reframes the artefacts, from a product-centric view to the adoption of product-service, paradigm and process innovations. Finally, the roadmap focuses on reducing uncertainty and building capability, as circular innovation is not only about technological development.

The roadmap scope may also vary according to the stakeholders participating in the transition (Phaal et al., 2010). The Circular Innovation roadmap can cascade from ecosystem roadmaps into more actionable operational ones, uncovering roadmaps specific to the different stakeholder roles in the ecosystems and even the different areas within companies. More aggregate roadmaps orchestrate partial roadmaps to-wards sector-wide transitions. A well-designed circular innovation roadmap will help to detail the connection between goals and development efforts, identify critical gaps in knowledge and capabilities, and avoid contradictory incentives at the ecosystem and company levels.

Experimenting your way towards circularity

Experimentation is the sustaining perspective of a circular innovation roadmap, which permits engaging in hypothesis-testing activities to persevere or pivot towards the defined vision (Eisenmann et al., 2012). It serves to reduce uncertainty and risk through each of the innovation efforts, from problem-solution fit to scale. Experimentation is critical in circular transitions because the final solutions might not be evident from the outset (Bocken & Snihur, 2020). In addition, companies are often resource-constrained, and experimentation enables maximising validated learning through minimum viable products (MVPs), i.e., artefacts holding the smallest set of features capable of falsifying hypotheses sustaining the innovation. The term Lean Start-up refers to such hypothesis-testing activities, as they avoid wasting time and resources in developing innovations that will not be adopted.

Apart from reducing uncertainty in innovation, the experimentation path is a platform for engaging with internal and external stakeholders in the innovation ecosystem, to promote collective learning towards economic, societal and environmental impact (Bocken & Snihur, 2020). It introduces the mindset for collectively and continuously moving towards circular innovation, where companies, customers and other critical stakeholders engage in the circular transformation. Moreover, continuous testing enables the ones involved in innovation to make sense of the operational capabilities needed for scaling and sustain commitment to the required changes.

The experimentation lane also reinforces the Circular Innovation roadmap as a living document. Experimentation will help discover if the general direction of the transformation path is correct, and validate with solid evidence that each of the ideas will work in practice (Bland & Osterwalder, 2020). Sufficient evidence for the hypotheses that support the fit between innovation adoption and societal needs allows stakeholders to persevere in the designed roadmap. If there is evidence that the initial hypotheses do not hold and additional opportunities emerge, pivot decisions might change the direction towards more promising paths. Finally, it might be the case that evidence discourages entire pathways, and they might perish, freeing resources to investigate more promising ones. In any case, adjustments on tactics, strategy and vision will occur based on knowledge gained from experimentation.

Doing it systematically through a circular innovation framework

In this chapter, we argued that companies must engage in circular business transformation on the portfolio level, in coordination with the innovation ecosystems they are part of, to create a timely enough transition. Companies should include in their strategic radar the rising odds of a circular transition (and the rising uncertainties

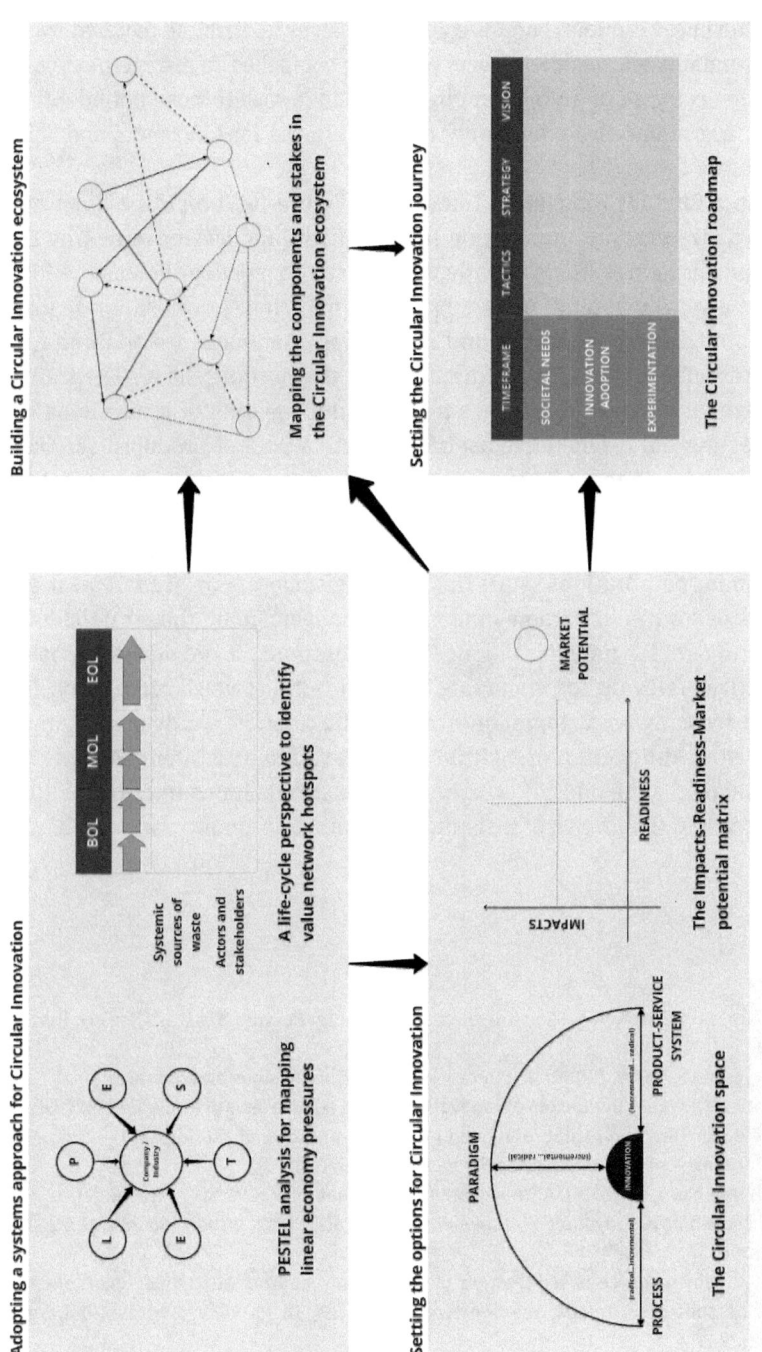

Figure 10.5: A Circular Innovation framework.
Source: Authors.

of insisting on the linear economy) and prepare themselves to drive the required transformation. Corporations and regime players might be compelled to engage in circular innovation, as the collapse of an old paradigm is relentless with those not adapting. On the flipside, transitions also offer plenty of opportunities for innovators and start-ups playing in niche fields.

We propose a Circular Innovation framework (Figure 10.5) building upon and adapting known strategic and innovation management practices governed by the CE principles and strategies. Tools already widely used in practice such as the PESTEL analysis, life cycle mapping, innovation space mapping, prioritisation matrix, ecosystem mapping and governance and innovation roadmaps are adapted and combined for Circular business transformation on the portfolio level. The framework is thoroughly conceived to balance planning and steering of innovations capable of introjecting the regenerative aspirations into a pool of products, services, processes and paradigms that will sustain the company's position in the (near) future while de-risking its circular transformation.

Such a framework and the complementary discussions enabled by it are valuable for academia, corporations, start-ups and the society. For academia, it addresses the lack of studies that frame innovation management on the portfolio level when addressing circular transformation. For corporations, it provides actionable tools and practices relevant for managing their portfolio towards circularity. For start-ups, often focusing on a single innovation, it helps strategically identify their roles as connectors and facilitators within the innovation ecosystem. Finally, for the society, it increases the odds of bottom-up circular transitions in critical industries to timely react to the threats of insisting on the linear economy rationale.

References

Almquist, E., Senior, J., & Bloch, N. (2016). The elements of value. *Harvard Business Review*, (no. September).

Andriopoulos, C., & Lewis, M. W. (2009). Exploitation-exploration tensions and organizational ambidexterity : Managing paradoxes of innovation linked references are available on JSTOR for this article : Exploitation-exploration tensions and organizational ambidexterity : Managing Paradoxes of In. *Organization Science*, *20*(4), 696–717.

Antikainen, M., & Valkokari, K. (2016). A framework for sustainable circular business model innovation. *Technology Innovation Management Review*, *6*(7), 5–12. https://doi.org/10.22215/timreview/1000

Barney, J. B. (2018). Why resource-based theory's model of profit appropriation must incorporate a stakeholder perspective. *Strategic Management Journal*, *39*(13), 3305–3325. https://doi.org/10.1002/smj.2949

Bergek, A., Jacobsson, S., Carlsson, B., Lindmark, S., & Rickne, A. (2008). Analyzing the functional dynamics of technological innovation systems: A scheme of analysis. *Research Policy*, *37*(3), 407–429. https://doi.org/10.1016/j.respol.2007.12.003

Bland, D. J., & Osterwalder, A. (2020). *Testing business ideas*, Strategyzer.Com/Test

Bocken, N. M. P., de Pauw, I., Bakker, C. A., & van der Grinten, B. (2016). Product design and business model strategies for a circular economy. *Journal of Industrial and Production Engineering*, *33*(5), 308–320. https://doi.org/10.1080/21681015.2016.1172124

Bocken, N. M. P., & Snihur, Y. (2020). Lean startup and the business model: Experimenting for novelty and impact. *Long Range Planning*, *53*(4), 101953. https://doi.org/10.1016/j.lrp.2019.101953

Bogers, M., Chesbrough, H., & Moedas, C. (2018). Open innovation: Research, practices, and policies. *California Management Review*, *60*(2), 5–16. https://doi.org/10.1177/0008125617745086

Cainelli, G., D'Amato, A., & Mazzanti, M. (2020). Resource efficient eco-innovations for a circular economy: Evidence from EU firms. *Research Policy*, *49*(1), 103827. https://doi.org/10.1016/j.respol.2019.103827

Casadesus-masanell, R., & Ricart, J. E. (2010). From strategy to business models and onto tactics. *Long Range Planning*, *43*(2–3), 195–215. https://doi.org/10.1016/j.lrp.2010.01.004

Chesbrough, H., & Bogers, M. (2014). Explicating open innovation. In: *New frontiers in open innovation* (Vol. 11, pp. 3–28). Oxford University Press. https://doi.org/10.1093/acprof:oso/9780199682461.003.0001

Circle Economy. (2021). *The circularity gap report*.

Cooper, R., Edgett, S., & Kleinschmidt, E. (2001). Portfolio management for new product development: results of an industry practices study. *R and D Management*, *31*(4), 361–380. https://doi.org/10.1111/1467-9310.00225

de Jesus, A. D., & Mendonça, S. (2018). Lost in transition? drivers and barriers in the eco-innovation road to the circular economy. Ecological Economics, *145*, (July 2017), 75–89. https://doi.org/10.1016/j.ecolecon.2017.08.001

Eisenmann, T., Ries, E., & Dillard, S. 2012. Hypothesis-driven entrepreneurship : The lean startup, 1–26.

Ellen MacArthur Foundation. (2015). *Growth within: A circular economy vision for a competitive Europe*. UK: Cowes.

Esposito, M., Tse, T., & Soufani, K. 2015. Is the circular economy a new fast-expanding market ?, 9–14. https://doi.org/10.1002/TIE.21764.

Esposito, M., Tse, T., & Soufani, K. (2018). Introducing a circular economy: New thinking with new managerial and policy implications. *California Management Review*, *60*(3), 5–19. https://doi.org/10.1177/0008125618764691

European Investment Bank. (2019). The EIB circular economy guide. *Supporting the circular transition*. www.eib.org/circular-economy

Evans, S., Vladimirova, D., Holgado, M., Kirsten, V. F., Yang, M., Silva, E. A., & Barlow, C. Y. (2017). Business model innovation for sustainability: Towards a unified perspective for creation of sustainable business models. *Business Strategy and the Environment*, *608*(April), 597–608. https://doi.org/10.1002/bse.1939

Fink, L. (2021). *Larry Fink's 2021 letter to CEOs*. 2021. https://www.blackrock.com/corporate/investor-relations/larry-fink-ceo-letter

Geels, F. W. (2002). Technological transitions as evolutionary reconfiguration processes : A multi-level perspective and a case-study. *Research Policy*, *31*, 1257–1274.

Geels, F. W. (2011). The multi-level perspective on sustainability transitions: Responses to seven criticisms. *Environmental Innovation and Societal Transitions*, *1*(1), 24–40. https://doi.org/10.1016/j.eist.2011.02.002

Geissdoerfer, M., Savaget, P., Bocken, N. M. P., & Hultink, E. J. (2017). The circular economy – A new sustainability paradigm? *Journal of Cleaner Production*, *143*, 757–768. https://doi.org/10.1016/j.jclepro.2016.12.048

Gibson, C. B., Gibson, S. C., & Webster, Q. (2021). Expanding our resources: Including community in the resource-based view of the firm. *Journal of Management*, *47*(7), 1878–1898. https://doi.org/10.1177/0149206320987289

Gorissen, L., Vrancken, K., & Manshoven, S. (2016). Transition thinking and business model innovation – Towards a transformative business model and new role for the reuse centers of Limburg, Belgium. *Sustainability*. https://doi.org/10.3390/su8020112

Guzzo, D., Trevisan, A. H., Echeveste, M., & Costa, J. M. H. (2019). Circular innovation framework: Verifying conceptual to practical decisions in sustainability-oriented product-service system cases. *Sustainability*, *11*(12), 3248. https://doi.org/10.3390/su11123248

Hart, S. L., & Milstein, M. B. (2003). Creating sustainable value. *Academy of Management Executive*, *17*(2), 56–69. https://doi.org/10.5465/ame.2003.10025194

Hauser, J., Tellis, G. J., & Griffin, A. (2006). Research on innovation: A review and agenda for marketing science. *Marketing Science*, *25*(6), 687–717. https://doi.org/10.1287/mksc.1050.0144

Heiskanen, E. (2002). The institutional logic of life cycle thinking. *Journal of Cleaner Production*, *10*(5), 427–437. https://doi.org/10.1016/S0959-6526(02)00014-8

Hekkert, M. P., Suurs, R. A. A., Negro, S. O., Kuhlmann, S., & Smits, R. E. H. M. (2007). Functions of innovation systems: A new approach for analysing technological change. *Technological Forecasting and Social Change*, *74*(4), 413–432. https://doi.org/10.1016/j.techfore.2006.03.002

Hopkinson, P., Zils, M., Hawkins, P., & Roper, S. (2018). Managing a complex global circular economy business model: Opportunities and challenges. *California Management Review*, *60*(3), 71–94. https://doi.org/10.1177/0008125618764692

Howllet, A. (2019). From smarter machines to greener products. *Financial Times*. 2019. https://www.ft.com/content/d9074342-0797-11ea-a958-5e9b7282cbd1

Kelly, E., & Marchese, K. (2015). Supply chains and value webs. In *Business ecosystems come of age*, 55–66. https://www2.deloitte.com/content/dam/insights/us/articles/platform-strategy-new-level-business-trends/DUP_1048-Business-ecosystems-come-of-age_MASTER_FINAL.pdf

Kirchherr, J., Piscicelli, L., Bour, R., Kostense-Smit, E., Muller, J., Huibrechtse-Truijens, A., & Hekkert, M. (2018). Barriers to the circular economy: Evidence from the European Union (EU). *Ecological Economics*, *150*(April), 264–272. https://doi.org/10.1016/j.ecolecon.2018.04.028

Kirchherr, J., Reike, D., & Hekkert, M. (2017). Conceptualizing the circular economy: An analysis of 114 definitions. *Resources, Conservation and Recycling*, *127*(September), 221–232. https://doi.org/10.1016/j.resconrec.2017.09.005

Laukkanen, M., & Patala, S. (2014). Analysing barriers to sustainable business model innovations: Innovation systems approach. *International Journal of Innovation Management*, *18*(6). https://doi.org/10.1142/S1363919614400106

Linder, M., Sarasini, S., & Patricia, V. L. (2017). A metric for quantifying product-level circularity. *Journal of Industrial Ecology*, *21*(3), 545–558. https://doi.org/10.1111/jiec.12552

Lüdeke-freund, F., Gold, S., & Bocken, N. M. P. (2018). A review and typology of circular economy business model patterns – Appendix II, 1–9.

Makov, T., & Vivanco, D. F. (2018). Does the circular economy grow the pie? The case of rebound effects from smartphone reuse. *Frontiers in Energy Research*, *6*(MAY), 1–11. https://doi.org/10.3389/fenrg.2018.00039

Milios, L. (2021). Overarching policy framework for product life extension in a circular economy – A bottom-up business perspective. *Environmental Policy and Governance*, *31*(4), 330–346. https://doi.org/10.1002/eet.1927

Nagji, B., & Tuff, G. (2012). Managing your innovation portfolio. *Harvard Business Review*, *37*(8), 30–42.

Norman, D. A., & Verganti, R. (2014). Incremental and radical innovation: design research vs. technology and meaning change. *Design Issues*, *30*(1), 78–96. https://doi.org/10.1162/DESI_a_00250

Pedersen, E., Gjerdrum, R., Lüdeke-Freund, F., Henriques, I., & May Seitanidi, M. (2021). Toward collaborative cross-sector business models for sustainability. *Business and Society*, *60*(5), 1039–1058. https://doi.org/10.1177/0007650320959027

Phaal, R. (2004). Technology roadmapping – A planning framework for evolution and revolution. *Technological Forecasting and Social Change*, *71*(1–2), 5–26. https://doi.org/10.1016/S0040-1625(03)00072-6

Phaal, R., Farrukh, C., & Probert, D. R. (2010). *Roadmapping for strategy and innovation: Aligning technology and markets in a dynamic world*. University of Cambridge.

Porter, M. E., & Kramer, M. R. (2011). Creating shared value. *Harvard Business Review*, *89*, 1–2. https://doi.org/10.32591/coas.ojss.0201.04037b

Ramkumar, S., Kraanen, F., Plomp, R., Edgerton, B., Walrecht, A., Baer, I., & Hirsch, P. (2018). Linear risks. *Circle economy*, 14. 'Linear Risks': How Business As Usual Is A Threat To Companies And Investors – Insights – Circle Economy (circle-economy.com).

Rastogi, N., & Trivedi, M. K. (2016). PESTLE technique – A tool to identify external risks in construction projects. *International Research Journal of Engineering and Technology*, *03*(01), 384–388. www.irjet.net

Ritzén, S., & Sandström, G. Ö. (2017). Barriers to the circular economy – Integration of perspectives and domains. *Procedia CIRP*, *64*, 7–12. https://doi.org/10.1016/j.procir.2017.03.005

Rovanto, I. K., & Bask, A. (2021). Systemic circular business model application at the company, supply chain and society levels – A view into circular economy native and adopter companies. *Business Strategy and the Environment*, *30*(2), 1153–1173. https://doi.org/10.1002/bse.2677

Sariatli, F. (2017). Linear economy versus circular economy: A comparative and analyzer study for optimization of economy for sustainability. *Visegrad Journal on Bioeconomy and Sustainable Development*, *6*(1), 31–34. https://doi.org/10.1515/vjbsd-2017-0005

Schaltegger, S., Freund, F. L., & Hansen, E. G. (2012). Business cases for sustainability: The role of business model innovation for corporate sustainability. *International Journal of Innovation and Sustainable Development*, *6*(2), 95. https://doi.org/10.1504/IJISD.2012.046944

Stahel, W. R. (2016). Circular economy. *Nature*, (no. March), 435–438. https://doi.org/10.1038/531435a

Stumpf, L., Schöggl, J. P., & Baumgartner, R. J. (2021). Climbing up the circularity ladder? – A mixed-methods analysis of circular economy in business practice. *Journal of Cleaner Production*, *316*. https://doi.org/10.1016/j.jclepro.2021.128158

Tidd, J., & Bessant, J. (2009). *Managing Innovation: integrating technological, market and organizational change* (4th. ed.). Chichester: John Wiley & Sons, Ltd.

Tidd, J. & Bessant, J. R. (2011). *Managing innovation: integrating technological, market and organizational change*. John Wiley & Sons.

Turnheim, B., Berkhout, F., Geels, F., Hof, A., McMeekin, A., Nykvist, B., & van Vuuren, D. (2015). Evaluating sustainability transitions pathways: Bridging analytical approaches to address governance challenges. *Global Environmental Change*, *35*(2015), 239–253. https://doi.org/10.1016/j.gloenvcha.2015.08.010

Veleva, V., & Bodkin, G. (2018). Corporate-entrepreneur collaborations to advance a circular economy. *Journal of Cleaner Production*, *188*, 20–37. https://doi.org/10.1016/j.jclepro.2018.03.196

Webb, A. (2019). How to do strategic planning like a futurist. *Harvard Business Review*.

Whalen, C. J., & Whalen, K. A. (2020). Circular economy business models: A critical examination. *Journal of Economic Issues*, *54*(3), 628–643. https://doi.org/10.1080/00213624.2020. 1778404

Yang, M., Evans, S., Vladimirova, D., & Rana, P. (2017). Value uncaptured perspective for sustainable business model innovation. *Journal of Cleaner Production*. https://doi.org/10. 1016/j.jclepro.2016.07.102

Zink, T., & Geyer, R. (2017). Circular economy rebound. *Journal of Industrial Ecology*, *00*(0), 1–10. https://doi.org/10.1111/jiec.12545

Paavo Ritala, Nancy M. P. Bocken, Jan Konietzko

11 Three lenses on circular business model innovation

Abstract: Circular business models help businesses to create and capture value in ways that are in line with circular economy strategies, while, at the same time, making feasible business cases. Identifying circular business cases is not simple, however, and calls for major innovation efforts in different aspects of the firm's business model. In this chapter, we identify and demonstrate three lenses to circular business model innovation. First, we explicate how firms innovate circular value propositions, that is, customer-facing promises of value creation that draw from circularity principles. Second, we focus on the firm-level business models, that is, on how firms innovate their value creation, delivery, capture schemes to operationalise circular strategies. Third, we adopt the ecosystem lens, by examining how firms can involve a variety of complementary actors, governed via coordination mechanisms such as contractual frameworks, digital platforms and around a shared value proposition. The three lenses are complementary and constitute an overarching view of circular business model innovation. We use examples to show how these lenses help companies implement the circular strategies of narrowing, slowing, closing and regenerating the loop. We also suggest directions for future research in this area.

Keywords: circular economy, sustainability, value proposition, business model innovation, value creation, ecosystems

Introduction

Circular business models – that is, business models that apply circular economy (CE) strategies – are a topic of increasing research interest (for reviews, see Ferasso et al., 2020; Geissdoerfer et al., 2020). At best, circular business models help businesses to create, deliver and capture value in ways that are aligned with economic as well as environmental goals of businesses and their stakeholders. In general, business model innovation is defined as "designed, novel, nontrivial changes to the key elements of a firm's business model and/or the architecture linking these elements" (Foss & Saebi, 2016, p. 201). Building beyond this general view, scholars have started to examine how firms can implement business models based on CE strategies of narrowing, slowing, closing and regenerating resource flows, with

Funding: Jan Konietzko and Nancy Bocken received funding from the European Union's Horizon 2020's European research Council (ERC) funding scheme under grant agreement no. 850159, Project Circular X (www.circularx.eu).

https://doi.org/10.1515/9783110723373-014

both intra-firm and inter-firm changes and reconfigurations (Bocken & Ritala, 2022; Konietzko et al., 2020). However, achieving such business models is all but easy. A CE creates new types of demands for firms and industries, and thus challenges the established business models that have been based on linear economy principles, requiring business model transformation (Frishammar & Parida, 2019; Hofmann & Jaeger-Erben, 2020). Identifying concrete business cases for circular business models, thus, calls for major innovation efforts at different levels. Furthermore, firms encounter multiple barriers in their circular business model innovation, including intra-firm as well as societal and institutional barriers (Guldmann & Huulgaard, 2020; Ranta et al., 2018).

This chapter consolidates and reconceptualises the rapidly cumulating literature and evidence, by highlighting three important lenses on circular business model innovation. First, we explicate how firms innovate circular *value propositions*, that is, customer-facing promises of economic, social and environmental value creation that draw from CE principles. Circular value propositions are important, since they allow firms to clearly demonstrate and communicate the benefits of their business models not only to the customers, but also the stakeholder benefits within the broader socio-environmental context (Ranta et al., 2020). Second, we adopt *firm-centric* lenses to circular business model innovation, that is, on how firms innovate their value creation, delivery and capture schemes to operationalise circular strategies. Focusing on the firm level is important, since many circularity-related innovations reside in the way a firm configures its own business processes in alignment with processes such as materials handling and resource efficiency (e.g. Cainelli et al., 2020). Third, we adopt the *eco-system* lens to circular business model innovation by examining how firms can utilise a variety of complementary actors to scale up the circular business models and related value propositions. The ecosystem lens – based on the concept of business and innovation ecosystem from management studies – helps to examine the interdependencies across different actors that can contribute to the realisation of a particular value proposition, and to take part in the focal firm's business model innovation (Adner, 2017; Snihur et al., 2018).[1] The ecosystem lens broadens the view from downstream value generation (i.e., customer interface) towards the system or upstream actors that enable circular business model innovation (Pieroni et al., 2019). In fact, circular business models often involve firm-external actors. For instance, in each of the circular business model innovations identified in a recent study by Frishammar and Parida (2019), each business model involved an ecosystem of complementary actors (e.g. industrial customers or partners), whose inputs and collaboration are needed for a circular solution to materialise. Furthermore, ecosystem lenses help understand the broad diversity of

---

1 We will be using the word "ecosystem" throughout this chapter when we refer to business and/ or innovation ecosystems. For discussion on the definitions of business and innovation ecosystems, please see (Thomas and Autio 2020; Autio 2021).

relevant societal and institutional actors for CE (Ranta et al., 2018), which are crucial to the legitimacy of new business and innovation ecosystems (Thomas & Ritala, 2021). Overall, these three lenses are complementary and constitute an overarching view of circular business model innovation.

In the remainder of this chapter, we first briefly discuss the big picture in terms of circular business model innovation. This is followed by an integrative discussion of the three lenses (value proposition, firm-centric and ecosystem). We end the chapter with a summary of concrete examples of circular business model innovation using the three lenses, as well as implications for further research.

Defining circular business model innovation

Businesses and societies will face increasing environmental risk, including the failure to mitigate climate change, biodiversity loss and resource scarcity (World Economic Forum, 2020). In fact, many advanced economies have a resource deficit – they import and consume more than they can generate. Thus, as environmental constraints tighten, competition for resources is likely to increase (Tukker et al., 2016). At the same time, global economies are struggling with waste management and processing (e.g. around plastic waste). Therefore, innovation across all aspects of the CE can help ensure not only environmental sustainability, but also effectiveness and longevity of resources, globally.

CE has been positioned as a driver for sustainability (Geissdoerfer et al., 2017) and a potential way to regenerate the economy in a sustainable way (Wuyts et al., 2020). The CE is related to innovations at the technical level of material, processes and products, the logistical level of value chains, and organisational level of value propositions, business models and ecosystems (Brown et al., 2019; Kalmykova et al., 2018; Lieder et al., 2017). While technical innovation is essential to create cleaner technologies and to ensure circular processes and practices, *business model innovation* is needed in order to fundamentally rethink the way business is done (Bocken et al., 2014). In this regard, sustainable and circular business models aim for a holistic approach that bridges economic, social and environmental value creation (Ritala et al., 2021; Stubbs & Cocklin, 2008) and, when designed as such, can have a significant positive impact on circularity (Tukker, 2004, 2015). Importantly, circular business model innovation can extend its impact far beyond individual businesses. Indeed, transformation of business models (e.g. from linear to circular ones) may be viewed as an important driver of sustainability transitions (Bidmon & Knab, 2018; Sarasini & Linder, 2018) and system-level sustainability-oriented innovation (Adams et al., 2016).

Due to the above-mentioned issues, circular business model innovation can help companies adapt to a new context shaped by environmental constraints and increasing governmental and societal demand for sustainable products and processes. The goal of circular business models is to maximise the economic value of products, components and materials, and to minimise absolute resource use, emissions, waste and pollution (Geissdoerfer et al., 2017). Such business models help firms create value from waste and underused resources (i.e., 'value from waste': Bocken et al., 2014), to develop stronger customer relationships by transforming to service models instead of products (Spring & Araujo, 2017), to develop resilience against volatile prices for critical raw materials and to prepare for upcoming tightening of environmental policies (Mathieux et al., 2017).

Distinct from generic business model innovation, circular business model innovation is based on a natural resource perspective (Bocken & Ritala, 2022), including questions such as how resources can be conserved, optimally used and reused and how natural environments can be improved rather than degraded. In this regard, the literature has started to consolidate along four key strategies that relate to different aspects of the circularity of natural resources (based on Bocken et al., 2016, 2021; McDonough & Braungart, 2002; Stahel, 1994):

1. **Narrowing:** use less resources through innovations and efficiencies in the production and design process, such as light-weight products and cleaner production or logistics processes
2. **Slowing:** prolong product and component lives (e.g. design for durability or design for remanufacturing) and avoid unnecessary consumption (e.g. propositions with premium or service offerings that promote long life)
3. **Closing:** reuse materials, that is, post-consumer recycling, enabled by innovations in product design (e.g. design for disassembly), supply chains (reverse logistics and procurement) and business models (e.g. take-back or rental)
4. **Regenerating:** use renewable material and energy where possible, and improve the natural environment (e.g. by collaborating to address biodiversity loss)

Business model innovation is defined as "designed, novel, nontrivial changes to the key elements of a firm's business model and/or the architecture linking these elements" (Foss & Saebi, 2016, p. 201). Building on this general definition, we can conclude that circular business model innovation involves a focus on the elements of a business model and its architecture to improve how tangible resources (e.g. energy, materials) are used, including the principles of narrowing, slowing, closing and regenerating.[2] Furthermore, as resources and their flows do not stop at the boundary of one organisation, an integrated perspective is needed to combine several different lenses to circular business model innovation.

---

2 The specific nature of circular business model innovation involves a focus on tangible resources and the efficiency in the use of those, and this is why we particularly highlight them here. However, also intangible resources such as data, knowledge, and financial resources are important for (any) business model innovation, including circular.

Three lenses to circular business model innovation

Figure 11.1 visualises three important lenses for understanding circular business model innovation. The 'narrowest' lens is that of a value proposition; it looks at the interface between the firm's business model and its customer offering (Osterwalder et al., 2014). However, starting from the customer value proposition is important, as, ultimately, all business models are built around some type of customer (e.g. consumer or an industrial customer), who assesses and appreciates different aspects of the value proposition (Payne et al., 2017). Then, we use the firm-centric lens to discuss a firm's value creation, delivery and capture schemes (Teece, 2010). Here, the focus is on how the firm designs the overall system of its activities that are required to create and deliver the value proposition, and also to capture value from doing this. Finally, the 'wide lens' view is that of ecosystem, where the focus is on the complementary actors and interdependencies among those actors in delivering and scaling value propositions and business models (Adner, 2017; Snihur et al., 2018).

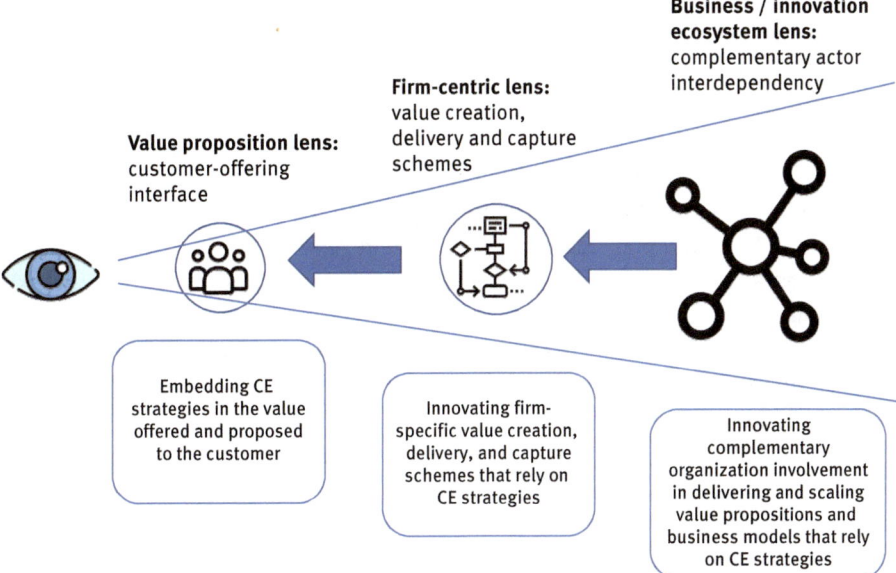

Figure 11.1: Three lenses to circular business model innovation[3].
Source: Authors.

---

3 Icon attributions: Icons made by Freepik, Eycalyp, and Kiranshastry from www.flaticon.com.

The three lenses we highlight are complementary as well as partially overlapping[4]. The value proposition has been viewed by some, as a part of the overall business model (Osterwalder et al., 2014), and also ecosystem scholars have viewed value proposition as an important facet of the ecosystem perspective (Adner, 2017; Thomas & Ritala, 2021). Furthermore, business models are often viewed not only as firm-centric constructs, but also including the external activity system where, for example, suppliers and complementors are included as part of the firm's business model (Zott & Amit, 2010). However, for the sake of analytical clarity and applicable implications, in this chapter, we use these three lenses as ways to look at different aspects of circular business model innovation.

Value proposition lens: Framing value for customers, environment and society

The value proposition refers to the customer-facing promise of the value that the provider creates, which is communicated by different means to the customer and the stakeholders – in essence, what value is delivered to whom (Payne et al., 2017). However, research has demonstrated that circular and sustainable value propositions also involve a promise of the social and environmental benefits (Manninen et al., 2018). Therefore, circular business model innovation should incorporate a clear value proposition for the customer, which, at the same time, explicates how CE strategies are embedded in this value proposition (Ranta et al., 2020). For example, product longevity and upgradability, or the regeneration of biodiversity, could be included as part of the value proposition and communication strategy towards the customer and stakeholders.

Some research has already identified examples or contents of value propositions building on CE strategies. For instance, Ranta et al. (2020) identify multiple B2B value propositions that communicate CE benefits to the industrial customer, while, often, also demonstrating a concrete business case. Furthermore, Bocken and Ritala (Bocken & Ritala, 2022) discuss six ways to build circular business models, including corresponding examples of value propositions. For instance, a value proposition that involves both customer value and CE strategies for slowing the loop includes the reuse of resources to broaden the offerings to the customer (e.g. vintage, second hand). In such a value proposition, the customer gets broader offerings due to the access to second-hand products, while the reuse aspect includes the environmental benefit. Similarly, apps that help restaurants sell food they would

---

[4] We realize that these three lenses might not be an overarching or the only way to represent circular business model innovation. Other useful lenses might include zooming in on the product or service, or zooming out to the institutional and regulatory context.

otherwise throw away provide a clear customer benefit (affordable food), additional revenue for the restaurant, as well as environmental benefits by reduction of food waste.[5] On the other hand, organic foods focus on the regenerative aspects: when consuming this type of food, the consumer can 'feel good' about buying a product that does not cause environmental harm, and instead pursues to regenerate and preserve nature and achieve personal health benefits (Anisimova, 2016; Davies et al., 1995). Such findings demonstrate the multidimensional nature of the value proposition concept: while customers almost always assess the economical and functional benefits, they also increasingly appreciate the emotional and symbolic benefits of value propositions (Rintamäki et al., 2007), many of which relate to CE and environmental issues, more broadly.

In summary, the aim of the value proposition lens to circular business model innovation is to design a clear value proposition that fulfils a customer need, while, at the same time, address environmental and social issues related to the resource use and efficiency.

Firm-centric lens: a scheme for creating, delivering and capturing value

In circular business model innovation, the starting point at the firm level is the CE vision that drives the innovation, in addition to an appealing or innovative offering to the customer (Brown et al., 2021). Thus, the firm-centric lens focuses on how a firm structures its processes and practices to enable value creation, delivery and capture in the CE context.

As for narrowing the loop, a good example is the Toyota Production System, focused on continuous improvement and editing out waste throughout all types of processes. The Toyota example represents a firm-level innovation focused on value creation (Bocken et al., 2014; Lüdeke-Freund et al., 2016). Part of this environmental improvement may be communicated to the customer, but it, first and foremost, leads to cost savings in the final product (in this case, a car), providing both a possible customer benefit in terms of reduced prices and circularity benefit in terms of reduced material usage. To slow the loop, Swapfiets[6] is a good example that provides a bike membership model on an as-a-service basis, which incentivises the company to maximise the useful lives of the bikes and minimise maintenance and operating costs. As for closing the loop, companies such as MUD jeans[7] offer jeans on a lease basis: after usage, customers can return the products and the materials

---

5 See e.g. https://www.resq-club.com/.
6 https://swapfiets.nl.
7 https://mudjeans.eu.

are turned into new garments. This is embedded in a new offering (e.g. 'lease jeans') in addition to a new logistics model, and financial model, where the jeans are only being paid back to the company over an extended time through monthly instalments. An example of a firm-centric lens to regenerating is the use of renewable electricity in the production and supply chain, as exemplified by Apple with its efforts to help suppliers and manufacturers switch to renewable energy.[8] Business model innovations can also focus on firm-internal resource efficiencies. In fact, Bocken and Ritala (2021) argue that the most effective business model in terms of CE is the one where the material and resource flows are fully processed and circulated in-house, leading to maximum material efficiency. However, many circular business models require integrating inputs across diverse supply chains (Hofstetter et al., 2021), and therefore, firms need to be able to design business models that involve ways to improve the effectiveness of collaboration, coordination and alignment with external actors. In the next section, we will discuss this more extensively in adopting the "ecosystem" lens.

In summary, the firm-level lens involves a focus on the business and how it can contribute to environmental and societal issues, through its value proposition, value creation and delivery and value-capturing mechanisms, while making an innovative or lucrative offering to the customer.

Ecosystem lens: boosting the scalability of circular business models

The ecosystem lens to circular business model innovation recognises how inter-organisational flows of natural resources (e.g. industrial-side streams or recycled materials) link to the firm's business model, and the roles of different complementors and partners in achieving circularity. While existing business model research has recognised that individual business models incorporate not only internal processes but also economic activities and actors in the broader ecosystem (Zott & Amit, 2010), the CE context widens this lens to include an ecosystem perspective with not only economic flows, but also the flow of materials (Aarikka-Stenroos et al., 2021). This systemic perspective is inherently connected to the CE context, and widely recognised in the literature on industrial ecology (Lowe & Evans, 1995) and industrial symbiosis (Chertow & Ehrenfeld, 2012). Indeed, recent case studies on CE business models have identified the roles for different ecosystem partners (Frishammar & Parida, 2019).

Ecosystems, often (but not always), use digital platforms (Cennamo, 2021) as a way to coordinate various complementary actors around a system-level value

---

8 https://www.apple.com/newsroom/2021/03/apple-powers-ahead-in-new-renewable-energy-solutions-with-over-110-suppliers/.

proposition (Thomas & Ritala, 2021). Aligned with this development, circular business models are increasingly being innovated and scaled up via digital platforms. Indeed, Konietzko et al. (2020) suggest that firms should focus on circular ecosystem innovation to understand how inter-firm collaboration and the emergence of digital platforms can improve the viability and sustainability of circular business models. Platform owners have been shown to act as 'circularity brokers' (Ciulli et al., 2020) connecting material flows together, thus reducing waste and excess resources. More broadly, digital platforms make social and economic exchange more efficient (Konietzko et al., 2020). In the context of the CE, they help to better market and disseminate circular products, components and materials, operate product-service systems and enable crowdsourced innovation (Konietzko et al., 2019).

Recent evidence in business practice has demonstrated multiple ways of how ecosystems are built around CE strategies, using both contractual and partnership models, as well as platform-based models to enable business model innovation. For instance, Adaptive City Mobility[9] seeks to maximise the efficiency of using cars in cities, by enabling the shared use of cars among fleet operators like bakery chains, security and taxi companies. This leads to fewer cars in a city, thus narrowing resource flows on an ecosystem level. Another example of closing and slowing the loop on an ecosystem scale is the 'Loop platform'[10] by Terracycle, which brings together fast-moving consumer goods manufacturers and retailers in a platform that enables typical supermarket products to be delivered in a reusable and recyclable manner, where the reuse and recycle effort takes place mainly within the platform organisation – and not the customer. This type of innovation focuses simultaneously on convenience and sustainability and has the potential to disrupt the supermarket business, while solving plastic packaging issues, as at present, only 3% of packaging is reusable (UN Environmental Programme, 2019). Another recent example in the business-to-business space is the Excess Materials Exchange, a start-up that is organising a platform for matching the supply and demand of excess materials across different industries.[11] As for regenerative business models, companies such as Patagonia and its food business have been focusing on regenerative agriculture and reviving wild salmon in nature through collaborations with various experts such as biologists and NGOs.[12] As for closing the loop, the Networks initiative between Interface, Aquafil and NGO Zoological Society of London seeks to clean up the sea of discarded nylon fishing nets, prevent them from being discarded in the future (in collaboration with local communities, e.g. in the Philippines), and turn these wasted materials into new carpets (Bocken & Ritala, 2022). The design of the carpets is also

---

9 https://acm.city/home/.
10 https://loopstore.com.
11 https://excessmaterialsexchange.com/en_us/.
12 https://www.patagoniaprovisions.com/pages/sourcing-salmon.

inspired by the ocean to communicate the story to the customer, which links back to the value propositions aspects.

In summary, ecosystem lenses are needed to address systemic CE innovations and business models, involving new collaborations with likeminded stakeholders to address the institutionalised sustainability and circularity challenges.

Discussion and conclusions

In this chapter, we have provided a conceptualisation of three alternative lenses that can be used to examine and explain circular business model innovation. In practice, all these lenses are beneficial and often required to build feasible circular business models. However, they can also serve as different viewpoints or points of departure to study the phenomenon.

First, a solid **value proposition** is able to address a concrete need and to communicate its benefits (Payne et al., 2017). In CE context, the best value propositions help to communicate the circularity benefits to different stakeholders, in addition to the direct benefits to the customer (Ranta et al., 2020).The improved communication, then, could mobilise further attention and action towards CE initiatives related to the particular business model, but ever more broadly. Furthermore, customer value is increasingly intangible and embedded in emotional and symbolic meanings (Rintamäki et al., 2007). Accordingly, firms should pay attention to consumers' or business customers' perceptions of the higher-order moral, ethical and sustainable dimensions, in addition to focusing on communicating the direct economic or resource circulation benefits (see also Sipilä et al., 2021).

Second, holding a **firm-centric lens** to circular business models helps analyse the internal workings of the business model; for instance, some of the circularity innovations such as process innovations in resource efficiency might, in fact, be completely invisible for customers. Focusing on such improvements and innovations has the potential to significantly improve a firm's competitive advantage, given the firm-specificity of many productivity improvements. On the other hand, there are great opportunities to innovate on the value creation and capture aspect of circular business models that transform the value proposition. For instance, take-back models (Bocken & Ritala, 2022) and remanufacturing models (Spring & Araujo, 2017) change the logic of what resources are used in production and how, and they also help redefine how the firm interacts with its customers and captures value from its value propositions.

Finally, it often pays off to look broader and wider, and adopt the **ecosystems lenses**. A firm's business model is, in no way, isolated from the external environment; in fact, in circularity contexts, the full analysis of material and resource flows typically requires looking across a variety of ecosystem actors (Aarikka-Stenroos et al., 2021). Ecosystem-based business models in CE typically rely on complementors that provide different types of materials, resources or services to complement or deliver the value proposition. For instance, Ciulli et al. (2020) demonstrated that reductions in food waste require coordination across multiple actors and brokerage of those actors' food waste by dedicated platform organisations. Furthermore, the recycling of components and materials often requires many actors to coordinate their activities. The Circular Footwear Alliance, for example, brings competing safety footwear firms, recycling firms and technology providers together to generate sufficient volume to make recycling viable and to experiment with joint reverse logistics and different recycling technologies.[13]

Table 11.1 summarises the three lenses with examples from four alternative ways to innovate resource circularity (narrow, close, slow and regenerate loops). The examples used in the table draw mainly from existing work that has pursued to

Table 11.1: Summary and examples.

	Value proposition lens: customer-offering interface	Firm-centric lens: value creation, delivery and capture schemes	Ecosystem lens: complementary actor interdependency
Narrow the loop	Cost saving, environmental benefit: e.g. reduced price for customer due to lower material use (smaller budget cars)	Major efficiency improvements in firm's processes: e.g. Toyota Production System	Cost saving, resource savings on the ecosystem level: e.g. Adaptive City Mobility
Slow the loop	Durability, quality, service, feel-good factor: e.g. Vintage and second-hand models, such as Depop, Rent the Runway, Vinted	Service models, incentives to minimise operating cost: e.g. Swapfiets	Align and coordinate complementary actors to improve the system: e.g. Loop store
Close the loop	Convenience, feel-good factor: e.g. ResQ Club and food waste reduction	Reverse logistics and access to secondary raw materials: e.g. MUD Jeans	Aligning actors around the recovery of waste material: e.g. Interface Networks or Circular Footwear Alliance
Regenerate the loop	Feel-good factor, environmental and health consciousness: e.g. organic foods consumerism	Renewable electricity in the production and supply chain: e.g. Apple solar panel installation initiative	Aligning actors around regenerating natural ecosystems: e.g. Patagonia regenerative agriculture

Source: Examples partially adapted and further developed from Bocken and Ritala (2022) and Lüdeke-Freund et al. (2016).

---

13 https://www.cfalliance.eu.

conceptualise different aspects of circular business model innovation (Bocken & Ritala, 2022; Lüdeke-Freund et al., 2016). Here, we adopt and extend those examples to highlight the differences across the three lenses.

The insights presented in this chapter highlight the utility of adopting multiple lenses to circular business model innovation. The three lenses of value proposition, firm and ecosystem are mutually complementary. For instance, an effective value proposition that communicates the circularity benefits is directly related to what a firm does internally and in its business or innovation ecosystem. Furthermore, the firm's internal activities and those by its complementors and supply-chain partners are certainly interdependent. Thus, all three lenses are helpful when used in concert; however, for analytical purposes, scholars or practitioners might benefit from zooming in with a particular lens. The key takeaway is that researchers and practitioners should make informed choices on the lenses they adopt when analysing the phenomena. Furthermore, our chapter serves as an illustration of the diversity of the circular business model innovation. Most firms can improve the majority of the aspects we identified; however, the firm-specific business models are likely to differ, and therefore, major innovation should be focused on the places where the biggest business and circular impact is expected.

Going forward, we propose to extend the lenses explored in this chapter to include insights from consumer behaviour, and to include cultural, legal and institutional aspects. It is important to consider the people that are adopting the lenses, and the values, norms and beliefs that are driving action for the CE. Integrating these as the starting point of circular business model innovation can generate multiple benefits, including daring to question the status quo of a linear and consumerist society, and to start the innovation process with a questioning of what the people involved in the process care about (Breuer & Lüdeke-Freund, 2018). Furthermore, it is important to recognise the importance of environmental policy to scale up circular business model innovation. Urgently needed for the viability of circular business model innovation are, among others, a higher carbon price, lower taxes on labour, producer ownership and higher recycling rates (Calisto Friant et al., 2021).

References

Aarikka-Stenroos, L., Ritala, P., & Thomas, L. D. (2021). Circular economy ecosystems: A typology, definitions, and implications. In S. Teerikangas, T. Onkila, K. Koistinen, & M. Mäkelä (eds) *Research handbook of sustainability agency* (pp. 260–276). Edward Elgar Publishing.

Adams, R., Jeanrenaud, S., Bessant, J., Denyer, D., & Overy, P. (2016). Sustainability-oriented innovation: A systematic review. *International Journal of Management Reviews, 18*(2), 180–205. https://doi.org/10.1111/ijmr.12068

Adner, R. (2017). Ecosystem as structure: An actionable construct for strategy. *Journal of Management, 43*(1), 39–58. https://doi.org/10.1177/0149206316678451

Anisimova, T. (2016). Integrating multiple factors affecting consumer behavior toward organic foods: The role of healthism, hedonism, and trust in consumer purchase intentions of organic foods. *Journal of Food Products Marketing, 22*(7), 809–823.

Autio, E. (2021). Orchestrating ecosystems: A multi-layered framework. *Innovation: Organization and Management, 00*(00), 1–14. https://doi.org/10.1080/14479338.2021.1919120

Bidmon, C. M., & Knab, S. F. (2018). The three roles of business models in societal transitions: New linkages between business model and transition research. *Journal of Cleaner Production, 178*, 903–916. https://doi.org/10.1016/j.jclepro.2017.12.198

Bocken, N., de Pauw, I., Bakker, C., & van der Grinten, B. (2016). Product design and business model strategies for a circular economy. *Journal of Industrial and Production Engineering, 33*(5), 308–320. https://doi.org/10.1080/21681015.2016.1172124

Bocken, N. M., & Ritala, P. (2022). Six ways to build circular business models. *Journal of Business Strategy, 43*(3), 184–192. https://doi.org/10.1108/JBS-11-2020-0258

Bocken, N., Short, S., Rana, P., & Evans, S. (2014). A literature and practice review to develop sustainable business model archetypes. *Journal of Cleaner Production, 65*, 42–56. https://doi.org/10.1016/j.jclepro.2013.11.039

Bocken, N. M. P., Stahel, W., Dobrauz, G., Koumbarakis, A., Obst, M., Matzdorf, P. (2021), Circularity as the new normal. Future fitting Swiss business strategies. WWF Switzerland and PWC. https://www.wwf.ch/sites/default/files/doc-2021-01/Circularity-as-the-new-normal_whitepaper-EN.pdf

Breuer, H., & Lüdeke-Freund, F. (2018). Values-based business model innovation: A toolkit. In onder redaksie van Lars Moratis, Frans Melissen, Samuel Idowu, *Sustainable business models: Principles, promise, and practice* (pp. 395–416). Springer. https://doi.org/10.1007/978-3-319-73503-0_18.

Brown, P., Bocken, N., & Balkenende, R. (2019). Why do companies pursue collaborative circular-oriented innovation? *Sustainability, 11*(3), 1–23. https://doi.org/10.3390/su11030635

Brown, P., Von Daniels, C., Bocken, N., & Balkenende, R. (2021). A process model for collaboration in circular-oriented innovation. *Journal of Cleaner Production, 286*, 125499. https://doi.org/10.1016/j.jclepro.2020.125499

Cainelli, G., D'Amato, A., & Mazzanti, M. (2020). Resource efficient eco-innovations for a circular economy: Evidence from EU firms. *Research Policy, 49*(1), 103827. https://doi.org/10.1016/j.respol.2019.103827

Martin, C. F., Vermeulen, W. J. V., & Salomone, R. (2021). Analysing European Union circular economy policies: Words versus actions. *Sustainable Production and Consumption, 27*, 337–353. https://doi.org/10.1016/j.spc.2020.11.001

Cennamo, C. (2021). Competing in digital markets: A platform-based perspective. *Academy of Management Perspectives, 35*(2), 265–291.

Chertow, M., & Ehrenfeld, J. (2012). Organizing self-organizing systems: Toward a theory of industrial symbiosis. *Journal of Industrial Ecology, 16*(1), 13–27. https://doi.org/10.1111/j.1530-9290.2011.00450.x

Ciulli, F., Kolk, A., & Boe-Lillegraven, S. (2020). Circularity brokers: Digital platform organizations and waste recovery in food supply chains. *Journal of Business Ethics, 167* [Springer Netherlands]. https://doi.org/10.1007/s10551-019-04160-5

Davies, A., Titterington, A., & Cochrane, C. (1995). Who buys organic food? A profile of the purchasers of organic food in Northern Ireland. *British Food Journal, 97*(10), 17–23.

Ferasso, M., Beliaeva, T., Kraus, S., Clauss, T., & Ribeiro-Soriano, D. (2020). Circular economy business models: The state of research and avenues ahead. Business Strategy and the Environment, 29(8), 3006–3024

Foss, N. J., & Saebi, T. (2016). Fifteen years of research on business model innovation: How far have we come, and where should we go? *Journal of Management, 43*(1), 200–227. https://doi. org/10.1177/0149206316675927

Frishammar, J., & Parida, V. (2019). Circular business model transformation: A roadmap for incumbent firms. *California Management Review, 61*(2), 5–29. https://doi.org/10.1177/0008125618811926

Geissdoerfer, M., Pieroni, M. P. P., Pigosso, D. C. A., & Soufani, K. (2020). Circular business models: A review. *Journal of Cleaner Production, 277,* 123741. https://doi.org/10.1016/j.jcle pro.2020.123741

Geissdoerfer, M., Savaget, P., Bocken, N. M. P., & Hultink, E. J. (2017). The circular economy – A new sustainability paradigm? *Journal of Cleaner Production, 143,* 757–768. https://doi.org/10. 1016/j.jclepro.2016.12.048

Guldmann, E., & Huulgaard, R. D. (2020). Barriers to circular business model innovation: A multiple-case study. *Journal of Cleaner Production, 243,* 118160. https://doi.org/10.1016/j.jcle pro.2019.118160

Hofmann, F., & Jaeger-Erben, M. (2020). Organizational transition management of circular business model innovations. *Business Strategy and the Environment, 29*(6), 2770–2788. https://doi. org/10.1002/bse.2542

Hofstetter, J. S., De Marchi, V., Sarkis, J., Govindan, K., Klassen, R., Ometto, A. R., Spraul, K. S., et al. (2021). From sustainable global value chains to circular economy – Different silos, different perspectives, but many opportunities to build bridges. *Circular Economy and Sustainability,* (2). https://doi.org/10.1007/s43615-021-00015-2

Kalmykova, Y., Sadagopan, M., & Rosado, L. (2018). Circular economy – From review of theories and practices to development of implementation tools. *Resources, Conservation and Recycling, 135*(February 2017), 190–201. https://doi.org/10.1016/j.resconrec.2017.10.034

Konietzko, J., Bocken, N., & Hultink, E. J. (2019). Online platforms and the circular economy. In: N. Bocken, P. Ritala, & L. Albareda, R. Verburg (Eds.), (435 –450) Innovation for sustainability. Palgrave Macmillan: London UK.

Konietzko, J., Bocken, N., & Hultink, E. J. (2020). Circular ecosystem innovation: An initial set of principles. *Journal of Cleaner Production, 253,* 119942. https://doi.org/10.1016/j.jclepro.2019. 119942

Lieder, M., Asif, F. M. A., Rashid, A., Mihelič, A., & Kotnik, S. (2017). Towards circular economy implementation in manufacturing systems using a multi-method simulation approach to link design and business strategy. *International Journal of Advanced Manufacturing Technology, 93*(5–8), 1953–1970. https://doi.org/10.1007/s00170-017-0610-9

Lowe, E. A., & Evans, L. K. (1995). Industrial ecology and industrial ecosystems. *Journal of Cleaner Production, 3*(1–2), 47–53. https://doi.org/10.1016/0959-6526(95)00045-G

Lüdeke-Freund, Florian, et al. "Business models for shared value." Network for Business Sustainability: South Africa (2016).

Manninen, K., Koskela, S., Antikainen, R., Bocken, N., Dahlbo, H., & Aminoff, A. (2018). Do circular economy business models capture intended environmental value propositions?. *Journal of Cleaner Production, 171,* 413–422. https://doi.org/10.1016/j.jclepro.2017.10.003

Mathieux, F., Ardente, F., Bobba, S., Nuss, P., Blengini, G., Alves Dias, P., Blagoeva, D., Torres De Matos, C., Wittmer, D., Pavel, C., Hamor, T., Saveyn, H., Gawlik, B., Orveillon, G., Huygens, D., Garbarino, E., Tzimas, E., Bouraoui, F. & Solar, S.

McDonough, W., & Braungart, M. (2002). *Remaking the way we make things: Cradle to cradle.* New York: North Point Press. New York.

Osterwalder, A., Pigneur, Y., Bernarda, G., & Smith, A. (2014). *Value proposition design: How to create products and services customers want.* New York. John Wiley and Sons Ltd.

Payne, A., Frow, P., & Eggert, A. (2017). The customer value proposition: Evolution, development, and application in marketing. *Journal of the Academy of Marketing Science, 45*(4), 467–489. https://doi.org/10.1007/s11747-017-0523-z

Pieroni, M. P. P., McAloone, T. C., & Pigosso, D. C. A. (2019). Business model innovation for circular economy and sustainability: A review of approaches. *Journal of Cleaner Production, 215,* 198–216. https://doi.org/10.1016/j.jclepro.2019.01.036

Ranta, V., Aarikka-Stenroos, L., Ritala, P., & Mäkinen, S. J. (2018). Exploring institutional drivers and barriers of the circular economy: A cross-regional comparison of China, the US, and Europe. *Resources, Conservation and Recycling, 135,* 70–82. https://doi.org/10.1016/j.resconrec.2017.08.017

Ranta, V., Keränen, J., & Aarikka-Stenroos, L. (2020). How B2B suppliers articulate customer value propositions in the circular economy: Four innovation-driven value creation logics. *Industrial Marketing Management, 87*(November 2019), 291–305. https://doi.org/10.1016/j.indmarman.2019.10.007

Rintamäki, T., Kuusela, H., & Mitronen, L. (2007). Identifying competitive customer value propositions in retailing. *Managing Service Quality: An International Journal, 17*(6), 621–634.

Ritala, P., Albareda, L., & Bocken, N. (2021). Value creation and appropriation in economic, social, and environmental domains: Recognizing and resolving the institutionalized asymmetries. *Journal of Cleaner Production, 290,* 125796. https://doi.org/10.1016/j.jclepro.2021.125796

Sarasini, S., & Linder, M. (2018). Integrating a business model perspective into transition theory: The example of new mobility services. *Environmental Innovation and Societal Transitions, 27*(October 2017), 16–31. https://doi.org/10.1016/j.eist.2017.09.004

Sipilä, J., Alavi, S., Edinger-Schons, L. M., Dörfer, S., & Schmitz, C. (2021). Corporate social responsibility in luxury contexts: Potential pitfalls and how to overcome them. *Journal of the Academy of Marketing Science, 49*(2), 280–303. https://doi.org/10.1007/s11747-020-00755-x

Snihur, Y., Thomas, L. D. W., & Burgelman, R. A. (2018). An ecosystem-level process model of business model disruption: The disruptor's gambit. *Journal of Management Studies, 55*(7), 1278–1316. https://doi.org/10.1111/joms.12343

Spring, M., & Araujo, L. (2017). Product biographies in servitization and the circular economy. *Industrial Marketing Management, 60,* 126–137. https://doi.org/10.1016/j.indmarman.2016.07.001

Stahel, W. R. (1994). The utilization-focused service economy: Resource efficiency and product-life extension. In *The greening of industrial ecosystems* (pp. 178–190). National Academy of Engineering. 1994. The Greening of Industrial Ecosystems. Washington, DC: The National Academies Press.

Stubbs, W., & Cocklin, C. (2008). Conceptualizing a 'sustainability business model. *Organization & Environment, 21*(2), 103–127. https://doi.org/10.1177/1086026608318042

Teece, D. J. (2010). Business models, business strategy and innovation. *Long Range Planning, 43*(2–3), 172–194. https://doi.org/10.1016/j.lrp.2009.07.003

Thomas, L. D. W., & Autio, E. (2020). Innovation ecosystems in management: An organizing typology. In *Oxford encyclopedia of business and management.* Oxford University Press. https://doi.org/10.5437/08956308X5706003

Thomas, L. D. W., & Ritala, P. (2021). Ecosystem Legitimacy Emergence: A Collective Action View. *Journal of Management,* 1–27. https://doi.org/10.1177/0149206320986617

Tukker, A. (2004). Eight types of product-service system: Eight ways to sutainability? Experiences from suspronet. *Business Strategy and the Environment, 13,* 246–260.

Tukker, A. (2015). Product services for a resource-efficient and circular economy – A review. *Journal of Cleaner Production, 97,* 76–91. https://doi.org/10.1016/j.jclepro.2013.11.049

Tukker, A., Bulavskaya, T., Giljum, S., de Koning, A., Lutter, S., Simas, M., . . . Wood, R. (2016). Environmental and resource footprints in a global context: Europe's structural deficit in resource endowments. *Global Environmental Change, 40*, 171–181. https://doi.org/10.1016/j. gloenvcha.2016.07.002

UN Environmental Programme. (2019). *The new plastics economy global commitment 2019 progress report*. https://www.unep.org/resources/report/new-plastics-economy-global-commitment-2019-progress-report.

World Economic Forum. (2020). *The global risks report 2020*. https://www.weforum.org/reports/the-global-risks-report-2020/

Wuyts, W., Marin, J., Brusselaers, J., & Vrancken, K. (2020). Circular economy as a COVID-19 cure? *Resources, Conservation and Recycling, 162*(June), 105016. https://doi.org/10.1016j.rescon rec.2020.105016

Zott, C., & Amit, R. (2010). Business model design: An activity system perspective. *Long Range Planning, 43*(2–3), 216–226. https://doi.org/10.1016/j.lrp.2009.07.004

Aglaia Fischer, Diane Zandee, Marleen Janssen Groesbeek

12 Finance and accounting in the circular economy

Abstract: This chapter discusses finance and accounting in the circular economy transition. Strategic decision-making in organisations is determined mostly by financial ratios to maximise financial profits. When companies adopt a circular business model, creating a positive environmental impact becomes part of the strategy and needs to be accounted for. Unfortunately, accounting for circular business output and impact is at odds with classical accounting, which is based on linear business models. By using data from three business cases that are in the process of creating a viable circular business model, we show opportunities and bottlenecks of financial accounting and financing for companies with circular business models.

The inability of current accounting principles and structures to register the additional value of circular business models, coupled with the lack of understanding surrounding the risks involved, is a major hurdle for valuing and funding these business models.

Keywords: Circular Economy, Sustainable finance, circular accounting, business model innovation, circular business models

Introduction

Circular business models (CBMs) are characterised by their value creation through circular activities and strategies designed to close the loop by reusing resources and materials. Design for longevity, modularity, refurbishing and recycling are all examples of company strategies to create financial value without harming the planet and its inhabitants (Bocken et al., 2016). For a company in transition towards a CBM, it is important to know how to capture and report the multiple values it aims to create with its CBM (Lahti et al., 2018). Understanding all costs and benefits of economic, environmental and social value created, and connecting them with the decision-making process, will support businesses that aspire to a CBM (Dewick et al., 2020).

It allows companies to integrate economic, environmental and social value and report on their multiple value creation. However, current managerial and financial decision-making is primarily informed by financial profits. This results in a situation where companies delivering high financial returns are favoured by capital providers even if they generate negative impacts and cause social and environmental harm. This creates an imbalance where companies that generate positive social and

https://doi.org/10.1515/9783110723373-015

environmental but have relatively low financial returns have difficulties raising enough capital to grow or survive. In short, externalities of business activities (i.e. the positive and/or negative social and environmental effects) are not sufficiently considered in decision-making because their impact is not expressed in financial terms.

Understanding the differences in financial and accounting valuation of linear and circular business cases could bring us closer to creating a level playing field to assess all companies by the same standards. This could result not only in accounting for financial value, but integrating financial, environmental and social costs and benefits (Aranda-Usón et al., 2019). Adding climate change and other environmental and social risks into the economic context would require mapping of and accounting for the externalities of creating products and services. The next step would be to incorporate them into the production and selling costs. (Albuquerque et al., 2019). This way, companies that generate the highest overall positive impact will become most attractive for investors in their pursuit to receive return on their investment (Elkington, 2017; Garcés-Ayerbe et al., 2019).

Creating a level playing field is easier said than done. The financial sector focusses on optimising financial profits (Musinszki & Suveges 2019) and neglects positive environmental and social impact (Galletta et al., 2020). This paradigm is not a matter of bad intention: it finds its origin in the industrial era in which large amounts of natural and human resources were available. Currently, society is facing the consequences of human and natural exploitation, and the damage to biodiversity and the onset of climate change (Salvioni & Brondoni 2020). The effects of climate change on living conditions require society to move to a new paradigm in order to stay within our planetary boundaries (UNESCO, 2013). In this chapter, we make the case that financial profits can no longer be the measure of success, but that economic prosperity depends on balancing financial, environmental and social returns (Wijkman & Skanberg 2014), and on executives who steer on maximising positive overall impact instead of financial profits. The case studies presented in this chapter show how businesses aim to develop CBMs that balance these returns. At the same time, these cases illustrate the hampering effect of current financial and accounting rules and frameworks.

With all societal and environmental challenges, companies and financiers have become aware of the economic costs of externalities and are changing their strategies. Companies with circular strategies create value for society and are the type of companies in which we ought to invest in order to avoid more and larger damages in the future (Ghisetti & Montresor 2020). The lack of understanding of CBM value creation and the vacuum of investment strategies hinder recognising and valuing the full potential of companies with CBMs (Aranda-Usón et al., 2019). Financial institutions are inclined to finance business models that they are already familiar with. This means that financiers are more likely to invest in CBMs that are familiar, such as recycling (Djuric et al., 2017), and reluctant to invest in CBMs taking a more radical approach: redesigning entire products or selling services connected to products (i.e. Product Service Systems (Tukker, 2004, 2015) (Stumpf et al., 2020)). More radical CBMs can deliver more substantial environmental and social benefits. But because they are different in terms of how and what value is created, financiers and accountants have difficulties understanding the strategy and the financial incentives behind them (Ozili & Opene 2021). Depending on the type of revenue model of a CBM and how circular incentives and activities are included in the agreements, the financial reports of CBMs – the main informants of financial decision-making – will differ from the reports of their linear counterparts. Hence, CBMs will encounter major obstacles in attracting necessary funding.

Structure of this chapter

We have used empirical materials from three cases to explain the current perspective of financiers and accountants on (the value of) CBMs, and how the financial sector is challenged to rethink assumptions and principles needed to better understand value creation by and financial funding for CBMs.

Methodology

Empirical data have been gathered by the authors in the Coalition Circular Accounting (CCA) (Coalition Circular Accounting, 2020a, 2020b, 2020c). This coalition was initiated by Circle Economy and the Dutch association of chartered accountants (NBA) with the aim "to identify and overcome accounting-related challenges that hinder the transition to the circular economy." The coalition is characterised by its pre-competitive environment through which experts and scientists in the fields of finance, accounting and law work on practical cases. Our case study focuses on three CCA cases: (A) road-as-a-service (RaaS), (B) facades-as-a-service (FaaS) and (C) valorising residual resources (VRR), spanning from 2019 to 2020. The three CCA case trajectories were organised with an interactive format in which a real-life circular business case was the focal point. Workshops and thematic deep dives were organised to discuss specific topics such as the value proposition, valuation and reporting issues, risk perception and financeability issues of the CBMs.

The three case trajectories have been fully recorded, and afterwards were transcribed, coded and analysed. Also, financial models of the CBMs that were created and discussed during the case trajectories were used in our analysis. These data provide detailed accounts of how CBMs are developing their value propositions and why financiers and accountants experience difficulties correctly interpreting and valuing these CBMs within their current professional frameworks.

This chapter is written as a position paper. Based on the empirical materials, ideas and concepts that are relevant for scholars and practitioners interested in the interface of CBM creation and finance and accounting are explained and discussed. The next section provides key concepts regarding CBMs in relation to accounting and finance.

These concepts are further clarified by using an illustrative business case[1] to compare accounting and finance results for circular and linear BMs. Then, three empirical business cases are introduced, and results are presented that show difficulties regarding understanding, valuing and financing CBMs. The preliminary section discusses what changes can be considered to better understand and support CBMs and create a level playing field. The final section of this chapter provides a brief conclusion.

Circular business models, accounting and finance

Circular business and value creation

In the circular economy, institutions and companies take a more long-term perspective on value creation. Instead of the linear take-make-use-waste trajectory of a product, a CBM aims at using products and materials in continuous loops where products are managed in ongoing cycles (Geissdoerfer et al., 2017). To get the most out of the value of raw materials, each product life cycle phase should be extended as much as possible, by applying the circular principles of cascading, e.g. staying in the inner loops as long as possible and finally reaching the last loop of material recycling (Ellen MacArthur Foundation, 2013). Bocken et al. (2016) describe these fundamental circular strategies as slowing, closing and narrowing the loop. Figure 12.1 visualises the cascading activities that can be used to slow, close and narrow the loop.

One-directional linear supply chains must develop to multidirectional value chains with products and materials going back and forth. They will have many different interactions between supply-chain partners and have to be supported by IT and software and supporting services to organise and manage the system effectively (Genovese et al., 2015). In a circular value chain, companies can capture the value of products and materials not just by selling them once, but also by selling them and buying them back at a later stage or even by retaining total ownership. There are three general ways for a company to make their products or materials available to a client and generate revenue, sometimes referred as revenue models (Lüdeke-Freund et al., 2019). The revenue models that we distinguish are:

1. **Sale:** A sale is a transaction between two or more parties in which the buyer receives tangible or intangible goods, services or assets in exchange for money. A

---

1 Given the confidentiality of the financial figures of the three cases, which are based on business cases for actual investment proposals, the insights from these cases have been incorporated into an illustrative case study, washing machine-as-a-service case study. In this illustrative case, financial figures can be shown and the insights from a business case, financial and accounting perspective can be shared.

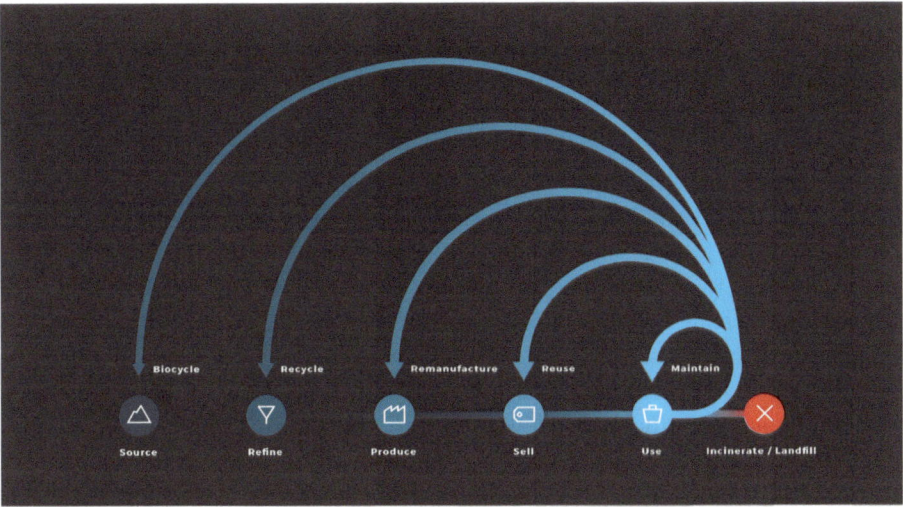

Figure 12.1: Circular 'cascading' activities (i.e. maintain, reuse, refurbish, recycle) can be used to slow, close and narrow the loop.
Source: Circle Economy (2016).

classic example is a company that sells a product to a client. By far, most economic transactions are sales transactions. In a sales transaction, the ownership of and the responsibility for a property is transferred (Ritzén & Sandström 2017).

2. **Sale and buy-back:** In a sale and buy-back transaction, the producer sells the product to the user with the aim (and potentially a formal agreement) to buy it back at end-of-use. There can be an agreement on a price upfront or the agreement can be made at the point of buy-back. Since the seller does not know what the state and value of the returned product will be, agreeing on a price beforehand can be a risk. However, if resource prices turn out substantially higher at the point of return, an upfront agreed price can be a benefit. When a sale and buy-back is formalised in a contract, this provides clarity and obliges both parties to adhere the contract. Without a formal agreement, a sale and buy-back becomes facultative. The sale can fall through in a situation where there is no formal agreement and the producer does not want to pay the price at the time of buy-back, or when there is another party that offers a higher price (Peace, 2014). In 2020, the Swedish furniture company Ikea started a large buy-back programme in 27 countries without a contractual agreement with its customers. They can get up to 50% back for their old Ikea furniture.

3. **Product-as-a-service (PaaS):**[2] In a PaaS business model, a product is not sold, but rather the performance of the product is sold. An example is the 'pay-per-

---

2 Note: There are many ways to combine products and services. Please see the extensive literature on product service systems (PSS) for a complete overview.

lux' business model from Philips. Philips maintains ownership of and is responsible for the fittings and light bulbs, while the client pays a periodic fee for enjoying the light (lux). PaaS is characterised by an ongoing contractual agreement – a service agreement – between the producer (and service provider) and the client. Moreover, since all products are returned to the producer after their use, the producer will benefit from creating a sustainable product that can easily be maintained, repaired, remanufactured and recycled (Romero & Rossi 2017).

These three revenue models each have certain effects on accounting structures of a company and are perceived differently by financiers. The longer the time horizon and the more control over products and materials (through buy-back agreements or service agreements), the more challenging for accountants and financiers to value circular companies. The reason behind this will be further explained below.

Circular business and accounting

The accounting process includes recording, summarising, analysing and reporting financial transactions pertaining to a business. The **financial statements** used in accounting are a concise summary of financial transactions over an accounting period, summarising a company's operations, financial position and cash flows. These financial statements include (1) the balance sheet, (2) the profit and loss statement (or income statement) and (3) the cash flow statement.

Balance sheet

The **balance sheet** shows the assets of the organisation, which must be in balance with the equity and debts of the company to finance the assets (Atrill & McLaney 2006). The **profit and loss statement** (P&L) shows the costs and revenues, and the profit (or loss) made.[3] The **cash flow statement** shows the cash flows of an organisation, consisting of profits and cash flows from operations, investments and financing. Transactions are recorded in the general ledger – the structure of accounting, each of which has a relationship with the balance sheet or P&L account (see Figure 12.2). Based on the relationships within this common structure, analysts and financial institutions can assess the vitality and the continuity of an organisation compared to

---

3 Definitions matter. Profit (loss), for instance, can relate to ongoing operations only, but can also include one-off transactions such as acquiring a firm or changing a firm's reserve position. Usually, profit (or loss) focuses on earnings before interest, taxes, depreciation and amortisation (EBITDA).

other organisations in the sector. The financial statements are subject to accountancy and other rules and guidelines set by financial and accountancy authorities such as the International Financial Reporting Standards Foundation (IFRS) or the Generally Accepted Accounting Principles in the United States (US-GAAP). Figure 12.2 presents a stylised picture of how the balance sheet, P&L and cash flow statements are related.

Figure 12.2: General accounting structure and relationships between the balance sheet, P&L and cash flow statement.
Source: Authors.

Accountants are responsible for the financial accounts and reports. The reports may include additional (non-financial) information to the financial data provided. In order to rightfully assess the activities and assets that a CBM represents, auditors have to understand the value created by companies that apply circular strategies and activities (Goretzki, 2013). Depending on the type of revenue model of a CBM and how circular incentives and activities are included in the agreements, these reports will differ from the reports of their linear counterparts. CBMs base their revenue on the multiple values created through circular activities (e.g. maintain, reuse, refurbish, recycle). Their aim is to use circular economy principles to create insight in the multiple value creation – optimising the use of products and materials and closing materials loops for continuous reuse of materials. Extending value creation beyond one product use cycle implies a long-term perspective. Moreover, this manner of managing products and materials in ongoing cycles implies a decrease in resource depletion, hence a decreasing environmental impact.[4] Exactly two benefits of circular business – (1) taking a long-term perspective, and (2) taking external benefits into account in production processes, and hence into price – are also visible in their accounting structures and are currently not well perceived by financiers. The reason for this is that when compared to reports from linear companies, the reports of circular businesses will show additional burdens due to holding on to

---

4 In reality, this is more complex. There is a trade-off between reducing resource depletion and increasing logistic movements. We assume that the impact of resource depletion is substantially more severe than the impact from increasing logistics. Moreover, organising logistics in a sustainable manner can further support this argument.

products and materials for a longer time span and performing circular activities that may yield return on investment only in the long run. Moreover, CBM activities that generate external benefits can entail extra costs that appear in the financial statements of the company.

Future financial results of CBMs are uncertain because they have strategies and revenue models that go beyond traditionally assumed time horizons and their control over products and materials. Because of their novelty, there are no long-term historical data available for CBMs, which makes it difficult for external auditors and financiers to assess their performance. Besides this lack of historical data, investment patterns often differ from their linear counterparts, because some CBMs choose to maintain ownership of, and responsibility for, products and materials over a longer period instead of buying materials and selling products following a rapid sequence, as is the characteristic of linear supply chains. Hence, companies with CBMs seem the odd ones out when they are benchmarked against their linear counterparts. Why and how financial reporting of CBMs is so different from linear ones will be discussed below.

Asset valuation and depreciation

Composing a balance sheet involves listing all the assets of a company. Assets contain economic value and/or future benefits and can be converted to cash. There are two broad asset categories: (1) Current assets can be converted into cash within one fiscal year. Examples are cash, accounts receivable and inventory. (2) Fixed assets are used to produce goods and services and have a lifespan of more than one year. Examples are land, buildings and machinery. Assets are depreciated throughout their useful life, meaning the asset value declines over time. This enables spreading the initial price of the asset over its useful life (i.e. economic lifespan). When an asset is no longer used, it is depreciated to zero, meaning it no longer has value for the company. In a circular economy, however, products and materials are managed in ongoing cycles. Therefore, one could argue that in a circular economy, we must take a different perspective on depreciation (Korse et al., 2016).

Circular business and financial assessment

Companies need financial capital to start and develop their business. Their financial structure reflects the mix of debt and equity that a company uses to finance its operations. Financial structures differ from company to company and from sector to sector. To increase the comparability of companies, financial ratios have been developed. Financial ratios are calculated with specific formulas and based on the numerical values taken from the financial statements. They provide a quantitative analysis to assess a company's solvency, liquidity, leverage, growth, margins, profitability, rates of return, valuation and more. Related ratios, explained in some detail below, have been developed over many centuries and are designed to give the capital providers insight into the risks they are taking by investing in the firm.

Financial ratios

Here the focus lies on three commonly used ratios that will turn out different for linear and circular business models: liquidity, solvency and profitability:
–   The *liquidity ratio* indicates whether a company has sufficient cash in hand to meet all its payment obligations in the short term and is calculated based on cash and cash equivalents and short-term borrowed capital. Quick ratio and current ratio are both indicators for cash flow and liquidity. Quick ratio is defined as the ratio between quickly available or liquid assets and current liabilities. The current ratio measures whether a firm has enough (financial) resources to meet its short-term obligations.
–   The *solvency ratio* indicates whether a company can repay all its debts if it ceases to exist or goes bankrupt. This ratio looks at the long-term debts and is the ratio of equity capital to loan capital or total assets. An example of a solvency ratio is the debt-to-equity (D/E) ratio.
–   *Profitability* ratios are used to assess a business's ability to generate earnings relative to its revenue, operating costs, balance sheet assets or shareholders' equity over time. The most-used ratio is the gross profit ratio. It is the profit a company makes after deducting the costs associated with making and selling its products, or the costs associated with providing its services.

A company that is financially healthy can (1) meet its short-term financial obligations with the liquid assets available to it (e.g. liquidity), (2) pay off its long-term debt (solvency) and (3) generate a profit (profitability). Historical data have been collected over many decades, resulting in clear brackets for ratios (for instance, with mining companies or consumer appliances companies). When a company that has traditionally been a manufacturing company wants to create a CBM and wants to increase its control over its assets (products) (for instance, in a PaaS model), this impacts the ratios. These ratios will likely fall outside the 'commonly accepted brackets', hence are difficult to accept within the current finance and accounting rules and regulations.

Risk

An important reason for applying the financial ratios is to calculate the risk profile of an organisation, or, if an organisation or investor is engaged in multiple activities, of an investment portfolio. One of the reasons management accounting was introduced is to monitor continuity perspective and therefore indicate the risk profile of a company (the continuity principle) (Vámosi, 2000). For investors and other financial stakeholders, the risk that a company will not continue to exist to meet its financial obligations is being assessed by financiers while using the commonly accepted accounting structures. When business models are altered to become more circular, the financial flows of an organisation will change as well (Larrinaga & Garcia-Torea 2022). The above-mentioned ratios will turn out differently in the financial reports of many circular businesses with a CBM. This results in a difficulty comparing these companies with their linear counterparts. From a financier's point of view, a company with a CBM is perceived to have a higher risk profile, since the upside benefits are long-term (often beyond financial assessment time horizon) and uncertain (due to the lack of historical data). Such a company will have to pay higher risk fee for its funding, if it's able to attract funding at all.

Application of concepts to an illustrative case

Having introduced basic and essential concepts related to accounting and finance for CBMs, we can now apply some of these concepts in a fictive case. This fictive case has been created based on the financial models that were developed during the real case trajectories and is a comparison of a linear and a circular scenario. For the sake of simplification and to protect sensitive company-specific data, the authors have created a dummy model for a sales model (the linear scenario) and a PaaS model (the circular scenarios) for washing machines.

Illustrative case: washing machines' business model – linear versus circular

For this illustrative case, we assume there is a washing machine producer that wants to move from selling the products (i.e. the linear scenario) to providing the washing machines in a PaaS model (i.e. the circular scenario). In the circular scenario, customers will pay a fee for every time they use their washing machine. The financial assumptions were modelled in a simplified balance sheet, income statement and cash flow statement for both scenarios to explain the differences between the two and to pinpoint the challenges to change from a linear to a circular business model.

The company currently sells 5,000 washing machines per year for a selling price of €750 each. These washing machines have a lifespan of approximately 10 years. In the proposed new circular model, washing-as-a-service (WaaS), the company will charge the customers for a unit price of €1.50 per wash, which amounts to a variable fee received by the producer every month. First, the balance sheet of the fictive linear and circular washing machine company is provided in Figure 12.3 (linear) and Figure 12.4 (circular).

Balance Sheet (linear)	Year 1	Income statement (linear)	Year 1
Assets	1,500,000	Turnover	3,750,000
Machines to produce washing machines	1,000,000	Costs overhead and retail	2,500,000
Cash	500,000	Depreciation	100,000
Liabilities	1,500,000	Interest	50,000
Equity	500,000	EBT	1,100,000
Debt	1,000,000		

Figure 12.3: Linear scenario balance sheet: Selling washing machines for a fixed price (*€).

Balance Sheet (PaaS)	Year 1	Income statement (PaaS)	Year 1
Assets	3,500,000	Turnover	1,260,000
Machines to produce washing machines	1,000,000	Costs overhead and retail	2,500,000
Rented washing machines	2,500,000	EBITDA	−1,240,000
Cash	0	Depreciation	350,000
Liabilities	3,500,000	Interest	150,000
Equity	500,000	EBT	−1,740,000
Debt	1,000,000		
Extra debt	2,000,000		

Figure 12.4: Circular scenario balance sheet in year 1: washing-as-a-service (WaaS) (*€).

In the balance sheet, the equipment (machinery) to produce the washing machines are reflected as fixed assets, which will be depreciated to zero within 10 years and fully financed by debt. In the case of a linear business model, the organisation will make a profit every year. Based on the CBM, the asset-base (the amount of assets owned by a company) of the company will be expanded by the amount of washing machines that are being 'serviced' to customers that pay a variable monthly fee (# wash cycles * €1.50). The circular company will make a yearly loss in the first few years and will need additional working capital to finance the production of the washing machines and to survive as a company in the first few years.

The washing machines, which have been produced by this company, are valued against the cost price on the balance sheet for €500 each, which results in a total of 5,000 washing machines, hence a valued asset-base of washing machines for €2,500,000. This extended balance sheet results in a solvency ratio of 14% (equity/total liabilities), which implicates a higher risk profile for this company than the linear business model, that has a solvency ratio of 33%.

The amount of rented washing machines will increase in the years to come, since per year 5,000 new units will be produced and provided to customers in a service agreement. In the second year, the company provides services of 10,000 units and the third year of 15,000 units. Based on the price per wash of €1.50 times 14 wash cycles per month, the business model will become more interesting over time. Figure 12.5 shows that the circular scenario starts to generate more revenue that the linear scenario in year 4.

Figure 12.5: Earnings before tax – linear and circular business model comparison washing machines (*€).

The financial results of the linear sales-business model will be more stable in the short run since sales transactions imply cash coming in quickly and this satisfies the short-term horizon of investors. However, the CBM has more future potential since the amount of assets increases over time and the revenue grows. Moreover, a circular asset base implies longevity of the assets; hence, more cash can be generated per asset. The perceived higher risk profile, based on a lower solvency rate and negative income in the first few years, marks an investment that results in a higher potential for the longer term.

Based on this simplified financial model, a cash flow forecast was generated to explain the financial differences between the linear and circular scenario. To start this CBM, additional cash is necessary to invest in an asset base. When revenue increases, based on a growing number of washing machines, the cash potential will strongly increase as well.[5] Figure 12.6 shows that the circular scenario will become more financially attractive than the linear scenario after year 5.

---

5 For simplicity, possible maintenance costs or other costs to support this business model are excluded from the financial model assumptions.

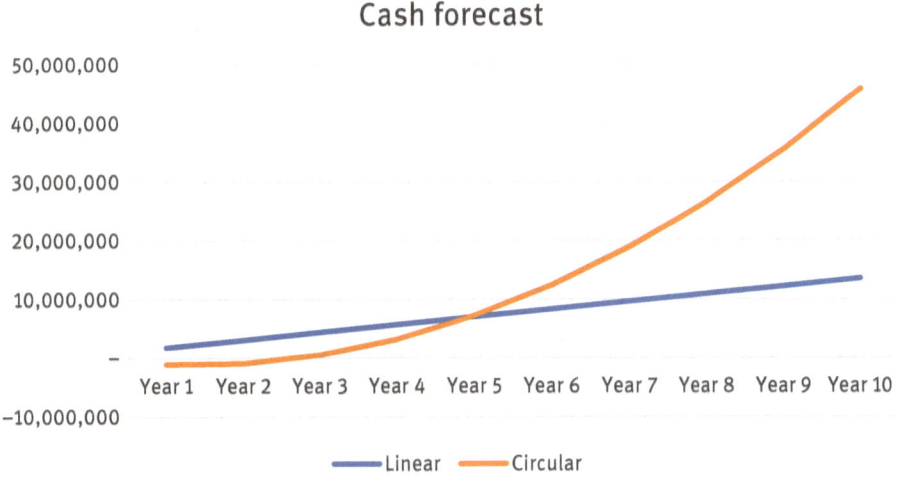

Figure 12.6: Cash forecast linear – circular business models washing machines (*€).

The profitability of the CBM will further increase by closing the loop and reusing old washing machine materials, since this saves on raw material costs. Using high-quality materials, which will lengthen the technical lifespan of the product, can result in an increased long-term profitability. Adding more services to the WaaS business model such as including the energy in this service model and therefore incentivising adding energy-reducing techniques to the washing machines, can further strengthen the network of sustainability and circularity businesses, provide incentives for both producers and users and can strengthen relationships between producers and users of the washing machines.

Financiers who were asked to invest in such a circular scenario were focussed on the solvency ratio (33% in a linear and 14% in a circular business case), which makes the linear case more attractive from a solvency point of view. Moreover, in years 1 to 5, the circular business case requires more cash, and the cumulative earnings will be higher in the linear business case. However, in the long run, the circular business case yields higher cumulative earnings.

Key accounting and finance challenges for CBMs

This section starts with an introduction of the three empirical cases that were analysed. The case introduction is followed by the analysis, supported with anecdotal examples of the difficulties these CBMs encountered regarding accounting and convincing financiers to invest.

Case A: Road-as-a-service

The road-as-a-service case revolved around the idea to (1) create a circular road, and (2) exploit this road as a service. The two main parties involved were a road building company (RBC) and a regional government (RG). The RBC has been experimenting with reusing materials from their old roads into new roads. This has resulted in a more sustainable road that can be used longer and can be taken apart to be reused. Under certain circumstances, the circular road could be demounted completely to be installed elsewhere. The RBC wants to explore the possibility of changing from a 'sales' revenue model to a 'product-as-a-service' revenue model. This implies that, instead of selling the road to the RG, the RBC stays owner of the road and remains responsible for the quality and safety. The RBC must maintain the road's optimal condition through monitoring, maintenance, and repair activities. The RG does not buy the road from the RBC, but instead pays a periodic fee for the services of the RBC (access to a road that is maintained to meet the standards of Dutch public roads). In this case, we focussed on the financial reality (accounting) and finance-ability (the ease of attracting funding) of a PaaS business model.

Case B: Facades-as-a-service

The facades-as-a-service case revolved around the idea to (1) create a circular façade, and (2) exploit this facade as a service. The main parties involved were a facade building company (FBC) and a real estate developer (RED). The FBC has developed a modular facade system that can be demounted and reinstalled. In this case, the focus lies on the contractual and financial structuring of a PaaS business model in the built environment. This case shows that the built environment is subject to a legal arrangement that complicates PaaS. In this case, so-called accession was a legal obstacle from real estate law that had to be overcome. Accession is a legal figure in which a smaller, independent physical object becomes part of a larger physical object. The FaaS case provides a clever contractual solution in the form of combining a rental agreement with a service contract. This provides a way to use contract law to bypass the issue of accession. From a financial perspective, however, no mortgage rights can be established in the case of a PaaS contract. This results in a misfit between contractual and financial structuring. Figure 12.7 shows the structure of the facades-as-a-service business model.

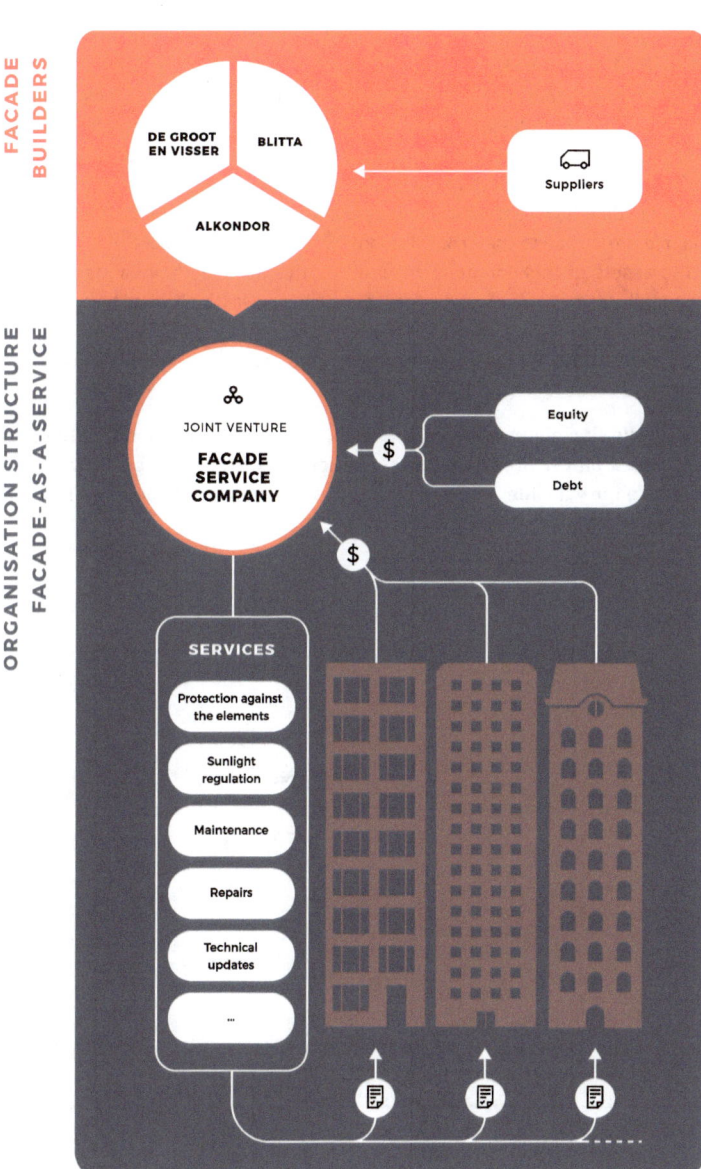

Figure 12.7: CBM structure of facades-as-a-service.
Source: The Circular Facade; Building a sustainable financial reality with facades-as-a-service[6] (2020).

---

6 https://www.circle-economy.com/resources/facade-as-a-service.

Case C: Valorising Residual Resources

This case revolved around a company that started a cooperative structure to use and valorise leftover resources from a food manufacturing process into new food products for human consumption. This enables higher value use of the leftover food that would otherwise be downcycled into animal feed. This company takes the lead in organising a supply chain that brings parties together with the aim of making new products from residual flows earlier in the supply chain. The profit is divided among the parties in the supply chain in proportion to their contributions. Specific challenges of this case were:

- How do we value the assets of this company when all parties involved have a stake in the cooperative and the main resource of the new food product is a waste product from another food production process of one of the parties involved?
- How do can we create incentives within this cooperative to pursue win-win situations where every collaborating party is rewarded fairly for its contribution?

This case led to a vibrant discussion about how to value residual resources, an important topic for circular economy that is based on cycling resources continuously. Various viewpoints were exchanged, leading to a range in the valuation of the residual stream. Figure 12.8 shows how residual resource flows are managed by the company, IntelligentFood.

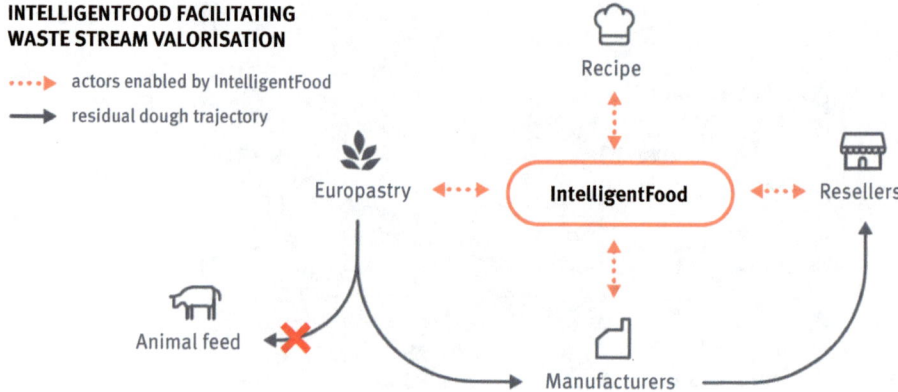

INTELLIGENTFOOD FACILITATING WASTE STREAM VALORISATION

····▶ actors enabled by IntelligentFood
——▶ residual dough trajectory

Figure 12.8: CBM structure of the Valorising Residual Resources case.
Source: Valorising Residual Resources; Mitigating food waste – How cooperatives can boost the circular economy[7] (2020).

Each case provided a different challenge in terms of finance and accounting. In Case A, the challenge was to create incentives for long-term usage of the materials within the CBM. Moreover, there were discussions on how to structure the service contract for the circular road on specific terms that reuse by the road building company would be guaranteed while staying within the brackets of what is acceptable

7 https://www.circle-economy.com/resources/valorising-residual-resources-mitigating-food-waste-how-cooperatives-can-boost-the-circular-economy.

for financiers. Case B zoomed in on the uncertainty in valuation of real estate when this real estate is being developed in by circular principles. The lack of historical data on separating material value from overall real estate value posed a challenge to valuation and to financial decision-making. Case C focussed on the pricing of residual (e.g. waste) streams, and on how to incentivise higher value reuse of residual resources. Creating a secondary market for residual resources resulted in discussions about how to correctly value these resources. Moreover, this CBM aimed at sharing profits by using a distribution key that was dependent on the value of these resources. A discussion about attributing value to resources before or after the sales of the final product was the result. The idea of sharing risks by dividing revenues in a later stage instead of paying a set price for resources and or services at the outset of a production process is interesting for incentivising circular supply chains, yet leads to discussions amongst accountants and financiers. This complexity translated in an additional identified risk.

Accountants, auditors and other financial professionals are trained to view organisations through the lens of risk and return. The illustrative case in the previous section has shown substantial differences between linear and circular CBMs. The structures and guidelines that financial professionals use to assess organisations are based on linear principles rather than circular ones. This section will show how linear accounting guidelines are an obstacle for a company to create a CBM. By using exemplary evidence from the three empirical cases, we illustrate that the current lenses are not equipped to fully understand and appreciate the value created with CBMs.

Financiers, investors and shareholders are informed on how a company is performing by a combination of non-financial and financial information. Financial assessment criteria presented above are the main data that inform financial decision-making. Financial ratios of circular businesses differ from their linear counterparts because of their circular business' strategies and activities and their different relationship with products, materials and the value chain. In CBMs, the value of resources is recognised and optimised before, during and after being used in products.

A company with a CBM aims to continuously reuse materials. For such a company, it is important to keep control over its assets, products and the embedded materials to ensure the continuity of a circular company. For a circular company, value lies in being less dependent on the availability of raw materials and less affected by price fluctuations of different resources. This goal of controlling the assets shifts the focus from assets to a combination of being in control of the assets *and* the management (controlling) of the value chain. The quality and continuity of the products and materials that flow back and forth through the value chain define partly the value of individual companies. Moreover, services become a more important part of the value proposition, especially when producers want to retain ownership of their products and make their use available in a PaaS model.

Balance sheet extension and ratios

In both PaaS cases (road-as-a-service and facades-as-a-service), the fact that companies wanted to remain owner of their products posed issues. Financial modelling of these cases showed that when a company owns a vast amount of assets, this results in an 'asset heavy' balance sheet. Based on the financial statements, financial ratios were calculated and discussed to assess the financial 'healthiness' of both projects. Perspectives of an accountant, a banker and a controller (i.e. internal accountant) provide insight in the perceived difficulties.

> Financing is difficult, balance sheet extension has been mentioned in this regard, including ratios that will be completely different [for PaaS]. (Accountant_PH_RaaS1)

> Preferably, I do not want it [the asset] on my balance sheet. But you notice that in the conversations we have with banks and other financiers that they are also struggling with this problem. Suppose 100 million [euro worth of assets] is added onto our balance sheet, then you see that it lowers the [solvency] ratio. Well then it probably wouldn't be financed by a bank. (Controller_AK_RaaS1)

> We are looking into how we, as a bank, can play a role in this and find a solution in the case of balance sheet extension. (Banker_IA_RaaS1)

CBMs and new securities

The revenue model behind a PaaS model is first and foremost about generating income by providing an ongoing service with an asset. It would be simple to compare selling a product with providing it as a service and earning a periodic fee. However, the activities that are taking place at the backend of a PaaS model – e.g. organising take back and logistics, monitoring software systems such as track and trace, maintenance, and repair, allocating products, elements and materials to the right value chain partners in order to add value again by circular cascading activities (refurbish, remanufacture, recycle) before products enter a new use cycle – create the *real* circular value potential. These elements also add substantial complexity to the system and to the exercise for accountants and financiers to judge its viability. The shift from selling products to PaaS models can be translated into financial structures – changing one single financial transaction (selling a product) into multiple and more frequent transactions (selling a service). The underlying contracts between companies and their customers on the one hand and between companies and their supply chain partners on the other hand can serve as new securities for financiers. These agreements secure ongoing cash flows (client contracts) and a smooth operation of the service (supply chain contracts). To guarantee cash flows with PaaS, the service provider must be sure there will be enough clients to use (and pay for) its service. The FaaS case shows how this uncertainty is perceived by financiers.

> The more certain the cash flow is, the better. We did talk about vacancy [of the apartments], but we also indicated that this risk is very limited. But they [financiers] do take that into account, so you have to show this in your scenarios and in your sensitivity analysis. (Controller of real estate developer_SV_FaaS3_model session)

Another bottleneck occurred relating to securities. Providing FaaS implied that mortgage rights – a well-known security used in real-estate – could not be established. A lawyer proposed to circumvent the issue altogether by creating an alternative contractual structure: a combination of a rental contract and a service contract. From a financial perspective, the underlying securities of this contractual structure are not mortgage rights, but step-in rights (in case of default, the bank can step in) and the right of removal (the FaaS company has the right to remove the facade if the user stops paying).

> We developed a rental structure that can be enforced by the owner of the façade. This is based on new case law, and it gives a legally enforceable right to take away the façade in case of non-payment [step in right]. Of course, this is undesirable, but it also gives the opportunity to say to the trustee in case of bankruptcy of the building owner, "are you going to fulfill the obligation? If not, then I will remove it, you are left with a building without facades." That would probably be enough pressure for the curator to continue to pay. Then you would have also secured those cash flows and could base financing on that. (WR Lawyer, CCA FaaS, 1)

However, current risk models assume mortgage rights, not step-in rights nor the right of removal. The financial sector currently has no framework or reference point for assessing this new contractual structure. This typically results in an increased risk premium, hence higher financing costs if such a CBM succeeds in obtaining financing at all.

> Establishing rights of superficies is from a legal point of view a somewhat less strong position than in case of mortgage rights, so you have to rely mainly on securing the cash flows as a security for the bank. (MO Banker, CCA FaaS, 1)

This example shows how circular businesses encounter many challenges due to the need to operate within the current linear financial metrics and ruleset. What is seen as a promising legal structure by pioneering businesses and legal experts can be interpreted as a liability by financiers. It shows how the conditions of the financial system and the (perception) of the risks in new CBMs lead to difficulties in valuing and financing CBMs.

Value, depreciation and risk allocation

Another challenge in the financial structures and valuation of CBMs is the uncertainty of estimating the value of a product during and after use phases. Roads (RaaS case) and buildings (FaaS case) will last for a long period and estimating the future value and application of materials is difficult because they depend on several variables such as future technical developments (e.g. new technologies such as self-driving cars) and geographic developments (e.g. how areas develop over the years). It is important to look at these issues from an accounting and valuation point of view, because in a circular economy, being able to retain and control the value of materials and resources is the main goal. Current accounting rules are designed to do the opposite – to depreciate their assets as quickly as possible.[8]

The introduction of a circular product may require more preparation, innovation and development. The purchase price can therefore be higher. On the flip side, being able to reuse components and materials harvested from used roads to compose a new road means production will require fewer virgin resources and potentially lower manufacturing costs if elements can be reused without much processing activities. After every use cycle, value is added again by circular activities such as maintenance, refurbishing and recycling.

During the RaaS case trajectory, it was discussed to stop *depreciation*, and start with the *appreciation* of materials. After all, these materials can be used indefinitely in continuous cycles. Current accounting rules do not allow for the full implementation of this residual value due to insufficient availability of practical examples and historical data. Due to insufficient data about the potential future value of products and materials in a CBM, depreciation stays the norm. This means the actual residual value remains 'nice to have' at the time of harvest (i.e. when the road is dismounted after, let's say, 35 years). However, to take this upside value into account in financing CBMs, we need to assume this 'future value' from the start. Several options of how to go about the value throughout the lifespan of the road were discussed.

> You could agree contractually that you have an annual or ten-yearly reassessment of the value of the entire property and then adjust the fee accordingly. We expect that the property will increase in value and thus the needed securities for the financier will decrease [over time]. (DZ_GaaS3)

Depreciation is still applied to these materials due to a lack of experience data of the value of raw materials over a long period and for reasons of prudence. Even if there are second-hand markets and a futures market for steel, concrete, bitumen, wood and other reusable materials in construction, valuing these materials 30 years from now is tricky because it needs insight in future economic developments. Moreover, assuming a higher residual value of products and materials in the future entails certain risks because of uncertainty. Contract parties must decide and agree beforehand how they will allocate these risks.

> Both [service provider] and [client] have to think very carefully, 'do I want certainty or am I also prepared to accept a minus in 30 or 40 years if the value turns out lower than expected. Or do I prefer to pay a little more periodically and have certainty that I do not have to pay

---

8 This is supported by current taxation schemes. Hence, tax is an important lever to discuss, yet remains out of scope in this chapter.

extra in 30 or 40 years in case that the value has collapsed.' That is what you should take into consideration. (RaaS_4)

In the Valorising Residual Resources (VRR) case, the aim was to use excess resources (for instance, from production processes) in high-value products. A resource flow that until recently was marked as a waste stream would become a valuable input for creating new products (in the case, the topic was making high-quality cookies from leftover dough that used to be sold against bargain prices to the animal feed industry).

A cooperative structure has been created to incentivise all stakeholders to collaborate and to benefit by sharing the profits from selling the new product. This inventive cooperative structure poses some challenges for valuing the residual resources ($t = 1$), since their value depends on the profit margin after sales ($t = 2$).

> For accounting we do not have many choices on valuation methods. It is regulated in the law and in accounting standards. Cost price or lower market value is applicable if you talk about inventory. If you receive the dough from the supplier, then there is the discussion on: what is the valuation on the balance sheet. When it is waste there is a valuation of 0, and if you must pay something then it's the cost price. If you feel that we might sell this product with the lower price, then you have a lower market value. This is not the case here I presume. However, the cost price is not known initially because the cost price can be considered as the 40% profit sharing which can be calculated only afterwards. That is a difficulty or challenge in this case study which is very interesting. (PH_VRR3)

Also, in the VRR case, there were discussions on how to take the value of the residual resources into account from the outset and not just as a nice to have after the sale of the final products. Being able to make this value explicit in the balance sheet signals a higher value of the business and makes it easier for a company to attract funding.

> The company really wants it [the value of the resources] on the balance sheet or at least somewhere to be visible because it provides a proof of actual value. When it is seen as 0 value it is seen as a waste stream. You want to signal that it does have a value towards financiers and other stakeholders. (AF_VRR2)

This section has shown that CBMs represent possibilities for adding and managing value through circular strategies and activities. The empirical materials from the three cases exposed the misfit between current finance and accounting practices and rulesets and CBMs. The following section presents solution directions for accounting and finance to better support of CBMs.

Finance and accounting for CBMs as the new default

A fresh perspective on accounting and finance practices and rulesets is needed to better understand and assess the value and risk profile of companies with a CBM. The current misfit has profound consequences for adopting and diffusing CBMs in both new and existing companies and for the wider circular economy transition. We argue that there is a need for an adjustment of the financial frameworks to acknowledge long-term sustainable value created by CBMs. Moreover, it is important to judge all companies with these adjusted frameworks in order to create a level playing field.

In the first section, key accounting and finance concepts were presented. Accounting is used to report the financial situation of a company in its financial statements. Financial ratios are calculated based on the financial statements. They enable comparing companies among one another and with historical data of similar companies. Moreover, financiers use these ratios to calculate a companies' risk profile – the likeliness a company will be able to fulfil its short-term and long-term financial responsibilities. The previous section showed that due to the newness of CBMs, historical data is often not available. Moreover, their tendency to generate value over a longer time horizon, and often by taking increased control (and potentially ownership) of their products and materials, causes the financial ratios to be outside the accepted brackets. This results in CBMs being perceived to be more risky than linear BMs, which makes it more difficult and more expensive to finance these CBMs, if possible.

How do we get out of this reinforcing cycle of not funding CBMs due to a lack of track record, which instigates the continuation of the lack of data? The answer lies in assessing the risks of not transitioning to a circular economy.

Mitigating climate risk and raw material risk

Financial professionals primarily assess organisations based on financial risk. Taking a long-term perspective, circular economy enables mitigating risks that occur at a different level than at the company level towards considering climate risk and the risk associated with scarcity of materials (Durán-Romero et al., 2020). Climate and resource risks will affect the continuity of businesses. Contributing to the circular strategies and activities can mitigate these risks and can support the long-term continuity of businesses.

Climate risk is increasingly recognised in annual reports, but has not yet found its way into the accounting guidelines. Since the year 2000, an increasing number of listed companies have started reporting on their carbon emissions, mostly according to the guidelines of the Greenhouse Gas (GHG) Protocol, with its scope 1, 2 and 3. The GHG protocol has become the reporting norm also in the legislation on corporate sustainability reporting and is used by governments to account for carbon emissions on a country level. Certain industries have already been subject to carbon pricing, but because of the complex calculations and thresholds, the financial costs of carbon pricing are not immediately recognisable in the financial accounts. The integrated annual reports of listed companies show their emissions, but not the related costs and not the direct and indirect risks related to carbon emissions. That is why, in 2015, the Task Force for Climate-related Financial Disclosure (TCFD) developed a framework for listed companies and financial institutions to disclose their physical and transition risks related to climate change.[9] Research shows that even though carbon accounting is still in its infancy, investors are already making their own assessments of the carbon risks of listed companies and pricing them in (Aswani et al., 2021). Companies that fail to take climate risk into account will likely drop in value when newly announced legislation such as carbon pricing becomes the norm. When this happens, the cost of carbon emission will show up in the financial accounts as a cost (as priced by the emissions trading system (ETS)).

Raw material risk. The impending scarcity of raw materials is not yet visible to many companies within the short-term horizon on which they manage their business. However, price volatility has increased over the years and there are prognoses of increasing material scarcity. Resource shortages and increasing prices will create a need to deal with raw materials differently. A company that anticipates this by creating circular strategies and activities that enable ongoing resource cycling will have an important competitive advantage over companies that fail to control their resource streams. These upsides of CBMs and risks of linear BMs are not sufficiently recognised in the accounting guidelines and remain out of scope for financial decision makers. These risks are not yet included in risk assessments, although they threaten the continuity of companies. The *continuity risk* in the longer term means that these risks do need to be clarified. CBMs do so by focusing on the long-term continuity of their own business, by preserving materials, reducing waste and environmental pressure (Gaustad et al., 2018).

Rethink depreciation and appreciate residual value

Rethinking depreciation in the circular economy can have an important positive effect on the financial attractiveness of CBMs. When a company owns assets that are

---

9 https://www.fsb-tcfd.org/.

no longer (fully) depreciated, this means they will end up having a higher *residual value* when compared to their linear counterparts that are fully depreciated. This is the case because in a CBM, products are designed so that they are easy to disassemble (modular), and they consist of durable materials that can be recycled. These products (and parts) are designed to be able to be used in multiple use cycles. A linear product made without these features is less easy to reuse and more likely to break down and be discarded. With its design for many use cycles, the circular product will likely have a higher *residual value* than the linear product. A higher residual value is beneficial to the owner of the product or materials. As explained above, the residual value of products and materials is currently seen as a 'nice to have' at the end of a product lifespan and cannot be taken into account as a real value from the outset of building the circular business case. If the residual value of the product or material can be priced in at the moment the product is created, it will likely have a positive effect on the value proposition of a company: lower costs, higher profits. At the same time, the future residual resource can be valued as an asset.

Valuation of materials and products is currently determined based on historical prices and knowledge. Because of physical and transitional changes in the economy, these historical data are no longer reliable or sufficient; it would require a review of calculation and valuation models, structures and methodologies currently used by financial professionals. At the same time, capturing ecological and social value of companies and insight into how companies capture value over different periods demand changes in risk management. Adapting to or mitigating climate change risks and continuously reusing raw materials would change the risk profile for a company and improve its license to operate and its continuity. A complicating factor that requires further research is the fact that value is determined by supply and demand, hence the existence of a market. When residual material markets are non-existent or supply and demand disconnected, the price of residual materials will be subject to uncertainty. The development of transparent (second-hand) markets is therefore a key leverage point for actual circular value creation and convincing financiers of accepting residual value as a security.

Longevity and increased control over assets as key risk mitigants

Above it was shown that a company that wants to transition towards a CBM in the form of a PaaS model encounters financial obstacles due to the growing amount of assets on the balance sheet. Companies with many assets are considered to be risky, because too much capital is tied up and its cash flow can be a problem. For financiers, that falls beyond their commonly accepted brackets.

However, if accountants and financiers would also consider the financial risks of the above-mentioned climate, raw material and continuity challenges and can assume a higher price and attribute less uncertainty to a future residual value, CBMs are mitigating risks for themselves and their financiers. As we have shown, CBMs are able to put value of reusing products and materials in multiple use cycles and make sure to keep control over their assets.

If we systemically start reusing products and materials and incentivise to produce and manage production in a sustainable way by scarcity and increasingly stringent policy measures, we can expect a higher market value for these products and materials in the future. With new policy measures in the pipeline especially in Europe, slowing, narrowing and closing the loop will increasingly become cost-effective. Companies that postpone the transition to CBMs will be confronted by extra costs, through carbon pricing, tax measures or scarcity. If this trend becomes apparent, companies with a CBM will turn out to be lower-risk investments than their linear counterparts.

Conclusions

CBMs focus on long-term continuity and limit negative environmental impact by optimising product and material reuse. Externalities are currently not considered in product prices, nor are they visible in the financial statements or risk assessments of a company. We propose accounting and finance practices, and frameworks must be adjusted to correctly value the upsides and risks of CBMs and Linear BMs.

This chapter has shown that the financial and accounting frameworks have been developed for an economic system, where depreciation is a leading principle and investment decisions are based on short-term financial profit rather than long-term value for society. Investors and financiers have become better in assessing environmental, social and governance performance of companies and have started changing their investment strategy and policy. But their assessments are still an overlay on old-fashioned financial assessments and fail to understand and value CBMs. The short-term focus and perceived higher risk profiles impede a rapid adoption of CBMs.

The current system of depreciation and reduced *residual value* is not reflecting the value of products and materials in CBMs that are maintained at their highest possible level throughout their lifespan. Depreciation costs reflect the diminishing value of materials and reduced profitability of the business case. A key element in CE is striving to keep the materials in a continuous loop of usage by recognising their usability and value. CE would require an alternative accounting standard for the value of materials used in a circular business model.

CBMs create new financial structures in balance sheet, cash flow and profit and loss statements. In the current system, this leads to a higher risk perception of financiers and investors, based on their interpretation of the financial ratios. At the same time, these new CBMs enable incorporating *climate change risks* and *raw material risks,* which are currently missing in the linear business models. Depending on the trade-off between the different risks, CBMs could therefore be perceived as lower-risk business models. New standards for accounting structures and financial ratios will be necessary to reframe and recalculate the risks and truly appreciate the long-term value of CBMs.

Understanding all aspects of CBMs and how value is created is key. Moreover, this requires multiple disciplines to tune in to one another's value proposition and collaborate in developing circular activities that optimise and strengthen the value chain. Understanding the potential upsides of CBMs regarding climate and raw material risk mitigation and increased residual value of products and materials is key; accountants must include this information in their financial statements, and financiers must recalibrate financial ratios and consecutive financial decision-making.

References

Albuquerque, T. L. M., Mattos, C. A., Scur, G., & Kissimoto, K. (2019). Life cycle costing and externalities to analyze circular economy strategy: Comparison between aluminum packaging and tinplate. *Journal of Cleaner Production*, *234*, 477–486. https://doi.org/10.1016/j.jclepro.2019.06.091

Aranda-Usón, A., Portillo-Tarragona, P., Marín-Vinuesa, L. M., & Scarpellini, S. (2019). Financial resources for the circular economy: A perspective from businesses. *Sustainability (Switzerland)*, *11*(3). https://doi.org/10.3390/su11030888

Atrill, P., & McLaney, E. (2006). *Accounting and finance for non-specialists*. 5th Edition. Pearson Education. https://books.google.nl/books?hl=nl&id=3JQ4z9xhwvcC&oi=fnd&pg=PR13&dq=principles+of+accounting+and+finance&ots=g24W7ghm9o&sig=-hq_U8s2oeKT1SdfxgyGs8feMZs#v=onepage&q=principlesofaccountingandfinance&f=false

Aswani, Jitendra and Raghunandan, Aneesh and Rajgopal, Shivaram, Are Carbon Emissions Associated with Stock Returns? (2022). Columbia Business School Research Paper Forthcoming, Available at SSRN: https://ssrn.com/abstract=3800193 or http://dx.doi.org/10.2139/ssrn.3800193

Bocken, N. M. P., de Pauw, I., Bakker, C., & van der Grinten, B. (2016). Product design and business model strategies for a circular economy. *Journal of Industrial and Production Engineering*, *33*(5), 308–320. https://doi.org/10.1080/21681015.2016.1172124

Circular 'cascading' activities; original figure is designed for presentation purposes and used in several Circle Economy presentations from 2016 onwards; Nicolas Rarpail for Circle Economy.

Coalition Circular Accounting. (2020a). *Pursuing financial reality of the circular road*. https://assets.website-files.com/5d26d80e8836af2d12ed1269/5e1de1e4a12b0a3cfa8999d4_Road-as-a-Service-Coalition-Circular-accounting-2020.pdf

Coalition Circular Accounting. (2020b). The Circular Facade. Facades-as-a-Service – Insights – Circle Economy (circle-economy.com).

Coalition Circular Accounting. (2020c). *Valorising residual resources*. https://assets.website-files. com/5d26d80e8836af2d12ed1269/5fd0f7fa8c0198c24fc7e3e3_CCA-Valorisingresidualresources-report-EN-compressed.pdf

Dewick, P., Bengtsson, M., Cohen, M. J., Sarkis, J., & Schröder, P. (2020). Circular economy finance: Clear winner or risky proposition? *Journal of Industrial Ecology, 24*(6), 1192–1200. https://doi.org/10.1111/jiec.13025

Djuric, S., Stosic-Mihajlovic, L., & Trajkovic, S. (2017). Circular economy and create new values: Recycling, renewable energy, ecology. *Journal of Process Management. New Technologies, 5* (3), 50–68. https://doi.org/10.5937/jouproman5-14306

Durán-Romero, G., López, A. M., Beliaeva, T., Ferasso, M., Garonne, C., & Jones, P. (2020). Bridging the gap between circular economy and climate change mitigation policies through eco-innovations and quintuple helix model. *Technological Forecasting and Social Change, 160* (August), 120246. https://doi.org/10.1016/j.techfore.2020.120246

Elkington, J. (2017). Saving the planet from ecological disaster is a $ 12 trillion opportunity. *Harvard Business Review*, 2–8.

Ellen MacArthur Foundation. (2013). Towards the circular economy: Economic and business rationale for an accelerated transition. *Ellen MacArthur Foundation*.

Galletta, S., Mazzù, S., Naciti, V., & Vermiglio, C. (2020). Sustainable development and financial institutions: do banks' environmental policies influence customer deposits? *Business Strategy and the Environment*, (no. March), 1–14. https://doi.org/10.1002/bse.2644

Garcés-Ayerbe, C., Rivera-Torres, P., Suárez-Perales, I., & Leyva De La Hiz, D. I. (2019). Is it possible to change from a linear to a circular economy? an overview of opportunities and barriers for european small and medium-sized enterprise companies. *International Journal of Environmental Research and Public Health, 16*(5). https://doi.org/10.3390/ijerph16050851

Gaustad, G., Krystofik, M., Bustamante, M., & Badami, K. (2018). Circular economy strategies for mitigating critical material supply issues. *Resources, Conservation and Recycling, 135* (January), 24–33. https://doi.org/10.1016/j.resconrec.2017.08.002

Geissdoerfer, M., Savaget, P., Bocken, N. M. P., & Hultink, E. J. (2017). The circular economy – a new sustainability paradigm? *Journal of Cleaner Production, 143*, 757–768. https://doi.org/ 10.1016/j.jclepro.2016.12.048

Genovese, A., Acquaye, A. A., Figueroa, A., & Koh, S. C. L. (2015). Sustainable supply chain management and the transition towards a circular economy: Evidence and some applications. *Omega, 66*(Part B), 344–357.

Ghisetti, C., & Montresor, S. (2020). On the adoption of circular economy practices by small and medium-size enterprises (SMEs): Does 'Financing-as-usual' still matter? *Journal of Evolutionary Economics, 30*(2), 559–586. https://doi.org/10.1007/s00191-019-00651-w

Goretzki, L., Strauss, E., & Weber, J. (2013). An institutional perspective on the changes in management accountants' professional role. *Management Accounting Research, 24*(1), 41–63. https://doi.org/10.1016/j.mar.2012.11.002

Korse, M., Ruitenburg, R. J., Toxopeus, M. E., & Braaksma, A. J. J. (2016). Embedding the circular economy in investment decision-making for capital assets – a business case framework. *Procedia CIRP, 48*, 425–430. https://doi.org/10.1016/j.procir.2016.04.087

Lahti, T., Wincent, J., & Parida, V. (2018). A definition and theoretical review of the circular economy, value creation, and sustainable business models: where are we now and where should research move in the future? *Sustainability (Switzerland), 10*(8). https://doi.org/ 10.3390/su10082799

Larrinaga, C., & Garcia-Torea, N. (2022). An ecological critique of accounting: Circular economy and COVID-19. *82*, 102320. https://doi.org/10.1016/j.cpa.2021.102320

Lüdeke-Freund, F., Gold, S., & Bocken, N. M. P. (2019). A review and typology of circular economy business model patterns. *Journal of Industrial Ecology*, *23*(1), 36–61. https://doi.org/10.1111/jiec.12763

Musinszki, Z., & Suveges, G. B. (2019). Strategic decision-making supported by traditional financial indicators. *Oradea Journal of Business and Economics*, *IV* (1): 41–57. http://real.mtak.hu/92482/1/OJBE_vol-41_29-37.pdf.

Ozili, P. K., & Opene, F. (2021). The role of banks in the economy. *Social Science Research Network*. https://doi.org/10.32609/0042-8736-2003-3-91-102

Peace, B. (2014). The Opportunities of a Circular Economy. *EG Magazine*, 2014. https://search.proquest.com/openview/1eca5e7d8d6e262a2854ef2b63a4d57f/1?pq-origsite=gscholar&cbl=2031817

Ritzén, S., & Sandström, G. Ö. (2017). Barriers to the circular economy – integration of perspectives and domains. *Procedia CIRP*, *64*, 7–12. https://doi.org/10.1016/j.procir.2017.03.005

Romero, D., & Rossi, M. (2017). Towards circular lean product-service systems. *Procedia CIRP*, *64*, 13–18. https://doi.org/10.1016/j.procir.2017.03.133

Salvioni, D., & Brondoni, S. (2020). Ouverture de 'Circular economy & new business models.' *Symphonya. Emerging Issues in Management*, (1), 1. https://doi.org/10.4468/2020.1.01ouverture

Stumpf, L., Schöggl, J.-P., & Baumgartner, R. J. (2020). Climbing up the circularity ladder? – a mixed-methods analysis of 131 circular economy business strategies. *Journal of Cleaner Production*, under revi.

Tukker, A. (2004). Eight types of product-service system: eight ways to sutainability? Experiences from suspronet. *Business Strategy and the Environment*, *260*, 246–260.

Tukker, A. (2015). Product services for a resource-efficient and circular economy – a review. *Journal of Cleaner Production*, *97*, 76–91. https://doi.org/10.1016/j.jclepro.2013.11.049

UNESCO. (2013). *World Social Science Report World Social Science Report*.

Vámosi, T. S. (2000). Continuity and change; management accounting during processes of transition. *Management Accounting Research*, *11*(1), 27–63. https://doi.org/10.1006/mare.1999.0122

Wijkman, A., & Skanberg, K. (2014). The circular economy and benefits for society: Jobs and climate clear winners in an economy based on renewable energy and resource efficiency. *Club of Rome*, 1–55. https://www.clubofrome.org/wp-content/uploads/2016/03/The-Circular-Economy-and-Benefits-for-Society.pdf

David Monciardini, Eléonore Maitre-Ekern, Carl Dalhammar,
Rosalind Malcolm

13 Circular Economy regulation: an emerging research agenda

Abstract: Circular Economy regulation is rapidly adopted at different levels of governance, including municipalities, private and industry standards, national law and supranational standard-setting. There are increasing demands from policymakers, businesses and civil society for theoretical insights and more knowledge on the effects of policies and laws ('policy learning'). However, despite this mounting interest, the literature still lacks a real strand of studies that systematically and comprehensively addresses this area of research. The chapter offers a review of the origins and limitations of the current model of regulation and the emergence of a law of the Circular Economy in Europe. We maintain that the policy and academic debate is currently too fragmented, depoliticised and undertheorised, highlighting existing gaps and new research directions.

Keywords: Circular Economy Action Plan, policy learning, Ecodesign Directive, EU law, waste regulation: sustainable product policy

Introduction

In the wake of an unprecedented level of interest in the risks of natural resource scarcity and waste pollution, and a consequent rise in related policies around the world, Circular Economy (CE) regulation is gaining attention in the political and regulatory agenda (Fitch-Roy et al., 2020; Milios, 2018; Monciardini et al., 2022). While many definitions of CE exist, central in all of them is the recognition that the resilience of ecosystems is no longer able to cope with the amount of natural resources extracted and of waste produced by our linear 'take-make-waste' economic model. Thus, the CE narrative proposes a future in which the concept of 'waste' is phased out "by transforming waste into biological and technical 'nutrients' circulating within ideally infinite loops, the extraction of natural resources and the waste produced to satisfy the needs of human societies will be significantly reduced"

https://doi.org/10.1515/9783110723373-016

(Borrello Pascucci & Cembalo, 2020). Policies[1] inspired by CE principles are rapidly adopted at different levels of governance including municipalities, private and industry standards, national law and supranational standard setting. This is not only happening in Europe – that is amongst the most active regions in promoting the CE – but also in China, in the United States and in many jurisdictions across the world. Regulation also features in academic CE debates as a key barrier or driving force in the transition towards a more circular and sustainable economy (Milios, 2018; Rizos et al., 2015). Clearly, 'circular business models' need support from policies and laws, as they need to compete in a 'linear economy,' where those laws are often adjusted to the needs of linear business models, which influences the socio-economic context (Milios, 2021). However, despite this mounting interest, the literature still lacks a real strand of studies that systematically and comprehensively addresses this area of research.

The chapter offers a review of the CE regulatory debate, starting from the following section that highlights the origins of the dominant model of environmental regulation that are historically rooted in the same linear paradigm that underpins economic thinking. It then introduces some of the main trends that have marked the emergence of a law of the CE in Europe, focusing in particular on the adoption of product-oriented regulations and what remains today the centrepiece of the European CE regulatory framework: the Ecodesign Directive. The following section provides a critical assessment of the current EU policy debate, and this is followed in turn by a section that identifies existing gaps in the literature on CE regulation and outlines new research directions. We conclude by calling for research on this area that is addressing the tough questions about what society we want and how CE regulation will contribute to achieving it.

---

1 *Public policy* can be understood as a system of laws, regulatory measures, targets, courses of action (e.g. roadmaps, strategies) and funding priorities concerning a given topic promulgated by a public body. Policies can be adopted at various levels of governance (international/EU/National/regional/local). The concept of 'policy' is thus broader than the concept of 'law'; *law* is a system of binding rules that a society or government develops in order to deal with crime, business agreements and social relationships. *Policy instruments* are specific governmental interventions aimed at achieving policy objectives, which provide actors with incentives to change behaviour. There are different classifications of policy instruments, but most of them differentiate between administrative, economic and informative instruments. *Laws relate to policy instruments in two main ways:* (1) Most policy instruments are implemented through laws (e.g. a national law that implements a carbon dioxide tax, or an EU law regulating content of hazardous substances in products); (2) laws influence the context for making use of policy instruments. For instance, EU law on public procurement does not regulate what kind of sustainability requirements public bodies in EU member states can use when purchasing goods and services for the public sector. However, EU member states must comply with the principles of EU procurement law – such as the principle of non-discrimination in order not to favour domestic manufacturers – when setting sustainability criteria.

Origins and limitations of the current model of regulation

This section aims to provide the historical background to the current discussion on how to establish a law of the CE. The model for environmental regulation of products in Europe and most high-income countries rests on a linear basis largely for historical reasons. Like the take-make-dispose model that underpins the linear economy and concentrates on the process rather than the result, the law has developed focusing mainly on impacts at key points instead of the source and overall occurrence of the harm. These key points of impact have been: quarrying and mining; manufacture; and waste disposal with fewer controls existing during the use phase of a product.

This linear regulatory model underpins the development of environmental legislation worldwide and is derived from the development of manufacturing industry. As the industrialisation of manufacturing processes occurred in earlier centuries accompanied inevitably by urbanisation, the problems of pollution became more acute and it was initially their relevance for public health, which was the driver for regulation. In Europe and other western economies, the movement of peoples from agricultural to urban settings was a feature of the shift from a pastoral economy to an industrial one where factories were built in towns and common lands were enclosed. With few controls on urban planning and public health in Western Europe, the early eighteenth and nineteenth centuries were marked by outbreaks of diseases such as typhoid and cholera resulting from inadequate sanitation, poor water supplies, bad air quality and squalid housing (Wohl, 1977). These patterns were and are still being replicated in low- and middle-income countries all over the world.

The expansion of industry was marked by a need to generate more energy and the source of this was, generally, water (Deemer et al., 2016) or coal, which resulted in increased atmospheric and other emissions. While energy sources may be changing in high-income countries, in large parts of the world, energy sources are still served primarily by non-renewable resources. Locally, this type of energy source can result, without technological refinements, in local or 'spot' pollution. Because of the highly visual and identifiable impacts of manufacturing, this became the focus for environmental as well as health and safety regulation.

Legislators reacted to a clear and perceived need to prevent harm to human health rather than the environment. Thus, in the nineteenth century, the UK passed sweeping public health legislation (1875) in response to the consequences of rapid industrialisation, such as air and water pollution, and waste disposal (Malcolm & Pointing, 2011; Marx, 1860). This was then followed first by powers then duties placed on municipal authorities to plan their neighbourhoods in ways that would alleviate such harm (Parliament of the United Kingdom, 1909; Sutcliffe, 1988).

These historical contaminations of land and water supplies linked to early industrialisation remain an intractable problem in many cases today with desultory attempts to continue developing regulatory frameworks designed to clean up the land and waterways (DEFRA, 1990).

As legislative efforts followed the event out of necessity they were mainly aimed at solving the problems of 'spot' pollution, national and regional regulatory frameworks developed in a piecemeal fashion. Environmental law followed a linear design that can be likened to the cutting of a salami sausage into segments where the segments have been developed in order to deal with the immediate and usually local effects of manufacturing activity. Thus, the current linear framework of governance focuses on the various stages of the manufacture of a product; from quarrying and mining (the extraction of the resource), to manufacture, and then to the disposal of waste. The resulting regulatory landscape is that of distinct legal instruments striking at key polluting points but ignoring the whole life cycle of the product and its lifetime impacts. Regulation addresses the spot pollution occurring at each slice of the salami; for example, at the factory, numerous controls relating to emissions and land use exist. But these controls are all focused on the process; none of them address the product or its life cycle.

In many respects, this development of environmental controls is understandable and justifiable, and advances were made. Process controls are important – the control of factories (European Commission, 2021) is critical and the progress made in this arena since the nineteenth century has been vital to public health and the protection of the environment. The current linear model of spot regulation has driven formidable changes, transforming former filthy industrial processes into low-emission manufacturing and significantly improving waste management. Yet, we still have an environmental problem of enormous concern with at least four planetary boundaries already trespassed, including climate change and biodiversity loss, which by themselves can tip us over the edge resulting in large, irreversible changes (Steffen et al., 2015). Amongst the new human activities creating this wrecking ball[2] is the production of waste arising from our addiction to consumption.

In the industrialised world, we can expect to live longer and in greater health than our ancestors, but what marks us out is our ability to buy more goods, which we dispose of after use. This gives rise to an array of environmental damage from the increasing amount of products in existence that, throughout their life cycle, require resources and create emissions, and the resulting waste to be disposed of. While we may be more or less successful in using law to moderate the 'spot' effects of industrialisation – the 'salami effect' – we now have a more diffuse problem: the life cycle

---

2 Current atmospheric problems in our cities result from fossil-fuel-powered cars and other vehicles, poor public transport policies, industry sited in the cities of industrializing countries, EEA 'Air Quality in Europe' – 2020 Report EEA Report No. 9/2020.

impact of products. For high-income countries, the problem has mostly gone elsewhere, as the impacts of our consumer lifestyles have been scattered across the planet through mining, delocalisation of manufacturing and waste shipments.

The problems of the past can be seen reflected in the emerging economies, where the development of manufacturing industries and the emphasis on economic growth as a way out of poverty have resulted in the sort of human and environmental impacts described previously. The problems are exacerbated by the trend of urbanisation and the growth in cities that leads to excessive consumerism.[3] History repeats itself.

The salami effect of spot pollution process controls has served a purpose in resolving many ills, and the linear approach has been inevitable when examining the historical perspective. Law is often used as a reaction to problems rather than as a preventive measure. Examples abound in the problems caused by industrial pollution: the dumping of waste,[4] polluted drinking water sources[5] and smell[6] are all problems that still occur. They represent the continuing and important basis for modern environmental law, which is still aimed at dealing with the problems of poor health, disease and injury that have arisen from industrial processes. While still important, the issue has become broader in that the consequences have changed and accelerated. It was relatively easy to regulate to deal with the immediate and local effects of a factory. What has now become more complex is the way in which the dynamics of society have changed with consequential effects. The problem has now changed to one that has a global effect in addition to its localised impact.

We need a change of paradigm – both in the economic model and in regulatory processes – from a linear to a circular approach. Current levels of consumption that are made to match our ability to produce are having an even greater impact than that of the early days of unregulated industrialisation. Thus, it no longer suffices to continue developing the piecemeal 'salami' approach to environmental regulation. It is paramount to shift the paradigm of regulation to one that is circular and limits our capacity to use natural resources and create waste while looking at the lifetime environmental impact of products. In the following section, we discuss how this shift from a regulatory framework mainly focused on manufacturing processes to one that is based on the life cycle of the product has started to take place in Europe. That is, we review the rise of a law for a CE.

---

3 Population trends in urbanisation in China continue: Jonathan Woetzel, Lenny Mendonca, Janamitra Devan, Stefano Negri, Yangmel Hu, Luke Jordan, Xiujun Li, Alexander Maasry, Geoff Tsen, Flora Yu, et al. (2009, February). Preparing for China's urban billion Report, McKinsey Global Institute.

4 Corby Group Litigation v. Corby Borough Council [2009] EWHC 1944 (TCC).

5 Cambridge Water Co v Easter Counties Leather plc [1994] 2 AC 264.

6 Barr and others v Biffa Waste Services Ltd [2012] EWCA Civ 312.

The emergence of a law of the Circular Economy in Europe

In this section, we introduce some of the main trends of regulating production and consumption in Europe towards the transition to a CE. We present in particular the rise of life cycle thinking that materialised most prominently in the EU with the adoption of the Ecodesign Directive. Although waste law remains a major part of current CE-related policies as the previous historical analysis revealed, we intentionally do not delve into this area in much detail. That is because, in our view, the CE is to be first and foremost about waste prevention and not waste management. Thus, we aim our attention at product design, product lifetime extension through maintenance, repair and re-use, and sustainable consumption. We recognise that waste law has played and will continue to play an important role in the policy mix for the CE, but it must nonetheless take a back seat. In addition, there is extensive literature about EU waste law that can be referred to for the purpose of this chapter (Maitre-Ekern, 20; Sachs, 2006; Scotford, 2007).

While there were EU regulations related to waste already in the 1970s, it was the early 2000s that saw the birth of many new product-related policies, such as the Waste from Electrical and Electronic Equipment (WEEE) Directive adopted in 2002. The Green Paper on Integrated Product Policy in 2001 outlined the contours of a more holistic product-oriented policy approach, aiming to regulate all the relevant life cycle phases of products.[7]

Following the IPP, the EU developed a number of strategies of relevance for the CE, such as thematic strategies for waste and resources (see Figure 13.1).

We will discuss some of the historical developments below.

The early days of EU environmental law saw the adoption of command-and-control regulations that would typically take the form of environmental standards that mandate a certain level of environmental quality (e.g. emission standards) (Field & Field, 2013). From the 1990s, however, there has been a trend – notably under the influence of the Chicago school (Coase, 1988) – of relying instead on economic policies and market-based regulation (environmental taxes, carbon markets, subsidies for electric cars, etc.). Later, information policies (eco-labelling, certifications and consumer information) also started to develop (Tietenberg, 1998). This explains why today we have regulations on industrial pollution,[8] the financial responsibility of producers,[9] and the energy labelling of products[10] – all at the same time.

---

7 It was followed by a Communication on IPP in 2003.
8 Directive 2010/75/EU of 24 November 2010 on industrial emissions (integrated pollution prevention and control), OJ L 334, 17.12.2010, p. 17.
9 Directive 2012/19/EU of the European Parliament and of the Council of 4 July 2012 on waste electrical and electronic equipment (WEEE), OJ L 197 24.7.2012, p. 38.
10 Regulation (EU) 2017/1369 of the European Parliament and of the Council of 4 July 2017 setting a framework for energy labelling and repealing Directive 2010/30/EU, OJ L 198, 28.7.2017, p. 1.

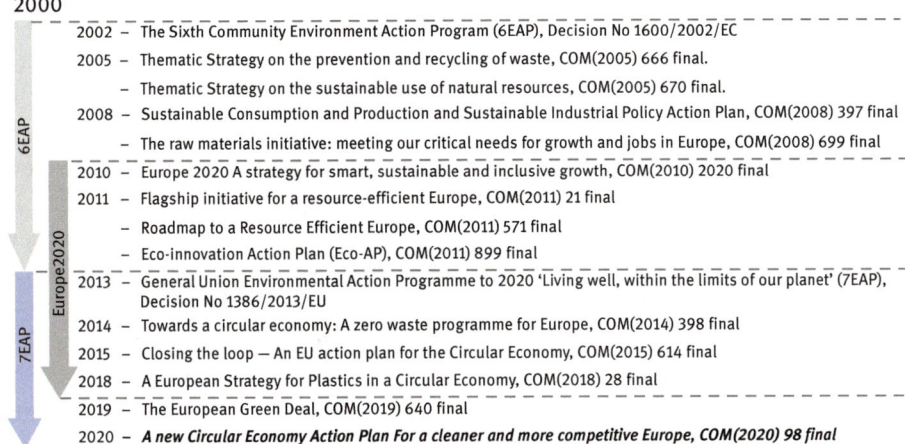

Figure 13.1: EU's strategies following the IPP.
Source: Milios, L. (2020). *Policy framework for material resource efficiency: Pathway towards a circular economy* [Dissertation, Lund University].

Throughout the years, the foci in policymaking have also widened. As presented in the previous section, environmental policy in its infancy focused on point of impacts and main sources of harm, such as hazardous substances (e.g. substances like DDT and lead), industrial and transport emissions to air and water (regulated, for example, through industrial permits and fuel standards), and polluted land. One of the tendencies in European environmental policy and law in the last few decades has been to shift the focus to the life cycle environmental impacts of products 'from the cradle to the grave' (Dalhammar, 2007; Maitre-Ekern, 2019; Malcolm, 2011). The EU and its member states have thus developed an increasing number of policies to address those environmental issues at every stage of the life cycle and with more focus on prevention of damage at the source.

The product-oriented regulations can be divided into four main categories:
1. Making products more energy- and resource-efficient (Regulated through the Ecodesign Directive and Energy labelling)
2. Banning hazardous substances (e.g. through REACH, the RoHS Directive and the General Product Safety Directive)
3. Ensuring a product is disposed of in an appropriate way at its end-of-life stage, including rules on collection and material and waste management (e.g. through producer responsibility rules)
4. Increasing product lifetimes and product reparability (latest trend discussed in more detail later)

These rules are complemented by various other instruments, such as mandatory and voluntary labelling schemes (e.g. the Energy Star and the EU Ecolabel), and public procurement (that is Directive 2014/24/EU on public procurement, and Directive 2014/25/EU on procurement by entities operating in the water, energy, transport and postal services). On the other hand, green taxes and charges, and the envisioned Green tax reforms – initiated in several European nations in the 1990s – have stalled (see e.g. Albrecht, 2006; Beuermann &Santarius, 2006; Deroubaix & Lévèque, 2006). Some countries (e.g. the Nordic European countries) have put in place significant taxes related to climate and energy. When it comes to resources, however, taxes are primarily aimed at waste management (i.e. landfilling, incineration and recycling), and there have been no successful attempts to directly tax resources, with the exception of specific products such as plastics bags.[11] Extended Producer Responsibility (EPR)_ rules allocate the costs of waste handling to the producers, and this has led to more economically efficient markets for collecting and recycling waste, but EPR rules have failed to incentivise a product design for longevity or recyclability (Dalhammar, 2018).

Where we have seen important developments towards a regulatory framework for CE is in the area of product-oriented laws. In the 1980s and 1990s, an increasing number of regulatory tools emerged, including eco-labelling, rules on chemicals in products and materials and producer responsibility rules. The first attempt to devise a larger, product-oriented strategy at the EU level – aiming to develop a more comprehensive 'policy mix' – was the Integrated Product Policy (IPP).[12] IPP aimed at a market transformation, in which environmentally sound and resource-efficient products become the mainstream and led to 'green' mass markets, improving resource efficiency in manufacturing and consumption activities in the EU and beyond (Rubik & Scholl, 2002). IPP supported the 'breakthrough' of new ideas in environmental policy, including 'life cycle thinking' as a guiding principle in environmental law. It also supported the more widespread use of green public procurement, and ideas expressed in the IPP strategies which also influenced important laws such as the Ecodesign Directive.

'Life cycle thinking' can be understood as a certain way to view the world (Heiskanen, 2002): each product "casts a shadow" as it is the embodiment of all environmental and social life cycle impacts during its life cycle (including extraction of raw materials, production of components, transport of materials and products and ultimately the waste phase when the product is thrown away). Life cycle thinking has influenced environmental law in several ways (Dalhammar, 2015). Most notable is the development of rules on the design of a product that target the product at one

---

11 In France.
12 Commission Communication – Integrated Product Policy – Building on Environmental Life-Cycle Thinking, COM (2003) 0302 final.

of the earliest and most determinant stages of the life cycle in terms of subsequent environmental impacts. It is estimated that up to 80% of the environmental impacts of a product are determined at the design stage. The Ecodesign Directive currently covers only energy-related products (that is, products that either use energy or affects energy use, such as windows), but the scope of the law is likely to be broadened in upcoming legislative review. Moreover, these design requirements have progressively covered more product groups,[13] and more features of the product, from energy efficiency to durability, and repair.

Another example is that of public procurement, where environmental and social aspects are increasingly part of the process – from selection criteria to technical specification, and award criteria (Sjåfjell & Wiesbrock, 2016). These standards have so far been voluntary; however, the European Commission announced in its 2020 Action Plan on the CE the revision of the public procurement rules to include mandatory green requirements. Lastly, there are new EU rules related to 'conflict minerals,' which refer to minerals, the trade of which can be used to finance armed groups, fuel labour and other human rights abuses, and support corruption in various regions of the world.[14] These rules have implications for supply-chain management because they effectively limit access to minerals (such as gold and tin) that are crucial to many manufacturing processes (e.g. mobile phones and cars). As a result, producers have to find alternative sources of supply, notably in the recovery of secondary raw materials.

With the emergence of the CE, and the two Action Plans of the European Commission, we are starting to see a new 'paradigm' emerge. The CE has a number of characteristics that make it stand out from prior initiatives, including:

-   *A focus on taking care of resources, not just waste management* – considerations of circularity have highlighted the need for product regulation that prolong product lifetimes through product design and more consumer opportunities for repairs. It has also highlighted the need for other practices that prolong the lifetime of products and components such as re-use and remanufacturing. For an overview of some practices and related emerging European policies, see Table 13.1.
-   *The need for a more disruptive approach* – one of the mantras of the IPP has been to 'work with the market.' On the contrary, many policies related to the CE consist of 'challenging' the market. Studies have increasingly established that market actors are less likely to design more long-lived products or to increase consumer opportunities to repair their products in the absence of regulation

---

13 31 product groups as of June 2021; https://ec.europa.eu/info/energy-climate-change-environment/ standards-tools-and-labels/products-labelling-rules-and-requirements/energy-label-and-ecodesign/ energy-efficient-products_en.

14 Regulation (EU) 2017/821 of 17 May 2017 laying down supply chain due diligence obligations for Union importers of tin, tantalum and tungsten, their ores, and gold originating from conflict-affected and high-risk areas, OJ L 130, 19.5.2017, p. 1.

(see e.g. Maitre & Dalhammar, 2016). Thus, changing the paradigm towards longer product lifetimes and repair represents a quite radical approach to policy-making. In particular, policies for more sustainable consumption patterns may contribute to changing the 'rules of the game' given the weak nature of other consumption policies (Dalhammar, 2019). It is also worth noting that such policies appear to have the support of the general public, and of consumer and environmental NGOs (Svensson et al., 2021). As discussed in other chapters in this handbook, there is a need also for innovation policy to contribute to new innovative products and solutions. Further, policymakers need to challenge the linear economy and support emerging circular business models which may be perceived as disrupting the current linear model with supportive policy packages; these should include both general and sector-specific policies (see Milios, 2021).

Table 13.1: Emerging policies for the CE in Europe.

Practice	Definition	Examples of laws and policies promoting the practice
Longer product lifetimes through design	Extending the technical lifetime through product design, e.g. using more durable materials or adopting design changes to make the product easier to repair.	Ecodesign regulations have regulated minimum lifetimes for some products/ components Changes in mandatory consumer warranties Modulated fees in producer responsibility schemes Eco-labels with criteria that aims to prolong product lifetimes (e.g. TCO) Proposals to provide durability information on products at point of sale
Repair	Extending the life of a product during its first use by retaining or restoring its functionalities with minor repairs that can be done by manufacturers or professional service providers.	Right-to-repair obligations in Ecodesign Directive (e.g. provision of spare parts, ease-of-disassembly) Reparability index (France) Repair fund linked to producer responsibility schemes (France) Lower taxes for the repair sector (several countries Local and regional initiatives, such as Repair Network Vienna (includes repair vouchers)

Table 13.1 (continued)

Practice	Definition	Examples of laws and policies promoting the practice
Re-use	Extending the life of a product or part by having a second-hand user utilise it for the same original purpose with no or only minor enhancements and changes; can be combined with refurbishing.	Re-use parks and re-use malls Quality labels for re-used products Support re-use in waste laws
Remanufacture	Enabling a full new service life of a product via a standardised industrial process that takes place within industrial or factory settings, in which cores are restored to original as-new condition and performance, or better. The remanufacturing process is in line with specific technical specifications, including engineering, quality and testing standards, and typically yields fully warranted products (and per agreement of global industry members).	Support to Remanufacturing networks Public procurement of remanufactured products such as laptops and furniture Changing trade agreements to support trade in of remanufactured products

Source: Dalhammar et al. (2021, May). Enabling re-use in extended producer responsibility schemes for white goods: Legal and organisational conditions for connecting resource flows and actors. *Circular Economy and Sustainability*.

One challenge going forward relates to the coordination of different levels of regulation and modes of governance, for instance between the EU and its Member States. Countries like France have signalled that they aim to go on with national policies, in the absence or even in anticipation of EU policies. This includes a 'reparability index' implemented in 2021 and a proposed product 'durability index' from 2024, and a repair fund that will be handled via the producer responsibility schemes.[15] Such national policies can provide dynamics to the policy process, but the adoption of national schemes brings the risk of fragmenting the market and creating legal loopholes.

---

15 French act of law against waste and for a circular economy [LOI n° 2020-105 du 10 février 2020 relative à la lutte contre le gaspillage et à l'économie circulaire], JORF n°0035 du 11 février 2020.

Existing policy debates: re-designing CE regulation

Having discussed some of the main trends in regulating the CE over the last decades, this section delves into what fundamental policy changes lawmakers may and should consider to enable a circular transition. We focus in particular on current EU regulatory debates. Our starting point is that circularity is not to be framed as an end in itself, but a necessary path towards a paradigmatic shift towards environmental and social sustainability. In other words, as the law of the CE is developing, it is crucial to ensure that it allows economic activities to prosper in keeping with the planetary boundaries framework and while achieving good living conditions for all people now and in the future (Raworth, 2017).

The Circular Economy Action Plan put forward in March 2020 by the European Commission makes some steps in this direction. However, as will be discussed, fundamental issues remain.

By establishing product sustainability principles as part of a 'sustainable product policy framework initiative,'[16] the Action Plan has introduced a new frame of reference to regulating for circularity that many had been calling for.[17] In particular, the Commission aims to establish overarching product sustainability principles and EU rules on producer responsibility that will not only prevent waste but also address social aspects throughout the product life cycle. The challenge of a new paradigm where a regulatory framework promoted a CE in preference to a linear one is reflected in the complexities surrounding the definition of 'waste.' Since the adoption of the first directive on waste in 1975, the definition of 'waste' has been established and refined through legislative reforms and jurisprudence.[18] However, there is no common framework or definitions for products. In the absence of clear definitions of key concepts such as 'resource,' 'material' or 'product,' the concept of 'waste' lives in a vacuum and effectively captures a very wide range of products and materials not least due to the vague notion of 'discard' that underpins it. This means that under the current regime, too many valuable resources fall under waste legislation not favouring circularity (Den Hollander et al., 2017). Once legally labelled as waste, refurbished,

---

16 Commission Communication, 'A new Circular Economy Action Plan For a cleaner and more competitive Europe', COM(2020) 98 final.

17 For example, E. Maitre-Ekern, M. B. Taylor, M. van der Velden. (2020). *Towards a sustainable circular economy. SMART reform proposals.* SMART report, available at SSRN: https://ssrn.com/abstract=3596076.

18 In 1975, 'waste' was defined as "any substance or object which the holder disposes of or is required to dispose of pursuant to the provisions of national law in force" (Article 1(a) of Council Directive 75/442/EEC of 15 July 1975 on waste). Following the revision of the directive, waste is currently defined as "any substance or object which the holder discards or intends or is required to discard" (Article 3(1) WFD) [emphasis added].

remanufactured or recycled materials do not have an easy access back to the market. The 'end-of-waste' process is conditioned by compliance with specific criteria, including the usage, the lawfulness, the demand for, and the risks linked to a particular substance or object.[19] In this context, the development of a product policy framework appears essential not only for the purpose of establishing good legislation, but also to reclaim some of the areas that have de facto fallen on waste law, not least activities of prevention and re-use.

In addition, the EU product policy framework remains fragmented, not just in terms of regulatory measures, but also and significantly in relation to its objectives, priorities and governance. Some regulations – like waste laws – prioritise environmental objectives, while others – like consumer protection and public procurement – are part of the EU's efforts to consolidate the internal market. Moreover, CE-related policies do not abide by clear guiding principles or specific definitions of key concepts like 'resource' or 'product.' Achieving a coherent set of policies for sustainable production and consumption requires taking a holistic approach not only to the product's life cycle, but also to the objectives that underpin that set of policies. In other words, a law of the CE needs to be based on clear policy objectives, common concepts and a coherent set of regulations.

The sustainable product policy framework, put forward by the European Commission in March 2022, appears therefore to be an important step in the right direction. As part of the initiative, a new Ecodesign Regulation Directive[20] is to be adopted that will widen the scope of the scheme beyond energy-related products. This broadening of the scope to new types of products has been long called for., and constitutes a crucial reform that may bring about much needed improvements, notably to the textile industry, the environmental impacts of which are considered and yet largely under-regulated. Now, the Commission is proposing to extend the eco-design scheme not only to textiles, but to any 'physical good', with the exception of food, medicinal products and plants.[21] The ecodesign requirements will also be extended from their current focus on energy efficiency of products in use, to include a wider range of issues such as a products' reusability, upgradability and reparability (Bohstrom and Micheletti (2016). If CE policies are to reduce overall life cycle impacts of products and, more importantly, contribute to achieving broader sustainability, they must first and foremost aim to extend the lifetime of products. These proposed new ecodesign requirements based on broader sustainability and circularity aspects offer important new perspectives, but determining those aspects

---

19 Articles 6 (1) and (2) of the Waste Framework Directive.

20 B. Bauer et al. *Potential ecodesign requirements for textiles and furniture.* Report for the Nordic Council of Ministers, TemaNord 2018:535; available at norden.diva-portal.org/smash/get/diva2:1,221,509/FULLTEXT01.pdf.

21 European Commission, Sustainable products initiative, available at https://ec.europa.eu/info/law/better-regulation/have-your-say/initiatives/12,567-Sustainable-products-initiative_en.

may prove intricate, in particular when it comes to defining durability or reconciling different objectives, such as reparability and recyclability. EU's future sustainable product policy is to focus the circular transition on products whereas existing policies have so far mainly targeted waste. The success of the ecodesign scheme has convinced policymakers not just to extend its scope and content but also to make it the central piece of the CE regulatory framework, and the core of a fast-developing product law. The proposed Ecodesign Regulation contains important definitions that are to be relevant for the entire field, such as 'durability', 'repair', 'maintenance', and 'reliability'. The ecodesign scheme might, however, prove to have limitations. The focus on design, as important and relevant as it is, does not offer avenues for adopting other measures to ensure a more sustainable production. In particular, the development of infrastructure to collect and process waste has been driven by the concept of extended producer responsibility that requires producers to bear at least the financial costs of waste management. If maintenance, repair and re-use are the driving force of tomorrow's markets, similar investments must be made, and systems developed to ensure that consumers have effective access to repair. It would also seem that re-use is no longer a niche.

Policies promoting longer product lifetimes must be underpinned by standards that make it possible to measure and regulate a product's lifetime and reparability. 'Lifetime' and 'reparability' are context-dependent. For instance, the reparability of a broken mobile phone is influenced by several different parameters, including the possibility to access repair (e.g. accredited or independent repairers, and cost of repair), the availability and cost of spare parts (such as screens and batteries), and the ease of disassembly (is it glued or screwed together). Legal requirements need to refer to standards to be relevant for an entire product group, like mobile phones. For this reason, the European Standardization Organizations (CEN, CENELEC and ETSI) have developed standards for product lifetimes and reparability upon request by the EU legislators.[22] Standardisation request to the European standardisation organisations as regards ecodesign requirements on material efficiency aspects for energy-related products in support of the implementation of Directive 2009/125/EC of the European Parliament and of the Council.

For instance, the standard on reparability (EN 45554:2020)[23] EN 45554:2020 General methods for the assessment of the ability to repair, re-use and upgrade energy-related products allows for the development of a scorecard where the reparability of a given product is measured and evaluated based on different criteria. These criteria relate in particular to the availability of spare parts, whether special tools are needed to open the product, and whether manufacturers provide repair manuals to independent repairers. The standard then offers a method to aggregate all relevant criteria into one reparability score that could pave the way for a European reparability scoring system similar to the French index.

The purchase of more durable products often comes with economic constraints as those products are more expensive up front Proske et al. (2016). Convincing consumers of the opportunity of such purchase requires notably providing information. It also requires that consumers are able to trust the information they receive and that they will indeed be able to keep and use their products over a longer period of time. Even the most durable products may encounter failures or break. In such cases, the extension of product lifetime is an issue of access to repair: is it feasible? Is it costly? Is it attractive? These are critical aspects in the discourse around policies for promoting repair (see Table 13.2) and the emergence of a 'right to repair' Svensson-Hoglund et al. (2021).

Table 13.2: Policies for promoting repair.

Objective	Approach	Policies	Main stakeholders
Reparability	Qualitative: is it feasible?	Ecodesign; producer obligations (spare parts availability); reform of IP and antitrust laws	Producers; government
Affordability	Economic: is it costly?	Extended (in time and scope) consumer guarantees; tax rebates	Consumers; government
Mainstreaming	Psychological: is it attractive?	Repair scores and labels; grassroots initiatives (e.g. repair cafés); education and awareness raising campaigns	Independent repairers and DIY; government

Hopefully, the current legal developments will be implemented quickly, and we will move towards a situation where more manufacturers compete on producing products with longer lifetime and better reparability. Equally important is that bad products are removed from the market: consumers should only be able to choose high-quality products. This process has already started. For instance, we can notice how EU product regulations affect producers outside the EU that want to ensure their products can be exported to the EU Bradford (2020).

A research agenda to address unanswered CE regulatory questions

Taking stock of some of the issues highlighted so far, this section focuses on three major gaps in the extant literature on CE regulation: the academic debate is currently under-theorised, depoliticised and fragmented. We maintain that trying to address these gaps would offer important areas for future research.

First, CE ideals have enjoyed a rapid and widespread success in the institutional and public debate. Nonetheless, CE remains ill-defined and under-theorised. As it has been often pointed out, "the scientific and research content of the CE concept is superficial and unorganised Korhonen et al. (2018a)." As lawmakers are currently debating the translation of CE principles into policy and legal instruments, this weakness has become more apparent, attracting growing criticism. Tensions exist in the definition of the respective roles and possible complementarity of private and public actors in the CE transition. Many large corporations have taken a leading role in adopting CE practices, but also various governments and public authorities, such as the European Commission, are taking a strong regulatory role. This is likely to trigger a reconsideration of the traditional boundaries between private and public spheres, particularly around areas such as property rights and contract relationships as certain CE ideals challenge established and conventional legal frameworks Ballardini et al. (2021). Tensions and challenges exist also when it comes to advancing both economic growth and societal goals. The CE approach has been often framed as a 'win-win-win' practical and technical solution able to contribute to all three dimensions of sustainable development – economic, environmental and social (EMF, 2016). However, recent research has started to question whether these benefits can actually be delivered and how Zink and Geyer (2017). As CE ideals turn into reality, conflicts are likely to emerge within and between business organisations as well as in relation to the regulatory demands of various social constituencies and groups of stakeholders Valenzuela and Bohm (2017). Thus, a new strand of studies is needed that systematically and comprehensively addresses CE regulatory and legal issues. We call for future research that can expand the current narrow and under-theorised approach to CE regulation to bridge more technical conceptualisations of the CE, stemming from engineering, design, technology, biology, etc., with perspectives coming from social sciences such as regulatory studies, law, sociology, economics and politics. In order to become a "unified narrative able to inspire policy changes" (Borello et al., 2020), the promoters of CE have to establish what it is really able to do in relation to the economic, social and environmental grand challenges that public policies are currently facing, such as limits of planetary and social boundaries, food, water and energy security, global warming, mass unemployment and precarity of work, global social inequalities.

Second and relatedly, despite the concept of CE being deeply contested (Korhonen et al., 2018b), the literature dealing with regulation tends to take a de-politicised and under-socialised approach to regulation. While there is a general recognition that technical and socio-political issues and economic issues are intertwined, CE regulation is often taken for granted and treated mainly as a technical matter where social dimensions have to adapt to the technical solutions offered by CE models. Murray et al. (2017) This is evident, for example, in the burgeoning literature on barriers to CE, which typically includes regulatory and institutional barriers (DeJesus & Mendonca, 2018) In line with a system theory approach to social dynamics, the latter are treated as independent variables to be adjusted and modified based on technical considerations that are often detached from the socio-economic and historical context (Kovacic et al., 2019). The risk is that the revolutionary and radical intuitions that have driven early CE debates (i.e. Stahel performance economy; Benyus innovation inspired by nature) is transformed into a "policy legend" that merely normalises and protects today's intensive model of consumption and production Giampetro & Funtowicz, 2020) It is important prevent the CE regulatory project being captured by powerful vested interests and becoming a covert way to depoliticise the sustainability debate and the demand for social and environmental justice.

Third, due to the lack of a coherent and ambitious socio-economic and political vision, the CE policy and regulatory debate is currently fragmented. We have already referred to the 'salami effect' by which the legislative efforts follow the events out of necessity and environmental regulation addresses the spot pollution occurring at each phase of the life cycle of a product. As van Eijk noted, "we tend to be late and reactive in our adjustments of regulation which is frustrating new initiatives (van Eijk, 2015)" Rizos and colleagues mention the need to "develop the supportive policy frameworks that address both supply and demand" (Rizos et al., 2015). In the European context, Hughes warns that "the EU has multiple laws that relate to life cycle thinking, but they do not yet form a coherent whole" (Hughes, 2017). The result is that the regulatory intervention on CE is focusing on certain areas, leaving others behind. For instance, key economic policy areas for the success of CE transitions such as taxation, sustainable finance or sustainability reporting have received so far very limited attention by policymakers as well as CE researchers.

Drawing on the gaps identified here, several questions are yet to be answered and represent fertile ground for future academic research in the emergent study of CE regulation. There is a long list of unanswered questions, for instance:

- How can CE reforms effectively support the organisational or societal transition towards a more just and sustainable economy?
- How do different levels of governance interplay (e.g. organisational, sectorial, regional, national)?
- In terms of effectiveness in promoting systemic change towards a more CE, which policies should be implemented at which level of governance and by whom?
- What are the right incentives to nudge organisations and individuals towards more circular behaviour?
- How do we design policies that deal with resistance to change and prevent potential rebound effects?
- How can we move from existing empirical research on CE regulatory barriers to a more comprehensive theorisation?
- Which socio-economic forces are driving the adoption of CE policies? Which ones are opposing them? Why?
- What are the regulatory preferences of companies, consumers, workers and other groups affected by CE regulation?

- How are CE policies actually implemented (as compared to the initial policy objectives)? What are the real effects of specific CE policies on businesses, workers or other key parties?
- What kinds of struggles take place in the emerging multi-level CE policy arena?
- Who benefits from the adoption of certain CE policies?
- What are the (often unintended) implications in developing countries of the adoption of CE policies by the developed world?
- Do CE regulations in the global south differ from the global north? How?
- How does CE policy diffusion take place across developed and developing countries?
- Do different CE regulatory initiatives present contradictions and inconsistencies? How can greater policy coherence be achieved?
- In which instances do economic, humanitarian, health or environmental concerns clash in the adoption of CE regulation?
- How is the legal responsibility associated with the implementation of CE policies distributed across different actors (e.g. producers, consumers, intermediaries) in complex supply chains?

Conclusion

CE regulation is gaining attention in the political and regulatory agenda. Although much remains to be done to achieve a law for the CE, the arena is not hollow. On the contrary, several regulations have been adopted and revised to deal with the increasingly negative repercussions of our industrial and post-industrial activities. In this chapter, we have examined how the traditional regulatory approach focusing on points of impact has created a scattered and disjointed landscape. This 'salami approach' is particularly ill-equipped to deal with the diffuse pollution created by the throwaway society. In fact, the CE requires a holistic approach to regulating products whereby their entire life cycle is examined, and policies are fitted to minimising damage throughout the life cycle rather than targeting a particular phase.

EU regulation originally centred on impact points – that is, industrial pollution and waste. Since the early 2000s, however, we have seen a progressive shift towards more comprehensive policies. The adoption of the IPP anchored the concept of 'life cycle thinking' into product-related law-making, and eventually led to the adoption of what remains today the centrepiece of the CE regulatory framework: the Ecodesign Directive. The eco-design scheme was soon completed with an energy labelling regulation. These two policies were innovative for two main reasons: they were mandatory and directly applicable in the entire EU market; and they targeted the use phase that had been largely neglected by environmental legislators so far.

The transition to a CE that can contribute to achieving a sustainable future requires going further than eco-design, however. We need to shift away from a consumerist and productivist paradigm to one that places value in longer product lifetime and the human capital. The 2020 CE Action Plan of the EU sanctioned a new phase of policymaking that had started blooming in Europe, notably through grassroots repair movements and national strategies combatting premature obsolescence. The 2022 Sustainable Product Initiative will take this further. But any future law of the CE will have to address these issues not only through the lens of ecodesign, but also through more intricate policies such as right to repair, tax reforms and socio-economic progress.

We found that research on the regulation of the CE remain very de-politicised and under-socialised, but it is more than time to address the tough questions about what society we want to create and how CE regulations will contribute to achieving it.

References

Albrecht, J. (2006). The use of consumption taxes to re-launch Green tax reforms. *International Review of Law and Economics, 26*(1). 88–103.https://doi.org/10.1016/j.irle.2006.05.007

Ballardini, R. M., Kaisto, J., & Similä, J. (2021). Developing novel property concepts in private law to foster the circular economy. *Journal of Cleaner Production, 279*, 123747.

Beuermann, C., & Santarius, T. (2006). Ecological tax reform in germany: Handling two hot potatoes at the same time. *Energy Policy, 34*(8), 917–929. https://doi.org/10.1016/j.enpol.2004.08.045

Borrello, M., Pascucci, S., & Cembalo, L. (2020). Three propositions to unify circular economy research: A review. *Sustainability, 12*(10), 1–22. https://doi.org/10.3390/SU12104069

Boström, Magnus, and Michele Micheletti. "Introducing the sustainability challenge of textiles and clothing." Journal of Consumer Policy 39, no. 4 (2016): 367–375.

Bradford, A. (2020). *The Brussels effect.* New York: Oxford University Press.

Coase, R. H. (1988). *The firm, the market, and the law. The firm, the market, and the law.* University of Chicago Press. https://doi.org/10.7208/chicago/9780226051208.001.0001.

Dalhammar, C. (2007). Product and life cycle issues in european environmental law: A review of recent developments. In *Yearbook of European environmental law.* Oxford, Oxford University Press.

Dalhammar, C. (2015). The application of 'life cycle thinking' in European environmental law: Theory and Practice. *Journal for European Environmental & Planning Law, 12*, 97–127.

Dalhammar, C. (2018). Extended producer responsibility. In L. Krämer & E. Orlando (Eds.), *Principles of environmental law.* Cheltenham, UK Edward Elgar Publishing.

Dalhammar, C. (2019). It is never too late to give up, or is it? Revisiting policies for sustainable consumption, Chapters. In O. Mont (Ed.), *A research agenda for sustainable consumption governance* (Vol. 9, pp. 137–155). Cheltenham Edward Elgar Publishing.

De Jesus, A., & Mendonça, S. (2018). Lost in transition? Drivers and barriers in the eco-innovation road to the circular economy. *Ecological Economics, 145*, 75–89; Kirchherr, J., Piscicelli, L., Bour, R., Kostense-Smit, E., Muller, J., Huibrechtse-Truijens, A., & Hekkert, M. (2018). Barriers to the circular economy: Evidence from the European Union (EU). *Ecological Economics, 150*, 264–272.

Deemer, B. R., Harrison, J. A., Li, S., Beaulieu, J. J., Delsontro, T., Barros, N., . . . Vonk, J. A. (2016). Greenhouse gas emissions from reservoir water surfaces: A new global synthesis. *BioScience, 66*(11), 949–964. https://doi.org/10.1093/biosci/biw117

DEFRA. (1990). Department for environment food and rural affairs. 2012. Environmental Protection Act 1990: Part 2A Contaminated Land Statutory Guidance, no. April, 2–62. https://www.gov.uk/government/uploads/system/uploads/attachment_data/file/223705/pb13735cont-land-guidance.pdf

Den Hollander, M. C., Bakker, C. A., & Hultink, E. J. (2017). Product design in a circular economy: Development of a typology of key concepts and terms. *Journal of Industrial Ecology, 21*(3), 517–525. https://doi.org/10.1111/jiec.12610

Deroubaix, J. F., & Lévèque, F. (2006). The rise and fall of French Ecological Tax Reform: Social acceptability versus political feasibility in the energy tax implementation process. *Energy Policy, 34*(8), 940–949. https://doi.org/10.1016/j.enpol.2004.08.047

EMF, Towards the circular economy. (2013). *Growth within: A circular economy vision for a competitive Europe (2016).*

European Commission. (2021). *Industrial emissions directive.* https://ec.europa.eu/environment/industry/stationary/ied/legislation.htm

Field, B. C., & Field, M. K. (2013). *Environmental economics: An introduction* (6th ed., pp. 206). New York: McGraw-Hill Education.

Fitch-Roy, O., Benson, D., & Monciardini, D. (2020). Going around in circles? Conceptual recycling, patching and policy layering in the EU circular economy package. *Environmental Politics, 29*(6), 983–1003.

Giampietro, M., & Funtowicz, S. O. (2020). From elite folk science to the policy legend of the circular economy. *Environmental Science & Policy, 109,* 64–72; Corvellec, H., Böhm, S., Stowell, A., & Valenzuela, F. (2020). Introduction to the special issue on the contested realities of the circular economy, 97–102.

Heiskanen, E. (2002). The institutional logic of life cycle thinking. *Journal of Cleaner Production, 10,* 427–437.

Hughes, R. (2017). The EU circular economy package–life cycle thinking to life cycle law? *Procedia CIRP, 61,* 10–16, p 14.

Korhonen, J., Honkasalo, A., & Seppälä, J. (2018a). Circular economy: The concept and its limitations. *Ecological Economics, 143,* 37–46.

Korhonen, J., Honkasalo, A., & Seppälä, J. (2018b). Circular economy: The concept and its limitations. *Ecological Economics, 143,* 37–46; Corvellec, H., Böhm, S., Stowell, A., & Valenzuela, F. (2020). Introduction to the special issue on the contested realities of the circular economy, 97–102.

Kovacic, Z., Strand, R., & Völker, T. (2019). *The circular economy in Europe: Critical perspectives on policies and imaginaries, Routledge explorations in sustainability and governance.* London: Routledge, Taylor & Francis Group. ISBN 978-0-429-06102-8. http://dx.doi.org/10.4324/9780429061028

Maitre, E., & Dalhammar, C. (2016). Regulating planned obsolescence: A review of legal approaches to increase product durability and reparability in Europe. *Review of European, Comparative & International Environmental Law (RECIEL), 25*(3), 378–394.

Maitre-Ekern, E. (2019). *Towards a circular economy for products.* University of Oslo.

Maitre-Ekern, E. (2021). Re-thinking producer responsibility for a sustainable circular economy from extended producer responsibility to pre-market producer responsibility. *Journal of Cleaner Production, 286.* https://doi.org/10.1016/j.jclepro.2020.125454

Malcolm, R. (2011). Ecodesign laws and the environmental impact of our consumption of products. *Journal of Environmental Law, 23*(3), 487–503. https://doi.org/10.1093/jel/eqr029

Malcolm, R., & Pointing, J. (2011). *Statutory nuisance: Law and practice* (2nd ed.). New York, Oxford University Press.

Markard, J., Raven, R., & Truffer, B. (2012). Sustainability transitions: An emerging field of research and its prospects. *Research Policy, 41*(6), 955–967; George, G., Howard-Grenville, J., Joshi, A., & Tihanyi, L. (2016). Understanding and tackling societal grand challenges through management research. *Academy of Management Journal, 59* (6), 1880–1895.

Marx, K. (1860). The state of British manufacturing industry. *New-York Daily Tribune*, No. 6016, in Marx and Engels Collected Works (Vol. 17, pp. 410–420). Moscow: Progress Publishers, 1980.

Milios, L. (2018). Advancing to a circular economy: Three essential ingredients for a comprehensive policy mix. *Sustainability Science, 13*(3), 861–878. https://doi.org/10.1007/s11625-017-0502-9

Milios, L. (2021). Overarching policy framework for product life extension in a circular economy – A bottom-up business perspective. *Environmental Policy and Governance, 31*(4), 330–346. https://doi.org/10.1002/eet.1927

Monciardini, D., Dalhammar, C., & Malcolm, R. (2022). Introduction to the special issue on regulating the circular economy: Gaps, insights and an emerging research agenda. *Journal of Cleaner Production, 350*, 131–341.

Murray, A., Skene, K., & Haynes, K. (2017). The circular economy: An interdisciplinary exploration of the concept and application in a global context. *Journal of Business Ethics, 140*(3), 369–380.

Parliament of the United Kingdom. n.d. Public Health Act 1875.

Parliament of the United Kingdom. (n.d.). The Housing and Town Planning in Great Britain Act 1909.

Proske, M., Winzer, J., Marwede, M., Nissen, N. F., & Lang, K. D. (2016, September). Obsolescence of electronics-the example of smartphones. In 2016 Electronics Goes Green 2016+(EGG) (pp. 1–8). IEEE.

Raworth, K. (2017). *Doughnut economics: Seven ways to think like a 21st-century economist.* Random House, London, UK.

Rizos, V., Behrens, A., Kafyeke, T., Hirschnitz-Garbera, M., & Ioannou, A. (2015a). *The circular economy: Barriers and opportunities for SMEs.* CEPS Working Documents No. 412/ September 2015, no. 412.

Rubik, F., & Scholl, G. (2002). Integrated product policy (IPP) in Europe – A development model and some impressions. *Journal of Cleaner Production, 10*(5), 507–515. https://doi.org/10.1016/S0959-6526(02)00016-1

Sachs, N. (2006). Planning the funeral at the birth: Extended producer responsibility in the European Union and the United States. *Harvard Environmental Law Review, 30*(1), 51–98.

Scotford, E. (2007). Policy tensions in EC waste regulation. *Journal of Environmental Law, 19*(3), 367–388. https://www.jstor.org/stable/44248616

Sjafjell, B., & Wiesbrock, A. (Eds). (2016). *Sustainable public procurement under EU law.* Cambridge: Cambridge University Press. https://doi.org/10.1017/CBO9781316423288

Steffen, W., Richardson, K., Rockström, J., Cornell, S. E., Fetzer, I., Bennett, E. M., Biggs, R., Carpenter, S.R., de Vries, W., de Wit, C.A., Folke, C., Gerten, D., Heinke, J., Mace, G.M., Persson, L.M., Ramanathan, V., Reyers, B., Sorlin, S. (2015). Planetary boundaries: Guiding human development on a changing planet. *Science, 347*, 6223. https://doi.org/10.1126/science.1259855

Sutcliffe, A. (1988). Britain's first town planning act: A review of the 1909 achievement. *Town Planning Review, 59*(3), 289–303. https://doi.org/10.3828/tpr.59.3.x275553ul404k728

Svensson-Hoglund, S., Richter, J. L., Maitre-Ekern, E., Russell, J. D., Pihlajarinne, T., & Dalhammar, C. (2021). Barriers, enablers and market governance: A review of the policy landscape for repair of consumer electronics in the EU and the U.S. *Journal of Cleaner Production, 288* (March), 125488. https://doi.org/10.1016/j.jclepro.2020.125488

Tietenberg, T. (1998). Disclosure strategies for pollution control. *Environmental and Resource Economics, 11*(3–4), 587–602. https://doi.org/10.1023/a:1008291411492.

Valenzuela, F., & Böhm, S. (2017). Against wasted politics: A critique of the circular economy. *Ephemera: Theory & Politics in Organization, 17*(1), 23–60; Corvellec, H., Böhm, S., Stowell, A., & Valenzuela, F. (2020). Introduction to the special issue on the contested realities of the circular economy, 97–102.

van Eijk, F. (2015). *Drivers towards a circular economy.* Available at: http://www.circulaironderne men.nl/uploads/e00e8643951aef8adde612123e824493.pdf

Wohl, A. S. (1977). *The eternal slum: Housing and social policy in victorian london.* London: Edward Arnold.

Zink, T., & Geyer, R. (2017). Circular economy rebound. *Journal of Industrial Ecology, 21*(3), 593–602.

Steffen Böhm, Chia-Hao Ho, Helen Holmes,
Constantine Manolchev, Malte Rödl, Wouter Spekkink

14 Circular society activism: prefigurative communities in everyday Circular Economy action

Abstract: Circular Economy (CE) is predominately approached through a technical and engineering paradigm, which aims to radically reduce waste by redesigning resource flows. This often ignores the CE's social dimension. While the entrepreneurship and business model literatures do recognise the importance of people in CE transitions, this chapter goes a step further by understanding CE through an activism lens. Our argument builds on social movement perspectives of societal transitions, showing that change is often enacted by grassroots communities in everyday settings. We provide three examples of what we term circular society activism, illustrating our argument. We contribute to the CE literature by conceptualising circular society as a form of prefigurative action that can be enacted by communities in the here and now.

Keywords: activism, communities, circular society, social movements, prefiguration

Introduction

The Circular Economy (CE) discourse has been growing exponentially in recent years. Policymakers, such as the European Commission (2021), have been promoting the CE as a vital framework for achieving 'clean growth,' 'biodiversity loss reduction' and 'carbon neutrality.' As part of the Ellen MacArthur Foundation's (EMF) CE100 network, many large, multinational companies now have CE action plans in place, which have resulted in circular products and services being launched. While this commitment to the CE by these influential actors is laudable, there are two key limitations, which provide starting points for our chapter. First, these 'top-down' CE discourses tend to emphasise technical and engineering solutions, highlighting the material aspects of the CE transition (Geng et al., 2019). This is largely due to the fact that significant parts of the CE discourse are driven by a concern for the availability of so-called 'critical materials,' such as rare-earths and other materials, which are essential for a transition to low-carbon economies (Foxon, 2011). While environmental pollution and sustainability in general are clearly key concerns, dominant CE discourses often focus on waste and resource efficiency, driven by a concern for resource scarcity (Gregson et al., 2015).

https://doi.org/10.1515/9783110723373-017

Second, the social dimension of the CE is often ignored or under-emphasised (Padilla-Riveraet al., 2020). While there is an increasing focus on 'mid-range' dimensions of the CE, including skills and jobs (Schroeder et al., 2019), entrepreneurship (Cullen & De Angelis 2021; Millette et al., 2020; Vecchio et al., 2020), circular business model innovation and experimentation (Aminoff & Pihlajamaa 2020; Konietzko et al., 2020) as well as city and regional transformations (Lekan & Rogers 2020; Petit-Boix & Leipold 2018; Palm & Bocken 2021), the role of grassroots communities is not sufficiently understood. That is, what is missing is a 'bottom-up' perspective of the CE.

This chapter highlights the role of community groups in the CE transition, understanding their everyday activities through a social movement activism lens. This bottom-up perspective appreciates the prefigurative action taken by organised community groups, enacting the CE in the here and now. In this chapter, we provide three illustrative examples, showing how communities engage in reuse, repair and provisioning activities on a self-organised, daily basis.

Building on the recent conception of 'circular society' (Jaeger-Erben et al., 2021), we contribute to the CE literature by arguing that social movement activism plays a crucial role in bringing about CE transitions at a wider societal scale. The CE movement should not be understood, however, as a unified people that speak with one voice. Instead, CE is enacted and diffused by people who have different interests and agendas (Corvellec et al., 2020), who, nevertheless, engage in a form of prefigurative 'sustainable materialism' (Schlosberg, 2019), bringing about change in the here and now.

CE's social dimension

The CE is often referred to as a 'triple win.' EU Environment Commissioner Sinkevicius, for example, identified CE to provide wins "for people, for the planet and for prosperity" (European Commission, 2021). In its most recent educational introduction, the EMF, which works largely with large, multinational companies, claims that the CE "gives us the tools to tackle climate change and biodiversity loss together, while addressing important social needs." It continues: "It gives us the power to grow prosperity, jobs, and resilience while cutting greenhouse gas emissions, waste, and pollution" (EMF, n.d.). For another major actor, the Chinese state, CE is a tool to improve resource efficiency and drive green technology innovations, while reducing carbon emissions and water usage (Chipman Koty, 2021). A key focus for China is the establishment of eco-industrial parks and elaborate recycling systems to achieving these goals (Zhao et al., 2017).

The EU, China and EMF have been driving the CE approach from the top-down, working with large industrial actors. While this has involved wider social goals, it is fair to say that the CE is dominated by technical, engineering and materials-based approaches. This is mirrored in the academic CE literature; a Web of Science search on the topic 'Circular Economy' reveals that more than 90% of academic articles on the CE are published within the fields of environmental sciences, technology, engineering and materials sciences.

Only a minor fraction of academic literature and professional reports on CE acknowledge the social dimension (Kirchherr et al., 2017). In fact, its explicit inclusion might be a rather recent development: Padilla-Rivera et al. (2020) identify social concerns in only 60 academic CE publications (out of literally thousands). They name four larger clusters of social issues, namely 'Labor Practices and Decent Work,' 'Human Rights,' 'Society' and 'Product Responsibility.' In later work, they develop these concerns into a set of indicators for a 'social CE' (Padilla-Rivera et al., 2021).

Entrepreneurship is another focus for CE scholars that involves the social dimension (Cullen & De Angelis 2021; Millette et al., 2020; Vecchio et al., 2020). Particularly, social and ecological entrepreneurship often involves creative CE approaches by ecopreneurs, social enterprises and community groups who want to make a difference (Dentchev et al., 2016; Stratan, 2017). These entrepreneurial approaches are often based on radically different business models, compared to those suggested or implemented by top-down actors (Aminoff & Pihlajamaa 2020; Konietzko et al., 2020), providing opportunities for grassroots innovation (Charter & Keiller 2014; Ziegler, 2019) as well as city and regional transformations (Lekan & Rogers 2020; Palm & Bocken 2021; Petit-Boix & Leipold 2018).

Such CE approaches could be termed 'mid-range,' involving not only large companies or policymakers (the top-down dimension identified previously), but also a broad constituency of societal actors. This is in line with calls for a 'circular society' "to provide an alternative framing that is going beyond growth, technology and market-based solutions" (Jaeger-Erben et al., 2021, 1), which would enable a renewed focus on societal problems (Leipold et al., 2021) and highlight that all societal actors need to be involved in a socio-ecological transformation towards circularity (Jaeger-Erben et al., 2021). Along similar lines, scholars have called for a 'sustainable circular society' (Velenturf & Purnell 2021), and they have sought to integrate CE's technological-material focus with the concept of 'circular human sphere,' originating in the social sciences and development studies (Schröder et al., 2020).

All these studies highlight the social dimensions of the CE, emphasising the fact that it is people, and often community activists, that perform an essential role in propagating, diffusing and implementing CE approaches. Such a bottom-up view of CE allows us to move beyond largely top-down analyses that tend to focus on economic (e.g. company or industry) and institutional analyses (e.g. governmental, regulation, national level approaches). While mid-range perspectives on entrepreneurship, social enterprises and urban transitions have more fully acknowledged the importance of people, they are often rooted in ecological modernisation (Hobson & Lynch 2016) and unlimited growth narratives (Kovacic et al., 2019) more in line with traditional business and innovation models.

We contend that the transition towards a CE is often 'messy' and non-linear, involving a range of multiple, grassroots stakeholders who will often have conflicting aims and approaches. In fact, there is often a "lack of a clear pathway for action, and no single entity in charge" (Kanter, 2015, 129). Hence, the CE field is a multiplicity and a space for grassroots activism – to which we turn now.

Circular society as activism

For many decades, scholars have discussed the role of activists, which we understand here as individuals or collectives that seek to influence organisational or governmental conduct bottom-up to drive social and environmental change (Carter, 2018; Dono et al., 2010). Civil society activism (Diani, 2015), consumer activism (Cherrier et al., 2011) and environmentalism (Mirvis, 1994) are examples of activism that is done at an individual level, within community groups but also in social movements and other larger actor networks (Battilana & Casciaro, 2012; Farla et al., 2012). We suggest that the CE can be viewed as another domain that demands attention by activists. We now discuss different types of activism – namely, 'activism from without' and 'activism from within' – through which actors enact the CE directly and indirectly from below. We conclude the section by combining these categories into a new conceptual dimension, that of 'everyday, prefigurative activism,' which reflects the messy, organic and emergent nature of activism from below.

Activism from without

Traditionally, activists tend to challenge organisations from the outside (Briscoe & Gupta 2016). Hence, such activists, who normally organise themselves within civil society-based community groups and social movements, are usually seen as a threat to companies, governments or any other institutions with conflicting interests. According to Diani (2015), the structure of civil society is made up through the links of voluntary organisations (either informal, partial or formal), acting on collective issues. These organisations may devote themselves to mobilising external resources and creating inter-organisational networks (Diani, 2015; Korhonen et al., 2018) towards social change such as new infrastructure or regulation. They could also target companies, pushing them to address environmental and social issues (Den Hond & De Bakker, 2007; Shrivastava & Scott, 1992).

The CE domain is no different in this regard. For example, the 'right to repair' movement has, for decades now, campaigned for the reparability of products, which are often badly designed or intentionally made to last only for a limited time (Bello & Aufderheide, 2021). Similarly, social movements have campaigned against plastics and chemical pollution and other environmental impacts of the linear, wasteful economy (Auyero et al., 2019; Sicotte & Brulle, 2017). Perhaps the biggest market failure has been climate change, which has been the focus of social movement activism for decades now (Askanius & Uldam, 2011; McAdam, 2017; North, 2011). Yet, activism does not only happen 'on the streets,' so to say.

Activism from within

Activists also aim to bring about change within, often incumbent, organisations (Briscoe & Gupta, 2016). These so-called internal activists are individuals who act as crucial change agents within organisations, pushing them to adopt better social and environmental practices from the inside. Management and organisational scholars have, for some time, studied how organisational change is brought about. Innovations and pressure to change organisational routines and cultures can be incubated endogenously (Friesl & Larty, 2013), a process that is influenced by organisational power and politics, capabilities, as well as organisational inertia (Greenwood & Suddaby, 2006; Greenwood et al., 2015). Individuals who feel misaligned with predominant institutional logics will mobilise internal resources against existing organisational strategies and systems (Meyerson & Scully, 1995).

Scholars have studied internal activists at the CEO level (Chatterji & Toffel, 2019) as well as other managerial levels (Bocken & Geradts, 2020). They have also analysed institutional entrepreneurs (Battilana et al., 2009), social intrapreneurs (Davis & White, 2015), tempered radicals (Meyerson & Scully 1995) and employees' environmental activism (Skoglund & Böhm, 2020). These internal activists often have unique knowledge and expertise, as they are often embedded in wider communities of action, epistemic communities and social movements (Skoglund & Böhm, 2022). They use their unique position to bring about institutional change, sometimes using resources and networks from outside the organisation (Den Hond & De Bakker, 2007). However, as we suggest next, activism in the CE domain can be much more fluid and transcend organisational and role boundaries. It can include and be led as much by employees, managers or social movement activists. This type of activism transcends existing conceptual duality and provides a canvass for unbound activity, which allows for a plurality of stakeholders to push their organisations towards adopting CE principles and processes, thus bringing about positive social and environmental change.

Everyday, prefigurative activism

Since organisational boundaries are often porous (Gulati et al., 2012), activism should not be seen as a dichotomy but, indeed, recognised as part of a multi-level process of change-making in, between and through organisations (DeJordy et al., 2020). Change-making activism should be seen as a hybrid concept where individuals may take both internal and external organisational positions, which they use to accelerate or control resource flows and change processes (DeJordy et al., 2020).

Social enterprises (Mair & Marti, 2006) and community-based enterprises (Hertel et al., 2019) are typical examples of business activism that is geared towards social and environmental agendas. Here, employees and managers do not simply have a narrow profit-oriented identity and purpose. Instead, they work across a multiplicity of domains that incorporate social, environmental and economic dimensions. As the CE approach makes cross-organisational collaboration inevitable and necessary, precisely because wider system change is often targeted, CE-adopting businesses could be seen as activists. Eco-industrial development (de Abreu & Ceglia, 2018), circular supply chain (Herczeg et al., 2018) and corporate-entrepreneur collaborations (Veleva & Bodkin, 2018) are examples of such a linkage between business-focused change to wider societal and environmental dimensions.

In her work, however, Hobson (2020, 99) suggests that people may commend CE approaches but are sceptical of its "new, resource-efficient business models in contexts of hyper-consumerism." Instead, Hobson emphasises an 'everyday CE' which permeates into private practices of resource use and recycling, but which must build on circular spaces and neighbourhoods that enable and make possible circularity beyond small-scale citizen initiatives. This points towards a CE perspective that is much more community-based, embedded in the wider social and cultural fabric of society.

While often small-scale, many community-based initiatives related to 'making, doing, and mending' can be seen as a form of everyday CE action. One frequently mentioned example is Repair Cafés (Cole & Gnanapragasam, 2017; Charter & Keiller, 2014; Spekkink et al., 2022), which aim to promote and help with repairing broken items in order to save them from becoming waste. Other examples that have been interpreted as CE-related communities are hackerspaces (Charter & Keiller, 2014) or makerspaces (Unterfrauner et al., 2019). We can understand these initiatives as locally replicable ideas connected through shared aims and skills, which have been analysed as translocal networks (Loorbach et al., 2020) or as social movements both on a local and global level (Pesch et al., 2019). Such bottom-up approaches to CE have also been applied to community development in rural settings with the aim to increase its resilience (Aguiñaga et al., 2018).

Here, communities engage in everyday CE activism, pro-actively changing social and material relations. This often involves ethical considerations of solidarity, reciprocity, cooperation, autonomy and participation, enabling everyone to partake in the CE, not only as a consumer, manager or business-person, but as a citizen (Gutberlet et al., 2017). Community CE activists, hence, set up alternative organisations, prefiguring the economic relations and societies they themselves would like to see (Monticelli, 2018). They engage in a material, pragmatic type of activism that is prefigurative and performative; that is, it is taking place in everyday settings in the here and now (Skoglund & Böhm, 2020).

Examples of everyday, prefigurative CE activism

In this section, we discuss everyday CE activism through three multilateral examples. In all of them, individuals work within communities to bring about the change they want to see taking place. They hence engage in prefigurative action on a daily basis, in order to bring about the CE transition in the here and now.

Cultivate Cornwall

UK community interest companies (CICs) are an example of organisations with 'porous' boundaries (Gulati et al., 2012), able to lead or participate in change-making initiatives (DeJordy et al., 2020). In the UK, a CIC is usually set up in order to serve wider social needs of a community. This does not necessarily require it to be a not-for-profit organisation, but it does mean that its *modus operandi* should include the provision of some social good, in addition to a return to its investors. One such CIC, based in the Duchy of Cornwall, South West England, is Cultivate. Originally founded by John Lakey and Harry Deacon in 2016, and joined by Lin Chapman in 2017, Cultivate has dedicated itself to the sole purpose of alleviating economic destitution in the Bodmin area of Cornwall. In this way, the company's business model can be described as an example of 'place framing' (Martin, 2003), whereby a number of communal problems are identified ('framed') and solutions sought.

Initially, Cultivate was motivated to address material waste as an economic aspect of business operation in the immediate Bodmin area. To do so, it collected waste from local companies and reused it, in order to produce hand-crafted utility items. This includes collecting textile materials, such as unsold t-shirts from local event organisers, and turning them into novelty clothing items, designed and made in-house. Cultivate expanded its product range and repurposed damaged boat sails, scaffolding and netting. As well as becoming the hub for a 'social movement scene' (Creasap, 2012), connecting like-minded individuals, business and community members, Cultivate have been able to create a growing range of sustainable products, such as the lunch box made from a torn wet-suit, shown in Figure 14.1.

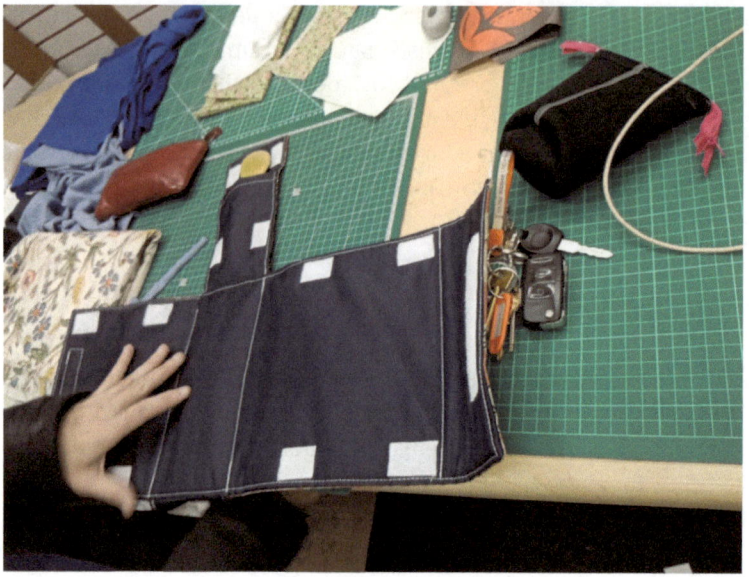

Figure 14.1: A lunchbox designed and made entirely from a repurposed wetsuit.
Source: Authors.

Note that the lunchbox opens out, so that the corners can be washed, and the item comes with sustainable cutlery (a wooden spoon is visible in the top pocket holder). Once opened, the lunchbox serves as a mat, so that lunch can be eaten directly from any surface, including a person's knees. Cultivate has also started producing its own range of backpacks, with a prototype, made from a damaged wetsuit, shown in Figure 14.2. Each product is hand-made at the company's Bodmin premises.

In this way, Cultivate operates within the fluid space of everyday activism, overcoming the separation between external organisational environments and internal decision-making (Skoglund & Böhm, 2020). As an example of what Skoglund & Böhm (2020) further refer to as 'prefigurative' merging of individual goodwill and professional knowhow, the reclaimed materials used in their own production are also offered to community members at the repair café-type sewing workshops Cultivate run for free, both from their Bodmin offices and in local schools. Cultivate also offers paid repair services, yet members of the public can bring their damaged items and use the sewing machines free of charge. Since 2019, their textile repairs offering has expanded and now also includes a music instrument library for young people in Cornwall.

The unequal impact of the coronavirus pandemic has pushed Cultivate to widen its social agenda (Hertel et al., 2019) and its communal 'framing' (Martin, 2003) even

Figure 14.2: Prototype of a backpack made from a damaged wetsuit.
Source: Authors.

further. In 2020, the CIC established an 'essential provisions in Cornwall' (EPIC) hub and has been working with over 20 local charities in the area. Although part of different domains, members of the EPIC hub are able to work together to deliver emergency food supplies across the South West, a collaboration which transcends individual boundaries in an effort to transform the local community.

Cultivate's Bodmin premises are the very embodiment of everyday activism. They are a ramshackle of space that acts as office, campaign HQ and warehouse at the same time (see Figure 14.3). Not a single corner, desk area or free space remains unutilised – every item, off-cut or pile serves a purpose and fulfils a business need, making it an example *par excellence* of 'sustainable materialism' (Schlosberg, 2019). There is nothing ornamental and no unnecessary decoration taking up valuable space. Not even a sign announces Cultivate's presence in the building – the CIC seems determined to serve the community and not dominate it. Instead, Harry, John and Lin rely on word-of-mouth to advertise their services and use their CIC as a vehicle to bring about positive social and environmental change for their local area.

Figure 14.3: Cultivate's main premises in Bodmin.
Source: Authors.

Repair Cafés

Repair Cafés are "free meeting places and they're all about repairing things (together)" (Repair Café, n.d.). Visitors of Repair Cafés bring broken items from home, ranging for electronics and mechanical devices to clothing, and then volunteer repairers help the visitors to fix these items (if still possible). This happens in public spaces that are typically set up in a way that facilitates social interaction between volunteers and visitors. Although Repair Cafés exist in multiple shapes and sizes, most Repair Cafés take place once a month, involve between 7 and 20 volunteers and have between 11 and 30 visitors in a typical session (Spekkink et al., 2020).

The first Repair Café was started by Martine Postma (who later became the chair of the Repair Café International Foundation (RCIF)) in Amsterdam in 2009. The initiative is a response to three developments:

1. people often throw away broken items too easily, even when they can still be repaired,
2. repairing is no longer a common practice and repair knowledge and skills are on the decline and
3. people that do have repair knowledge and skills are not sufficiently appreciated and their experience is underutilised.

The concept of Repair Cafés was quickly replicated in different parts of the Netherlands, as well as in other countries across the world. At the time of writing, more than 2,000 Repair Cafés exist across more than 35 countries. This quick diffusion is facilitated by, among other things, the existence of the RCIF, which was established in 2010, and which specifically focuses on international growth of the community since 2016.

Repair Cafés engage in a practical, material kind of activism. Rather than demanding from companies that they change their ways through adversarial action, Repair Cafés prefigure a society in which repairing is part of everyday life, a mundane activity through which anyone can make a tangible contribution to a circular society. Their vision of a circular society is one in which people are knowledgeable, skilled and active contributors to circularity, rather than the passive consumers that appear in the top-down visions that dominate CE debates. However, rather than trying to persuade people of this vision by talking about it, Repair Cafés are focused on enacting it. As the notion of 'Café' also suggests, bringing people from local communities together is a crucial part of their *modus operandi*, because it is through the social connections that Repair Cafés build that the repair knowledge and skills vital to a circular society.

Repair Cafés thus have a strong local orientation. At the same time, they form a translocal community with a common underlying vision (Loorbach et al., 2020), which has the potential to bring about broader social change by engaging in local activities in many different places at the same time. The direct environmental impact of Repair Cafés is likely to be modest (e.g. the RCIF roughly estimates that Repair Cafés prevented 400,000 kilos of waste in 2019), but they are 'playing the long game,' attempting to change how people see their possessions and the role of repair in their everyday life, which may have a larger indirect environmental impact. For Repair Cafés, it is not just important that their visitors take home a repaired item that would otherwise have been wasted, but also that visitors develop a different mindset.

Food groups

Long-term austerity has led to a rise in community-based food provisioning across the whole of Europe, with food banking now seen as the archetypal community food model. Grassroots food-based initiatives based on commoning, sharing and reciprocity are not new (Holmes, 2018). Indeed, the work of Pahl (1984) and Gibson and Graham (2008), amongst others, has illuminated the vast array of community and neighbour-driven informal economies of reciprocity and sharing of food and other items that have existed for centuries. Nonetheless, in 2021, in the UK alone there were over 2,000 food banks (House of Commons, 2021). With the pandemic severely affecting many people's livelihoods, the need for community and charity-based food provisioning has grown rapidly. Recent statistics illustrate that those accessing community-based food provisioning increased by up to 88% in October 2020 compared to the same period in 2019 (House of Commons, 2021). Yet, whilst the focus remains on food banking as the main means of community food provision, this often overlooks the numerous other endeavours that similarly bolster local communities whilst also engaging in everyday CE activism and practices.

Emerging work on community gardening schemes (Cumbers et al., 2018; Valle, 2021), community fridges (Morrow, 2019), food swaps (Schor et al., 2016) and cooking clubs (Marovelli, 2019) reveals a plethora of locally based, volunteer-led, grassroots initiatives using purpose-driven circular action to provide food for local people whilst also operating sustainably through a focus on redistribution and reuse.

Coffee Club is one such initiative. Based weekly in a community space on the high street of a former mill town in North West England, Coffee Club is an alternative community food provisioning model that exemplifies local everyday CE action. Like a food bank, its aim is to provide food to those in need in the local area. However, it attempts to avoid the stigma attached to such charity-based models by operating on a membership basis (Caplan, 2016). Anyone can join the group and for a nominal weekly fee receive approximately three bags worth of food. The food is sourced from a variety of suppliers – including surplus directly donated by local supermarkets and a weekly coordinated delivery from the food redistribution charity Fareshare. One critique levelled at many food banks is the removal of choice – in other words, recipients simply get a given set food parcels rather being able to choose the food they like and would want. This, of course, adds to the stigma attached to attending a food bank. At Coffee Club, members can choose the food they want, picking their own fruits and vegetables or tinned goods from food laid out on tables – not unlike a market stall.

What also sets it apart from most food banking models is its café and networking focus. Members do not just attend the group to collect food, they also attend to socialise with others. This often involves swapping recipes or tips of how to cook the produce on offer. Sometimes members will bring in things they have baked or surplus food they have grown to share with others. Thus, Coffee Club not only circulates and redistributes food from the wider and corporate food supply chain, but it also redistributes and encourages the circulation of food at much more micro and household levels (Holmes, 2018). In this respect, Coffee Club works with local businesses to achieve social and environmental action – providing low-cost food to those who need it, whilst ensuring food is not wasted. Furthermore, Coffee Club encourages and facilitates social relationships through such everyday circular practices – bringing local people together and connecting the community.

Discussion

The examples in the preceding section provide insight into how community groups simply get on with the CE transition in their everyday settings. They do not wait for companies to offer them circular products or services. They also do not wait for government to introduce CE legislation or regulation. They do what people have always done: self-organise to address social and environmental issues in their local setting (Heiskanen et al., 2010; Calvário et al., 2020). In this chapter, we have termed this 'circular society activism,' in order to provide an alternative, grassroots perspective of CE, whose discourse tends to be dominated by large, top-down actors, such as multinational companies, policymakers and influential thinktanks. We have argued that what is often forgotten or under-emphasised by most CE literatures is the vital role of grassroots communities who engage in CE activities in self-organised, everyday, place-based settings.

These bottom-up CE initiatives can be conceptualised as 'everyday, prefigurative activism,' which reflects the messy, organic and emergent nature of activism from below. This builds on recent scholarship that emphasises that perhaps we should talk less about the 'CE' but, instead, talk more about the 'circular society' (Jaeger-Erben et al., 2021). In this chapter, we have highlighted the role of grassroots activists in CE transitions. When we say that this type of activism is 'messy, organic and emergent,' we imply that a circular society should not be understood as a unified people. Instead, all social relations involve struggles, misunderstandings, divergent interests and power relations. The political theorist Mouffe (2014) calls this 'agonism.'

What is striking, however, is that communities, while embedded in agonistic spaces, nevertheless just get on with it. They engage in a form of prefigurative 'sustainable materialism' (Schlosberg, 2019), bringing about change in the here and now. While prefigurative social action has existed for a long time, what is perhaps distinct about CE prefigurative activism is its explicit engagement with materials and resources. All three illustrative examples deal directly with things we can touch and need on a daily basis: clothing, gadgets and food. What Schlosberg (2019) calls 'sustainable materialism' is a vital extension of existing forms of environmental activism that challenges dominant actors, often discursively, from without or within organisations. In our cases, the prefigurative communities do not necessarily shout too much about what they are doing; they simply get on with it, working with the everyday materials of their activist focus, creating change in the here and now (Chatterton & Pickerill, 2010; Maeckelbergh, 2011; Skoglund & Böhm, 2020).

While we think that circular society activism from below plays a vital part in bringing about social and environmental change, we do not claim that it can be seen in isolation of other forms of activism we discussed in this chapter (without and within), nor should we ignore the more mainstream top-down and mid-range CE discourses and practices by governments, corporations, large NGOs, consultancies and entrepreneurs. Social movement theory has understood the differentiated relationship between moderate and more radical actors through the lens of the concept of 'radical flank,' arguing that more radical actors often have a direct or indirect role in pushing mainstream actors in a particular direction of change (Fitzgerald & Rodgers, 2000; Rucht, 2004). While it goes beyond the purpose of this chapter to discuss this in detail, perhaps the type of circular society activism discussed here provides insight into the future direction of the whole CE transition movement.

We do claim, however, that place-based, everyday community activism is underemphasised in most CE scholarship as well as wider CE discourses and practices. There is an urgent need, we argue, to make more visible the thousands of grassroots initiatives that engage in concrete, material CE practices that are of ethical, cultural, social and economic value to communities. Without these actions from below, the CE runs the risk of being perceived as a top-down, elitist model of bringing about social and environmental change. History has shown that this can easily involve autocratic tendencies, while not creating sustained, inclusive change in the first place.

Conclusion

This chapter has conceptualised what we have called 'circular society activism' to provide an alternative, community-focused perspective of CE transitions. We have provided three illustrative examples to show how CE is enacted by community groups in an everyday, place-based setting, working with materials that are of relevance to communities: clothing, technical gadgets and food. We have understood these CE practices through the lens of prefiguration, which means that social and environmental change is brought about through concrete, material activities in the here and now. These grassroots practices from below are testament to the creative, innovative potential of the CE that can be enacted by anyone and at any time. In this way, the CE should not be seen as an economy to come in the distant future. Instead, the CE is already here, as it is being prefiguratively enacted by a circular society that does not want to wait for governments and large businesses to tell them what to do. The circular society is already actively engaged to bring about a CE.

References

Aguiñaga, E., Henriques, I., Scheel, C., & Scheel, A. (2018). Building resilience: A self-sustainable community approach to the triple bottom line. *Journal of Cleaner Production*, *173*, 186–196.

Aminoff, A., & Pihlajamaa, M. (2020). Business experimentation for a circular economy-Learning in the front end of innovation. *Journal of Cleaner Production*, *275*, 124051.

Askanius, T., & Uldam, J. (2011). Online social media for radical politics: Climate change activism on YouTube. *International Journal of Electronic Governance*, 4(1–2), 69–84.

Auyero, J., Hernandez, M., & Stitt, M. E. (2019). Grassroots activism in the belly of the beast: A relational account of the campaign against urban fracking in Texas. *Social Problems*, 66(1), 28–50.

Battilana, J., & Casciaro, T. (2012). Change agents, networks, and institutions: A contingency theory of organizational change. *Academy of Management Journal*, 55(2), 381–398.

Battilana, J., Leca, B., & Boxenbaum, E. (2009). 2 how actors change institutions: Towards a theory of institutional entrepreneurship. *Academy of Management Annals*, 3(1), 65–107.

Bello, B., & Aufderheide, P. (2021). The public interest and the information superhighway: The digital future coalition (1996–2002) and the afterlife of the digital millennium copyright act. *Information & Culture*, 56(1), 49–89.

Bocken, N. M. P., & Geradts, T. H. J. (2020). Barriers and drivers to sustainable business model innovation: Organization design and dynamic capabilities. *Long Range Planning*, 53(4), 101950.

Briscoe, F., & Gupta, A. (2016). Social activism in and around organizations. *Academy of Management Annals*, 10(1), 671–727.

Calvário, R., Desmarais, A. A., & Azkarraga, J. (2020). Solidarities from below in the making of emancipatory rural politics: Insights from food sovereignty struggles in the Basque country. *Sociologia Ruralis*, 60(4), 857–879.

Caplan, P. (2016). Big society or broken society?: Food banks in the UK. *Anthropology Today*, 32(1), 5–9.

Carter, N. (2018). *The politics of the environment: Ideas, activism, policy*. Cambridge, UK: Cambridge University Press.

Charter, M., & Keiller, S. (2014). Grassroots innovation and the circular economy. *The centre for sustainable design® University for the creative arts*.

Chatterji, A. K., & Toffel, M. W. (2019). Assessing the impact of CEO activism. *Organization & Environment*, 32(2), 159–185.

Chatterton, P., & Pickerill, J. (2010). Everyday activism and transitions towards post-capitalist worlds. *Transactions of the Institute of British Geographers*, 35(4), 475–490.

Cherrier, H., Black, I. R., & Lee, M. (2011). Intentional non-consumption for sustainability: Consumer resistance and/or anti-consumption? *European Journal of Marketing*. Vol. 45 No. 11/12, pp. 1757–1767

Chipman Koty, A. (2021). China's circular economy: Understanding the new five year plan. *China Briefing*, July 16, 2021. https://www.china-briefing.com/news/chinas-circular-economy-understanding-the-new-five-year-plan/

Cole, C., & Gnanapragasam, A. (2017). Community repair: Enabling repair as part of the movement towards a circular economy.

Corvellec, H., Böhm, S., Stowell, A., & Valenzuela, F. (2020). *Introduction to the special issue on the contested realities of the circular economy*. Culture and Organization, 26(2), 97–102.

Creasap, K.(2012). Social movement scenes: Place-based politics and everyday resistance. *Sociology Compass*, 6(2), 182–191.

Cullen, U. A., & De Angelis, R. (2021). Circular entrepreneurship: A business model perspective. *Resources, Conservation and Recycling*, 168, 105300.

Cumbers, A., Shaw, D., Crossan, J., & McMaster, R. (2018). The work of community gardens: Reclaiming place for community in the city. *Work, Employment and Society*, 32(1), 133–149.

Davis, G. F., & White, C. J. (2015). The new face of corporate activism. *Stanford Social Innovation Review*, 13(4), 40–45.

de Abreu, M., Sá, C., & Ceglia, D. (2018). On the implementation of a circular economy: The role of institutional capacity-building through industrial symbiosis. *Resources, Conservation and Recycling, 138*, 99–109.

DeJordy, R., Scully, M., Ventresca, M. J., & Douglas Creed, W. E. (2020). Inhabited ecosystems: Propelling transformative social change between and through organizations. *Administrative Science Quarterly, 65*(4), 931–971.

Den Hond, F., & Bakker, F. G. A. D. (2007). Ideologically motivated activism: How activist groups influence corporate social change activities. *Academy of Management Review, 32*(3), 901–924.

Dentchev, N., Baumgartner, R., Dieleman, H., Jóhannsdóttir, L., Jonker, J., Nyberg, T., . . . Tang, X. (2016). Embracing the variety of sustainable business models: Social entrepreneurship, corporate intrapreneurship, creativity, innovation, and other approaches to sustainability challenges. *Journal of Cleaner Production. 113*(1), 1–4.

Diani, M. (2015). *The cement of civil society.* Cambridge, UK: Cambridge University Press.

Dono, J., Webb, J., & Richardson, B. (2010). The relationship between environmental activism, pro-environmental behaviour and social identity. *Journal of Environmental Psychology, 30*(2), 178–186.

Ellen MacArthur Foundation. *"Introduction" What is a circular economy?* Accessed Nov 19, 2021. https://ellenmacarthurfoundation.org/topics/circular-economy-introduction/overview

European Commission. (2021). *Commissioner sinkevicius speech at EP plenary debate – new circular economy action plan*, 08 February 2020, Brussels. https://ec.europa.eu/commission/commissioners/2019-2024/sinkevicius/announcements/commissioner-sinkevicius-speech-ep-plenary-debate-new-circular-economy-action-plan-08-february-2020_en

Farla, J. C. M., Markard, J., Raven, R., & Coenen, L. E. (2012). Sustainability transitions in the making: A closer look at actors, strategies and resources. *Technological Forecasting and Social Change, 79*(6), 991–998.

Fitzgerald, K. J., & Rodgers, D. M. (2000). Radical social movement organizations: A theoretical model. *The Sociological Quarterly, 41*(4), 573–592.

Foxon, T. J. (2011). A coevolutionary framework for analysing a transition to a sustainable low carbon economy. *Ecological Economics, 70*(12), 2258–2267.

Friesl, M., & Larty, J. (2013). Replication of routines in organizations: Existing literature and new perspectives. *International Journal of Management Reviews, 15*(1), 106–122.

Geng, Y., Sarkis, J., & Bleischwitz, R. (2019). *How to globalize the circular economy.* Nature *565*, no. 7738 (2019): 153–155.

Gibson, K., & Graham, J. (2008). *Diverse economies: Performative practices for'other worlds' and Diverse economies in geography online bibliography.*

Greenwood, R., Devereaux Jennings, P., & Hinings, B. (2015). Sustainability and organizational change: An institutional perspective. In: *Leading sustainable change: An organizational perspective* (pp. 323–355). Oxford, UK: Oxford University Press.

Greenwood, R., & Suddaby, R. (2006). Institutional entrepreneurship in mature fields: The big five accounting firms. *Academy of Management Journal, 49*(1), 27–48.

Gregson, N., Crang, M., Fuller, S., & Holmes, H. (2015). Interrogating the circular economy: The moral economy of resource recovery in the EU. *Economy and Society, 44*(2), 218–243.

Gulati, R., Puranam, P., & Tushman, M. (2012). Meta-organization design: Rethinking design in interorganizational and community contexts. *Strategic Management Journal, 33*(6), 571–586.

Gutberlet, J., Carenzo, S., Kain, J.-H., & Azevedo, A. M. M. D. (2017). Waste picker organizations and their contribution to the circular economy: Two case studies from a global south perspective. *Resources, 6*(4), 52.

Heiskanen, E., Johnson, M., Robinson, S., Vadovics, E., & Saastamoinen, M. (2010). Low-carbon communities as a context for individual behavioural change. *Energy Policy*, *38*(12), 7586–7595.

Herczeg, G., Akkerman, R., & Hauschild, M. Z. (2018). Supply chain collaboration in industrial symbiosis networks. *Journal of Cleaner Production*, *171*, 1058–1067.

Hertel, C., Bacq, S., & Belz, F.-M. (2019). It takes a village to sustain a village: A social identity perspective on successful community-based enterprise creation. *Academy of Management Discoveries*, *5*(4), 438–464.

Hobson, K. (2020). Beyond the consumer: Enlarging the role of the citizen in the circular economy. In *Handbook of the circular economy*. Edited by Miguel Brandão, David Lazarevic, and Göran Finnveden (pp. 479–490). Cheltenham, UK: Edward Elgar Publishing.

Hobson, K., & Lynch, N. (2016). Diversifying and de-growing the circular economy: Radical social transformation in a resource-scarce world. *Futures*, *82*, 15–25.

Holmes, H. (2018). New spaces, ordinary practices: Circulating and sharing within diverse economies of provisioning. *Geoforum*, *88*, 138–147.

House of Commons. (2021). *Food Banks in the UK. Research briefing*, 21st April 2021. Accessed via: https://commonslibrary.parliament.uk/research-briefings/cbp-8585/

Jaeger-Erben, M., Jensen, C., Hofmann, F., & Zwiers, J. (2021). There is no sustainable circular economy without a circular society. *Resources, Conservation and Recycling*, *168*, 105476.

Kanter, R. M. (2015). How purpose-based companies master change for sustainability: A systemic approach to global social change. In: *Leading sustainable change: An organizational perspective* (pp. 111–142).

Kirchherr, J., Reike, D., & Hekkert, M. (2017). Conceptualizing the circular economy: An analysis of 114 definitions. *Resources, Conservation and Recycling*, *127*, 221–232.

Konietzko, J., Bocken, N., & Hultink, E. J. (2020). Circular ecosystem innovation: An initial set of principles. *Journal of Cleaner Production*, *253*, 119942.

Korhonen, J., Nuur, C., Feldmann, A., & Birkie, S. E. (2018). Circular economy as an essentially contested concept. *Journal of Cleaner Production*, *175*, 544–552.

Kovacic, Z., Strand, R., & Völker, T. (2019). *The circular economy in Europe: Critical perspectives on policies and imaginaries*. London: Routledge.

Leipold, S., Weldner, K., & Hohl, M. (2021). Do we need a 'circular society'? Competing narratives of the circular economy in the French food sector. *Ecological Economics*, *187*, 107086.

Lekan, M., & Rogers, H. A. (2020). Digitally enabled diverse economies: Exploring socially inclusive access to the circular economy in the city. *Urban Geography*, *41*(6), 898–901.

Loorbach, D., Wittmayer, J., Avelino, F., Timo, V. W., & Frantzeskaki, N. (2020). Transformative innovation and translocal diffusion. *Environmental Innovation and Societal Transitions*, *35*, 251–260.

Maeckelbergh, M. (2011). Doing is believing: Prefiguration as strategic practice in the alterglobalization movement. *Social Movement Studies*, *10*(01), 1–20.

Mair, J., & Marti, I. (2006). Social entrepreneurship research: A source of explanation, prediction, and delight. *Journal of World Business*, *41*(1), 36–44.

Marovelli, B. (2019). Cooking and eating together in London: Food sharing initiatives as collective spaces of encounter. *Geoforum*, *99*, 190–201.

Martin, D. G. (2003). "Place-framing" as place-making: Constituting a neighborhood for organizing and activism. *Annals of the Association of American Geographers*, *93*(3), 730–750.

McAdam, D. (2017). Social movement theory and the prospects for climate change activism in the United States. *Annual Review of Political Science, 20*, 189–208.

Meyerson, D. E., & Scully, M. A. (1995). Crossroads tempered radicalism and the politics of ambivalence and change. *Organization Science, 6*(5), 585–600.

Millette, S., Hull, C. E., & Williams, E. (2020). Business incubators as effective tools for driving circular economy. *Journal of Cleaner Production, 266*, 121999.

Mirvis, P. H. (1994). Environmentalism in progressive businesses. *Journal of Organizational Change Management. 7*(4), 82–100.

Monticelli, L. (2018). Embodying alternatives to capitalism in the twenty-first century. *TripleC: Communication, Capitalism & Critique. Open Access Journal for a Global Sustainable Information Society, 16*(2), 501–517.

Morrow, O. (2019). Sharing food and risk in Berlin's urban food commons. *Geoforum, 99*, 202–212.

Mouffe, C. (2014). Democratic politics and conflict: An agonistic approach. In: *Political power reconsidered: State power and civic activism between legitimacy and violence* (pp. 17–29).

Lakitsch, M. (Ed.). (2014). Political Power Reconsidered: State Power and Civic Activism between Legitimacy and Violence. Peace Report 2013 (Vol. 66). Münster: LIT Verlag.

North, P. (2011). The politics of climate activism in the UK: A social movement analysis. *Environment & Planning A, 43*(7), 1581–1598.

Padilla-Rivera, A., Telles Do Carmo, B. B., Arcese, G., & Merveille, N. (2021). Social circular economy indicators: Selection through fuzzy delphi method. *Sustainable Production and Consumption, 26*, 101–110.

Padilla-Rivera, A., Russo-Garrido, S., & Merveille, N. (2020). Addressing the social aspects of a circular economy: A systematic literature review. *Sustainability, 12*(19), 7912.

Pahl, R. E. (1984). *Divisions of labour*. Blackwell Oxford, UK: Blackwell.

Palm, J., & Bocken, N. (2021). Achieving the circular economy: Exploring the role of local governments, business and citizens in an urban context. Energies, *14*(4), 875.

Pesch, U., Spekkink, W., & Quist, J. (2019). Local sustainability initiatives: Innovation and civic engagement in societal experiments. *European Planning Studies, 27*(2), 300–317.

Petit-Boix, A., & Leipold, S. (2018). Circular economy in cities: Reviewing how environmental research aligns with local practices. *Journal of Cleaner Production, 195*, 1270–1281.

Repair Café International Foundation. "About". What is a Repair Café? Accessed Nov 19, 2021. https://www.repaircafe.org/en/about/

Rucht, D. (2004). Movement allies, adversaries, and third parties. In: *The Blackwell companion to social movements* (pp. 197–216).

Snow, D. A., Soule, S. A., & Kriesi, H. (Eds.). (2008). The Blackwell companion to social movements. John Wiley & Sons. Location: Hoboken, New Jersey.

Schlosberg, D. (2019). From postmaterialism to sustainable materialism: The environmental politics of practice-based movements. In: *Environmental politics*. https://doi.org/10.1080/09644016.2019.1587215

Schor, J. B., Fitzmaurice, C., Carfagna, L. B., Attwood-Charles, W., & Poteat, E. D. (2016). Paradoxes of openness and distinction in the sharing economy. *Poetics, 54*, 66–81.

Schröder, P., Lemille, A., & Desmond, P. (2020). Making the circular economy work for human development. *Resources, Conservation and Recycling, 156*, 104686.

Schroeder, P., Anggraeni, K., & Weber, U. (2019). The relevance of circular economy practices to the sustainable development goals. *Journal of Industrial Ecology, 23*(1), 77–95.

Shrivastava, P., & Scott, H. I. (1992). Corporate self-greenewal: Strategic responses to environmentalism. *Business Strategy and the Environment*, *1*(3), 9–21.

Sicotte, D. M., & Brulle, R. J. (2017). Social movements for environmental justice through the lens of social movement theory. In: *The routledge handbook of environmental justice* (pp. 25–36). London: Routledge.

Skoglund, A., & Böhm, S. (2020). Prefigurative partaking: Employees' environmental activism in an energy utility. *Organization Studies*, *41*(9), 1257–1283.

Skoglund, A., & Böhm, S. (2022). *Climate activism*: How Communities Take Renewable Energy Actions Across Business and Society. Cambridge, UK: Cambridge University Press.

Spekkink, W., Rödl, M., & Charter, M. (2022). Repair Cafés and Precious Plastic as translocal networks for the circular economy. Journal of Cleaner Production, 380, 135–125. https://doi.org/10.1016/j.jclepro.2022.135125

Stratan, D. (2017). Success factors of sustainable social enterprises through circular economy perspective. *Visegrad Journal on Bioeconomy and Sustainable Development*, *6*(1), 17–23.

Unterfrauner, E., Shao, J., Hofer, M., & Fabian, C. M. (2019). The environmental value and impact of the Maker movement – Insights from a cross-case analysis of European maker initiatives. *Business Strategy and the Environment*, *28*(8), 1518–1533.

Valle, G. R. (2021). Learning to be human again: Being and becoming in the home garden commons. Environment and Planning E: Nature and Space, 4(4), 1255–1269.

Vecchio, P. D., Ndou, V., Passiante, G., & Vrontis, D. (2020). Circular economy innovative entrepreneurship: A conceptual foundation. In: *Innovative entrepreneurship in action* (pp. 129–144). Springer.

Velenturf, A. P. M., & Purnell, P. (2021). Principles for a sustainable circular economy. *Sustainable Production and Consumption*, *27*, 1437–1457.

Veleva, V., & Bodkin, G. (2018). Corporate-entrepreneur collaborations to advance a circular economy. *Journal of Cleaner Production*, *188*, 20–37.

Zhao, H., Zhao, H., & Guo, S. (2017). Evaluating the comprehensive benefit of eco-industrial parks by employing multi-criteria decision making approach for circular economy. *Journal of Cleaner Production*, *142*, 2262–2276.

Ziegler, R. (2019). Viewpoint– water innovation for a circular economy: The contribution of grassroots actors. *Water Alternatives*, *12*(2), 774.

Esther Goodwin Brown, Marijana Novak, Constantine Manolchev, Sharon Gil, Esteban Munoz

15 Circular economy jobs: risks and opportunities in the labour market

Abstract: The social dimension of the Circular Economy (CE) is often under-represented in research, policy and corporate discourse. As a result, little is known about what labour market impacts will be felt locally as a result of adopting CE strategies. This chapter outlines recent work on bringing the impacts of circularity on employment to the surface. It considers international research and commentary on the potential for the CE to generate employment and what the adoption of circular strategies could mean for skills development, job quality and the inclusivity of labour markets. The chapter provides three snapshots of analysis on jobs that contribute to the CE in three territories, using a method co-developed by Circle Economy and the United Nations Environment Programme. It concludes on the importance of data and metrics for achieving a just CE transition, alongside tackling climate change and resource scarcity.

Keywords: Circular Economy, circular jobs, upskilling, decent work, employment, metrics

Introduction

A means to an end, the Circular Economy (CE) enacted on a global scale has the potential to drive us towards a net-zero, sustainable and just future. In addition, if managed well, its adoption can generate multiple societal benefits while simultaneously tackling climate change and resource scarcity and closing the emissions gap (Circle Economy, 2021c; United Nations Environment Programme, 2020). The environmental benefits of such goals are obvious, but what are the implications of the CE transition for workers and employees? Can it reverse lasting economic trends such as labour market polarisation and income inequality, creating employment in areas of underemployment? Can it improve job quality and create 'decent work' as one of the cornerstones of sustainable development (International Labour Organization, 2021)? This chapter focuses on the job market dimension of the CE, as it is workers who will and are driving the transformations needed, and who will also be impacted most by the changes in business models and supply chains that come with the adoption of the CE.

Strategies that prioritise regenerative resources in production increase the usage intensity of products and components and close material cycles are at the heart of the CE (Circle Economy, 2021b). In turn, the benefits of sustainable development have generated traction with a number of supranational bodies and are slowly

https://doi.org/10.1515/9783110723373-018

giving rise to tangible changes in legislation. For instance, the United Nations Sustainable Development Goals (SDGs), and the Nationally Determined Contributions (NDCs) outlined under the Paris Agreement (2015),[1] are guiding countries to develop policies that reduce resource scarcity, protect livelihoods and tackle climate change. The CE is gaining increasing attention as a means of addressing these challenges (United Nations Development Programme and United Nations Environment Programme, 2020).

However, the social dimension of the CE transition is often ignored or underrepresented in research, policy and corporate discourses (Schröder et al., 2019, 2020). As a result, the environmental and economic impacts of circular practices have received the most attention to date (Merli et al., 2018). Walter Stahel discusses the job-creation benefits of moving away from linear industries, and towards an economic framework that maintains the value of key resources (Stahel, 2021). Human labour is an example of such a resource. Human capital – that is, the skills, experience and general know-how of workers – Stahel argues, is at the heart of the CE (Lovins & Braungart, 2014). He asserts that people are a unique resource in the CE because their value can grow through training and education, but at the same time this value can be lost if un- or underemployment prevents workers from applying their acquired knowledge and skills (Stahel, 2013).

The overemphasis on the technocratic rather than social elements of CE has led to less clarity about whether circular strategies can also alleviate unemployment, particularly at the local level, or how they can bridge the gap between well-paid, high-skilled jobs at the top of the labour market and those more labour-intensive roles, with often limited opportunities for progression and job security commonly found downstream (Brynjolfsson, 2011). This is particularly important given the widening gap between productivity and wages (Webster, 2021); with the risk being that the new jobs created in the context of automation and digitisation will be created in the upper tier of the labour market. There is also uncertainty over whether all those currently in the labour market will be provided with sufficient opportunities for up-skilling in order to take up new or transformed job roles (Levy & Murnane, 2004; World Economic Forum, 2021).

The emergence of new and transforming occupations linked to circular business models presents an opportunity to challenge current norms within industries that we must harness (International Labour Organization, 2011). However, less than 40% of NDCs pledged under the Paris Agreement include any plans for training to support their implementation (International Labour Organization, 2021). To understand the opportunities and challenges of moving towards a CE, its social impacts

---

1 Adopted by 196 members in 2015, the Paris Agreement came into force in 2016 and required all signatories to submit their NDCs towards reducing global warming by 2020. Further information: https://unfccc.int/process-and-meetings/the-paris-agreement/the-paris-agreement.

require attention and research, and must be placed at the centre of policies and legislative provisions. In turn, to unlock the potential that circularity presents labour markets, we need better evidence of the relationship between the CE and the labour market in different contexts. Which jobs are already part of the CE? Where are these jobs located? How are these jobs distributed across sectors? Understanding this can not only help to place people at the heart of the CE, but open the door for much needed further scrutiny into how to promote job quality and skills development within the CE.

This chapter outlines recent work, bringing the impacts of circularity on employment to the surface. It considers international research and commentary on the potential for the CE to generate employment and what the adoption of circular strategies could mean for skills development, job quality and the inclusivity of labour markets. It then seeks to answer the question of employment generation by providing snapshots of analysis conducted by Circle Economy and the United Nations Environment Programme (UNEP) on the number and range of jobs that contribute to the CE in three territories in Europe. This analysis forms part of the two organisations' joint work on better understanding the social impacts of the transition to a CE, the cornerstone of which is methodology for mapping the progress of territories towards the CE, using employment as a proxy indicator.

Bringing the social impacts of the CE to the surface

In a CE, the value of products and components are preserved for as long as possible and waste is minimised, with the aim of being eventually designed out completely (European Commission, 2020). This alters how goods and services are produced and consumed, and consequently, the relationship between capital and labour (Laubinger et al., 2020). Large-scale adoption of CE strategies is expected to initially be labour-intensive (Willeghems & Bachus, 2018). This is because circularity relies heavily on activities and services that are focused on reusing materials and closing material cycles – strategies that require more secondary processes and more hands than many activities in the one-directional, linear economy (Rreuse, 2021). For example, products manufactured under a linear business model are made up of raw materials and generally designed for one use cycle. As such, their value is created by producing and selling as many products as possible (Brein, 2020). Under a circular business model, products are remanufactured, as much as possible, from secondary materials or pre-used components. This could have a range of implications for labour markets. It could, for instance, create demand for higher level skills and expertise on how to deploy secondary materials and design for reuse upstream. At the same time, technology-enabled efficiencies, for instance, in design and production, could have the undesired effect of job loss, since high-tech industries often require much lower workforce numbers (Ford, 2015).

Furthermore, even though more labour-intensive secondary processes may still be required downstream in order to procure, retrieve, sort, clean and prepare secondary materials for reuse, these practical jobs are often associated with precarious labour market participation, including in service and resource management industries (Manolchev et al., 2018).

Macroeconomic forecasts for the impacts of circularity on the labour market

The adoption of CE processes by companies, in combination with government policies that encourage CE production modes, demand patterns, trading and competitiveness, can have cumulative effects on the labour market. The Organisation for Economic Co-operation and Development (OECD) recently summarised four ways that the adoption of circular strategies impacts labour markets: (1) some additional jobs are created; (2) some jobs are substituted; (3) some jobs are lost; and (4) some jobs are transformed (Laubinger et al., 2020). It is estimated that the net increase in jobs related to the adoption of the CE will be approximately 700,000 jobs in the EU by 2030 (Cambridge Econometrics, 2018) and 7 to 8 million worldwide (International Labour Organization, 2021). The greatest gains will be seen in sectors that rely less on materials and more on workers and the services they provide. This shift will also see declining employment in material-intensive extractive industries, like mining and traditional manufacturing. However, research suggests that overall job losses from CE adoption are likely to be modest, and outnumbered by the jobs created in other areas of the economy (Laubinger et al., 2020). Evidence suggests that the distribution of these jobs, including their channelling towards areas of chronic unemployment, could be achieved through more ambitious policy making at the national level (Green Alliance, 2021).

Although recent studies, like those discussed here, have begun to quantify employment impacts on a macroeconomic level, much less is known about the quality of and skills needed for jobs that contribute to the CE and how those jobs will be distributed across social groups and labour markets in different cities, countries and regions around the world.

Making the CE work for everyone

Work is more than a means of providing for our basic needs. Work can also provide social value, agency and opportunities to develop our interests and skills (Egdell & McQuaid, 2014). The International Labour Organization (ILO) defines decent work as "productive work for men and women in conditions of freedom, equity, security and human dignity (Somavia, 1999)." In a fully CE, value chains should perform in optimal ways for both human capital and material resources. However, the ability to achieve positive social impact with the CE may be limited by the fact that labour markets do not currently offer equal access to decent work, nor present equal opportunities for skills development. For example, a meta-analysis of studies across OECD countries found members of ethnic and racial minority groups consistently face discrimination in hiring processes (Zschirnt, 2016). Furthermore, precarious working conditions are on the rise across global regions. Precarious work is typically short-term and lacks legal protection as well as progression and skill-development opportunities (Manolchev et al., 2018). In addition to being associated with low-paid or informal work in resource management sectors in the Global South (WIEGO, 2019), precarious work is also prevalent in the Global North with the rise of the 'gig-economy' and the spread of technology-enabled platforms that have helped to make precarious work a staple feature of global labour markets, including in the Global North due to declines in industry and the rise of the services industry (Karabarbounis & Neiman, 2014; Kessler, 2018).

Unequal market access is exacerbated by labour market polarisation, caused by the historic loss of mid-level jobs with the most well-remunerated jobs existing at the top end of labour markets, and continues to give rise to inequality and on-the-job poverty (Jacobs, 2015; Standing, 2011). Technological progress has been known to aggravate this (Brynjolfsson, 2011). Without a stable income or protection from unfair dismissal, workers at the bottom of the labour market are unlikely to be in a position to develop the skills that are demanded of them and as a result may be less able to progress. Furthermore, rising employment levels are often accompanied by an increase in underemployment (employees working less hours than they would like) and in-work poverty (Bell & Blanchflower, 2013; ILOStat, 2019).

As such, it is important that we acknowledge the structural issues that are embedded in our current labour markets, and that any new economic framework put forward, such as the CE, should strive to find ways to address and overcome such issues. We also need to acknowledge that jobs that contribute to the CE will not inherently be of better quality than jobs related to the linear economy. The labour conditions and social protections afforded to more practical, labour-intensive jobs associated with the CE, most notably those in resource management, have received the most criticism (Gregson et al., 2021). Issues have also been identified in the sourcing and manufacturing processes that repair, maintenance, resale and services business models rely upon, in the electronics and textiles industries in particular (Circle Economy, European Environmental Bureau and Fair Trade Advocacy Office, 2020).

The need to demonstrate how circular practices can create decent work has become even more pressing against the backdrop of the COVID-19 pandemic, with the CE being positioned, particularly within Europe, as a mechanism for achieving a transformed and more resilient and equitable economy (Euro Cities, 2020). As such, in promoting the CE, we need to understand and share best practices for mitigating the potential negative impacts, or blindspots, of circular practices. This is needed to ensure that the harmful side effects of linear value chains are not carried over into the CE and that the adoption of circular strategies does not exacerbate the existing inequalities in labour markets described here, often known as linear lock-in (Schröder, 2020; Sopjani et al., 2020).

Beyond net job creation

One of the main ways the CE can create positive social, as well as environmental, outcomes is through creating jobs. Focusing on job creation alone, however, is not enough. The CE needs to create decent work and good-quality jobs. In an ideal world, this work would also be meaningful and fulfilling, offering not only remuneration and safety from harm but a sense of purpose, promoting work that enables human capital to grow in value and contribute to positive individual, organisational and collective well-being. How can we take steps towards this goal? More to the point, do we have the means to measure both the quantity and quality of CE jobs? In the remainder of this chapter, we lay the foundation for monitoring the quantity and range of jobs that contribute to the CE in different territories, and set out the need for further work to combine measures of job quantity with indicators of job quality in order to fully understand and safeguard the nature of work in the CE.

Mapping the number and range of circular jobs across territories

This section is for researchers or policy makers that want to use the CE as a framework for tackling job creation and skills development alongside environmental issues. This chapter provides an overview of the Circle Economy and UNEP's Circular Jobs Methodology alongside three territorial snapshots – Cornwall and the Isles of Scilly, the Netherlands and Antwerp[2] – to illustrate the method and metric in action. A CE premises that material footprints and consequent impacts will be reduced, although cross-border trade and data availability make this difficult to measure. Standards for how to measure circularity in these terms are evolving. However, circular business models (such as repair) or enablers (such as digitalisation or design) are largely agreed upon in the literature. The Circular Jobs Methodology evaluates circular activity by looking at circular business models across the economy, and analysing interdependencies.

---

2 Please note that at the time of publishing this chapter, the methodology is still under peer review and these snapshots do not yet incorporate the material import dependency.

Evolution of the circular jobs monitor and methodology

Circle Economy first identified the need to monitor employment related to the CE through the organisation's work with city-level stakeholders. Stakeholders repeatedly expressed an interest in understanding how adopting the CE in their city would impact local employment. They were also interested in which instruments local governments have at their disposal to incentivise other governments, citizens and businesses to adopt circular strategies and ensure that circularity results in the highest possible level of societal value. In response to this need, Circle Economy first developed a method to define and measure employment in the CE with the Erasmus Happiness Economics Research Organisation (EHERO) in 2017, through a project funded by the Goldschmeding Foundation in the Netherlands (Circle Economy, 2020). This project resulted in a methodology that evaluated the extent to which the activities of a certain sector are circular, based on its interactions with other sectors. The King Baudouin Foundation then funded a baseline analysis of the Belgian context along with the development of an online dashboard to display the results of these baseline analyses: the Circular Jobs Monitor (Circle Economy, 2019; Circular Jobs Monitor, 2021).

As interest in the employment impacts of the CE and the potential for these insights to support policy makers in prioritising green recovery strategies grew, Circle Economy was commissioned to conduct similar baseline assessments for Cornwall and the Isles of Scilly, New York City, Amsterdam and Scotland through projects and partnerships with government agencies in those regions – all of which are now are represented on the Circular Jobs Monitor (Circle Economy, 2018, 2020; Circle Economy and New York Circular City Initiative, 2021; Circle Economy and Zero Waste Scotland, 2020). Next to this work, a dedicated programme for understanding the employment and social impacts of the CE was established at Circle Economy: the Circular Jobs Initiative.

UNEP has been working on circularity for some years from a material flow perspective (urban metabolism) through the Global Initiative for Resource Efficient Cities.[3] A series of pilots with cities[4] around the world from 2017 to 2018 and analysis of 2000 'CE' indicators (Dohmen), however, revealed gaps in measures that link circularity and human well-being. UNEP saw jobs as a key indicator to ascertain the well-being of people locally during a shift from linear to circular. In response to this, UNEP developed an independent methodology that tracked how a CE transition would impact job distribution across cities.

---

3 UNEP was a key driver of the Marrakech Process on sustainable consumption and production which led to One Planet Network and also pushed for SDG 12 on Sustainable Consumption and Production.
4 UNEP pilot cities for the Global Initiative for Resource Efficient Cities included Brussels, Belgium; Cape Town, South Africa; Dongguan, China; Recife, Brazil; and Sorsogon, Philippines.

To unite their shared interests in integrating the employment lens into measuring progress on the adoption of the CE in different localities, Circle Economy and the UNEP Cities team formed a partnership. Their joint work initially focused on integrating and harmonising the different methodologies the organisations had developed for measuring progress and analysing CE and activity contextually. The results of the harmonisation and application of these methodologies are described in the next section: Circular Jobs Framework. UNEP and Circle Economy continue to work closely on monitoring circular jobs across global regions and have now estimated the number of circular jobs in over 200 cities. These results can be viewed on the Circular Jobs Monitor (Circular Jobs Monitor, 2021).

Circular jobs framework

Circular strategies are suited to different economies depending on their composition, natural resources and current dominant industries (Laubinger et al., 2020). For example, automated recycling technologies may be more appropriate in highly industrialised economies and regions, whereas regeneration and preservation strategies may be more pertinent in environments rich in natural resources, in either context offering opportunities to reduce linear risks, generate new revenues and reduce costs (First Circular Economy Action Plan; Ogunmakinde, 2019; Assessment of the Current Status of the Circular Economy for Developing a Roadmap, 2018; Linear Risks, 2018).

In measuring the CE, Circle Economy and UNEP's joint methodology utilises Circle Economy's Key Elements Framework (see Figure 15.1, Table 15.7, Table 15.8), which outlines the full breadth of relevant strategies that give direction to the CE, developed on the basis of a literature review of the various terms and definitions that describe circular activity (The Key Elements of the Circular Economy).

The methodology also adheres to the following principles:

- **Materials**: In principle, the relationship between the economy and materials (virgin, or in components/products) should move towards the decoupling of economic growth from environmental degradation, and concern strategies that slow, narrow, close or regenerate material and energy loops (The Key Elements of the Circular Economy; Bocken, Bocken et al., 2015).
- **Activities**: In principle, activities that indicate progress towards a CE include 'core circular activities:' activities relating to handling materials according to the above principle. For example, repair strategies, upcycling and regenerative water systems all bring about a direct material impact (See Table 15.7). Then, there are 'enabling circular activities' that relate to resolving barriers for core circular actors. For example, rental business models *may* result in increased

Figure 15.1: 'Key Elements' framework.
Source: Circle Economy.
Note: This framework aims to make the CE practical and tangible. It maps CE terms and definitions and distils the common themes into fundamental strategies. The framework outlines core strategies (those that are related directly to material flows, and comparable to the RESOLVE framework) and enabling elements (those that remove obstacles to the implementation of core strategies) (Gower & Schröder, 2016).

usage intensity of products, and more durable products (See Table 15.8). All other activities not classified as core or enabling are referred to as 'indirectly circular activities,' for instance construction or recreational activities.

To prepare and structure the data, classification systems, datasets and metrics are utilised according to these principles. The analysis is organised according to economic sectors as traditionally classified:
- **Material framework**
  - Material data at a sector level is prepared using classifications as found in EORA, EXIObase and EUROSTAT depending on the region of the study (Lenzen et al., 2013; Wood et al., 2015; Eurostat).
- **Activity framework**
  - Sectors are classified as circular or non-circular based on Circle Economy's key elements framework.

The methodology utilises input–output (IO) analysis to derive the circular activities per sector. Once the sectors are classified as core, enabling or indirect, we estimate the proportion of demand for these sectors that is circular. We then use IO analysis to derive the circular proportion of economic output per sector. When conducting analyses on a subnational scale, downscaled tables are used.

The methodology (see Figure 15.2) utilises material import dependencies to proxy the material circularity of the sector, on the premise that if a sector has a low material import dependency then the uptake of 'regional economy' is higher, which is a proxy for circular (low impact, local or recovered) materials being prioritised in the region; or the material demand is just lower and there is no need/demand to import. This forms part of the estimate of circular demand for all sectors that are neither core nor enabling (i.e. indirect sectors).

Integrated Method

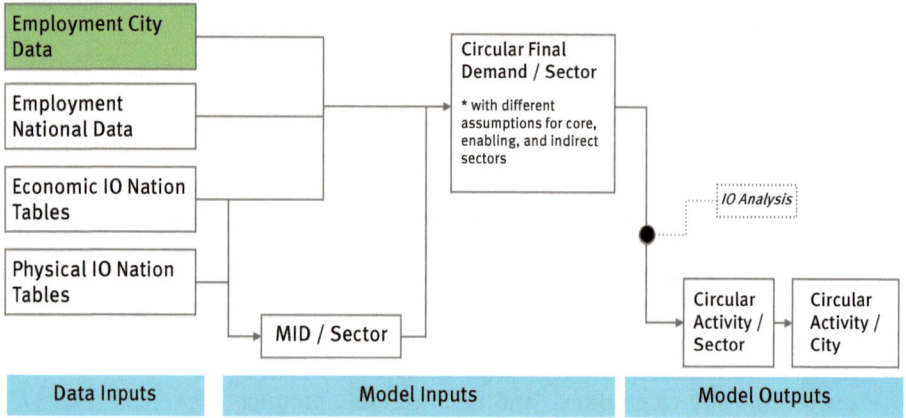

Figure 15.2: Schematic of the methodology.
Source: Circle Economy.

We will illustrate how this is a rich and multidimensional way of interpreting the economy of the region in the following section.

The following territorial snapshots display and interpret the results of this calculation in three different territories: a national study of the Netherlands, a regional study of Cornwall and the Isles of Scilly in the United Kingdom (UK) and an urban study of Antwerp, Belgium.

Results including the material indicator are currently under peer review, and are therefore not yet incorporated into these results.

Territorial snapshots

Cornwall and the Isles of Scilly, UK

Data description: The employment data was sourced from Nomis at the UK Office for National Statistics, is at a five-digit sector code classification and is from 2018. The number of direct jobs in renewable energy was obtained using national estimates from the IRENA database. The IO table was sourced from the UK Office for National Statistics and is also from 2018. The IO table is scaled down to the region using the same employment data.

Table 15.1: Headline indicator.

No. of circular jobs	17,886
No. of total jobs	210,424
Percent of circular jobs	**8.5%**

Table 15.2: Top five sectors – by volume of circular activity.

Sector code	Sector name	Total jobs in sector	Jobs as a % of all jobs in all sectors	Total circular activity in sector	Circular activity as a % of all employment in all sectors
J	Information and communication	3,634	1.73%	2,649	1.26%
M	Professional, scientific and technical activities	9,375	4.46%	1,950	0.93%
I	Accommodation and food service activities	34,545	16.42%	1,900	0.90%
E	Water supply; sewerage; waste management and remediation activities	1,705	0.81%	1,705	0.81%
G	Wholesale and retail trade; repair of motor vehicles and motorcycles	37,886	18.01%	1,667	0.79%

Conclusions on the stage of transition for this territory: Cornwall and the Isles of Scilly's job market is 8.5% circular (see Table 15.1) and has a services and tourism-based economy, and the Circular Jobs Methodology reflects this. The primary circular jobs arise in the enabling sectors of digital services and design, though these are not prominent sectors in the economy more generally.

Accommodation and food service activities employ 16% of the population, but only 5% of those pertain to circular activities. This indicates that there is the potential to increase circular activity in this sector, and create knowledge infrastructure to ensure a higher utilisation of circular inputs to produce the same, or better, economic output more efficiently and sustainably.

There seems to be an active repair industry for motor vehicles and motorcycles, which may call for a wider green transport strategy in the region. The waste management and remediation sectors make up a proportionately small portion of the economy, considering the country's high material and waste footprint. For more sector-specific figures, see Table 15.2.

The Netherlands

Data description: The employment data was sourced from LISA 2014, and the IO table from the same year was sourced from Centraal Bureau voor de Statistiek (Central Bureau of Statistics) (CBS) in the Netherlands.

Table 15.3: Headline indicator.

No. of circular jobs	686,337
No. of total jobs	7,977,635
Percent of circular jobs	**8.6%**

Table 15.4: Top five sectors – by volume of circular activity.

Sector code	Sector name	Total jobs in sector	Jobs as a % of all jobs in all sectors	Total circular jobs in sector	Circular jobs as a % of all jobs in all sectors
J	Information and communication	268,639	3.37%	184,018	2.31%
Q	Human health and social work activities	1,307,831	16.39%	77,162	0.97%
M	Professional, scientific and technical activities	2,444,095	30.64%	51,326	0.64%
R_S	Arts, entertainment and recreation, other services activities	356,683	4.47%	37,095	0.46%
E	Water supply; sewerage; waste management and remediation activities	39,354	0.49%	37,032	0.46%

Conclusions on the stage of transition for this territory: The Netherlands' job market is 8.6% circular (see Table 15.3). The circular activities appear to be oriented around, on the one hand, the high-tech digitisation of waste and green industries, and on the other hand, the development of labour-oriented business models such as architecture and design, repair and community services. The high employment rate in social work activities and professional activities in the national economy indicate the *potential* to enable these sectors to support circular activities in the economy more extensively. The waste management and remediation sectors make up a proportionately small portion of the economy, considering the country's high material and waste footprint. For more sector-specific figures, see Table 15.4.

Antwerp, Belgium

Data description: This 2016 data was sourced from Graydon, and the IO table was sourced from WIOD for the same year. The number of direct jobs in renewable energy was obtained using national estimates from the IRENA database.

Table 15.5: Headline indicator.

No. of circular jobs	14,516
No. of total jobs	226,143
Percent of circular jobs	**6.4%**

Table 15.6: Top five sectors – by volume of circular activity.

Sector code	Sector name	Total jobs in sector	Jobs as a % of all jobs in all sectors	Total circular activity in sector	Circular activity as a % of all employment in all sectors
M	Professional, scientific and technical activities	16,013	7.08%	2,209	0.98%
G	Wholesale and retail trade; repair of motor vehicles and motorcycles	35,121	15.53%	2,078	0.92%
J	Information and Communication	60,545	26.77%	2,021	0.89%
C	Manufacturing	23,609	10.44%	1,488	0.66%
N	Administrative and support service activities	44,364	19.62%	1,464	0.65%

Conclusions on the stage of transition for this territory: Antwerp's job market is 6.4% circular (see Table 15.5). Here, the transition is noticeably different from the others, in that the Waste Management sector does not present itself as a major circular sector. This is because of the low volume of people employed in waste management in Antwerp. Instead, we see that service sectors come up as the most prominent: for instance, architecture and digital services. Wholesale and retail trade come up as second most prominent; however, core circular jobs within this sector (i.e. direct repair) are very low, and rather the circularity arises from the utilisation of other core services and inputs across trade. Manufacturing is prominent for the same reason: because of its interaction with circular sectors. For more sector-specific figures, see Table 15.6.

Dealing with data

The methodology presented above aims to be globally applicable and relatively easy to produce as a baseline analysis in a new context. The principles meet these conditions, although in practice the methodology is applied through frameworks and datasets and these present contextual challenges and differences that need to be resolved each time the methodology is applied.

As with development of any metric, each region follows their own sector classification, and though local sector classifications may be matched to global standards, in doing so there is often data loss as sectors are aggregated or disaggregated accordingly. Global classification consequently will introduce bias in certain regions and this should be considered each time. There is an unavoidable trade-off between global benchmarking and comparison, and local results.

In relation to circularity, current global classification sets may not adequately represent circular strategies, such as urban mining, or innovation within an industry, such as alternative product design. Companies often list multiple sector classification codes, and the listing may not adequately reflect the activities (circular or otherwise) of the company. Alternative datasets and bottom-up data collection can resolve this.

Beyond frameworks, practices in data collection vary. Employment data is available with varying granularity, and sometimes not at all – especially in the realm of the informal economy which may contribute significantly to the CE transition across sectors. Nevertheless, employment data is relatively accessible and understood, both at the national and the company level.

The selection of material import dependency as a metric is partially due to global data availability, as opposed to a metric such as the recovery rate of materials per sector within a region, which is not readily available. Soon it will be necessary to incorporate the product life cycle and origin of materials relative to the context under study.

In the absence of data, alternative approaches to estimating the bottom-up activities were considered, such as data scraping and interpretation of website and GIS data to understand the location in more detail. In scaling down national datasets, even partial knowledge of how economic activities and natural resources are distributed can improve accuracy of scaling down datasets. Data gathered from technology (Internet of Things, mobile data, etc.) will become an important source, as will more elaborate coordination around collection standards and frameworks.

The growth of a dataset for CE modelling allows for the creation of typologies, both at the national and subnational level. Typologies allow regions to create transition paths that are based on preselected or statistical components of similar regions, and distinguish learnings or data from those regions as appropriate.

Applications of the methodology

The Circular Jobs Methodology allows the evaluation of the circular activity in a region, relatively quickly with relatively accessible datasets. The headline indicator can be compared year on year to give an indication of the delta in circular activity. This metric can be compared against other social, environmental or economic metrics to assess the extent and types of impact of circular activities becoming more widespread. The headline indicator would not ever – or need never – be 100%, nor should a region necessarily strive to increase the number for its own sake. It serves as a baseline assessment of current circular activity, and the basis to ask further questions.

Disaggregating the results by sector allows us to identify the pioneering sectors that are engaging with circular activities already, and conversely, allows for the identification of priority sectors that are ripe for redevelopment, as well as interdependencies between the two.

Bottom-up investigation is recommended to deepen and verify the analysis. Regional sector classifications need to be reviewed on a case-by-case basis. Yet, some circular activities happen within companies and are therefore not exposed via the sector classifications; for instance, optimised repair and maintenance schedules within a manufacturing company. Similarly, the material information is typically available only on a national level, alongside national IO tables. Effort must be made to gather regional material flow information and IO tables. Both the materials and the circular sectors must be evaluated in the local context before scenario modelling for policy making can be undertaken.

The Circular Jobs Methodology described here, despite some data gaps, does offer sufficient insight to establish the local link to the CE, by identifying sectors that are valuable to the CE that might not currently be the political priority, which could require skills development packages, have technological inefficiencies or be ripe for new market developments.

Challenges and opportunities: avoiding trade-offs

In this chapter, we have reconnected with early thinking on the subject of the CE. From early conceptualisations of industrial symbiosis and the 'performance economy' (Enberg, 1992; Stahel, 2021) to the Ellen MacArthur Foundation and Circle Economy's framing (Circle Economy, 2021a; Lovins & Braungart, 2014) there has been an urgency to reintroduce people – human capital – back into their rightful place at the heart of the CE. Within this chapter, we have tried to surface both the risks and the opportunities. This includes concerns over how to overcome deeply entrenched inequalities faced by workers on a global scale with the CE.

Our chapter includes data from Europe, where the incidence of labour market inequality is being observed in part through the so-called 'hollowing-out' of the labour market (Kalleberg, 2011). New jobs created on a global scale often continue to be situated in the upper segment of the labour market, requiring higher level skills and offering appropriate remuneration. Whilst this is good news both for national productivity and for those educated to a sufficiently high standard and possessing the relevant skills, there is a downside. Manual work is increasingly subject to automation in the Global North and the rise in productivity is not always accompanied by a rise in worker pay. While the jury is still out on whether we face a post-work future (Rifkin, 1996) where the majority of manual tasks are performed by machines, it is clear that work will retain its central place as not only a means of providing a better life, but also a source of meaning, fulfilment, status and – more broadly – a way of relating to others and participating in society (Coleman, 2017). Many of these issues are becoming increasingly prevalent in the Global South, where the issues of protection and long-term sustainability of livelihoods remain at the forefront. Changing environmental conditions, national environmental policies and corresponding legislation are driving demand for more green and circular jobs across global regions. This includes safeguarding many of the estimated 1.47 billion global jobs that depend on a stable climate (Schaller & Reitler, 2021).

At the local level, municipal governments are concerned with translating national ambitions into outcomes, including support for local businesses, workforce integration and the delivery of labour market policies and social assistance that support this. Conversely, circular strategies for utilising local resources and increasing local labour demand may result in reduced reliance on global supply chains at both the national and subnational scales, which may give way to an adjustment of the job mix. For example, services-based regions could begin to see a rise in primary and secondary activities, and vice versa.

In order for the CE to thrive and meet the needs of everyone, however, we do not simply need more jobs. We need good jobs. Jobs that offer protection from harm as well as opportunities for skills development. A distribution of jobs that enable equal labour market participation for different regions, age groups, genders and minority groups. History has shown us that a handful of well-meaning employers are not enough for this to happen. Indicators and policy drivers for improving working lives across the globe, such as the ILO's Decent Work Agenda, need to be developed and firmly embedded into the fabric of the CE. They should form part of the criteria for recognising and prioritising the circular practices of any organisation or government.

The CE is yet to prove it can overcome long-standing labour market challenges. Yet with better data and metrics, we will be in a better position to navigate these hurdles. In this chapter, we have shared the results of applying the Circular Jobs Methodology to three territories, with a national study of the Netherlands, a regional study of Cornwall, UK and a city study of Antwerp, Belgium. With this, we illustrated the circular activity indicator on three geographical scales in North-West Europe. Whilst the total circular activity across the scales is in the range of 6.4–8.6%, the sectoral breakdowns differ per economy. Prominent sectors are analysed in terms of their contribution to the CE, and therefore expose gaps and opportunities for further circular development. Understanding of the opportunities and risks will be expanded as this analysis is applied further by Circle Economy and UNEP to territories across global regions.

Through the Circular Jobs Metric, we aim to help quantify the circular activity in a given geographic context and industry sector. We believe this is an important start to putting people at the heart of the CE. Yet coordinated work on this mission is only at its beginning. We are committed to developing the Circular Jobs Metric further and bringing it together with supranational agendas to reflect on job quality and access, drawing on existing social indicators. This is not an easy task. The integration and sharing of metrics is important, but requires further research and collaboration to achieve, including to understand which social indicators are relevant for understanding who the CE creates jobs for and where. Thus, our organisations welcome the better integration of CE metrics, such as the Circular Jobs Metric, with other headline and performance indicators, in order to create alignment and action towards tackling the major issues we face, including reducing resource scarcity, protecting livelihoods and tackling climate change. This is possible only through further collaboration between researchers, think tanks, businesses and policy makers. If we wish to make the transition to a CE work for everyone, we must ensure it also works for workers. We feel a sense of urgency and responsibility, for which we make no apology. The time for complacency and implicit assumptions about the impact of changes in our economy on people and our environment is over. It is time to act.

Appendix

Table 15.7: Core elements.

Core Elements	Elements Description	Strategy Group	Strategy Group Description
Prioritise regenerative resources	Ensure renewable, reusable, non-toxic resources are utilised as materials and energy in an efficient way	Regenerative materials	Utilise bio-based, reusable, non-toxic and non-critical materials for products
		Regenerative water	Replace freshwater with rainwater or seawater and enact water efficiency measures
		Regenerative energy	More efficiently using energy that is ideally renewable and electric
Preserve and extend what's already made	While resources are in-use, maintain, repair and upgrade them to maximise their lifetime and give them a second life through take-back strategies when applicable	Maximise lifetime of products in-use	Upgrade, repair, and maintain products while they are still in-use
		Maximise lifetime of products after use	Producttake-backand giving products and parts another life after their end-of-use
		Maximise lifetime of biological products	Ensure that biological products are properly managed and preserved
Use waste as a resource	Utilise waste streams as a source of secondary resources and recover waste for reuse and recycling	Valorise waste streams—closed loop	Reuse, repurpose and recycle waste streams within the same industry
		Valorise waste streams—open loop	Reuse, repurpose and recycle waste streams within other industries
		Energy recovery from waste	Recover waste energy or generate fuels and energy from waste streams

Table 15.8: Enabling elements.

Enabling Elements	Element Description	Strategy Group	Strategy Group Description
Rethink the business model	*Consider opportunities to create greater value and align incentives that build on the interaction between products and services*	**Product business models**	Deliver products to customers through business models that ensure maximum value
		Service business models	Deliver services to consumers through business models that ensure maximum value
Design for the future	*Account for the systems perspective during the design process, to use the right materials, to design for appropriate lifetime and to design for extended future use*	**Design out waste**	Designing products to reduce waste during production and use
		Design for cyclability	Designing products to enable multiple uses and life cycles of a product and its materials
		Design for durability	Designing products that are made to last and to ensure longer use
Collaborate to create joint value	*Work together throughout the supply chain, internally within organisations and with the public sector and communities to increase transparency and create joint value*	**Industry collaboration**	Engage with industry peers to create joint value and identify synergies
		Customer/ consumer collaboration	Engage and guide customers and consumers to ensure circular use of products
		Government collaboration	Engage with the government on circular policies and programmes
		Internal collaboration	Engage internally to guide employees and facilitate greater knowledge sharing between internal divisions
		Community collaboration	Engage with the local community where facilities or offices are located
Incorporate digital technology	*Use digital, online platforms and technologies that provide insights to track and optimise resource use, strengthen connections between supply chain actors and enable the implementation of circular models*	**Data and insights**	Employ technologies to gather and analyse data to provide insights on resource use
		Digital platforms	Employ online platforms to connect and improve information sharing between stakeholders

Table 15.8 (continued)

Enabling Elements	Element Description	Strategy Group	Strategy Group Description
Strengthen and advance knowledge	*Develop research, structure knowledge, encourage innovation networks and disseminate findings with integrity*	**Research and development**	Research and facilitate new technological developments to aid the transition to a CE
		Education and curriculum	Integrate principles of circularity into primary, secondary and tertiary curriculum and conduct workplace trainings
		Knowledge management	Solidify definitions and create frameworks to support the understanding of the CE across contexts and maintain cohesive systems of data sharing, processing and storing
		Awareness and communication	Raise awareness and run information campaigns about strategies for and impacts of the CE in different contexts

References

Assessment of the Current Status of the Circular Economy for Developing a Roadmap. Climate Technology Centre & Network, 2018. Accessed 16 August 2021. https://www.ctc-n.org/technical-assistance/projects/assessment-current-status-circular-economy-developing-roadmap

Circular Jobs Initiative. Circle Economy. Accessed 16 August 2021. https://www.circle-economy.com/circular-jobs-initiative

Circular Jobs Monitor. Circle Economy. Accessed 16 August 2021. https://www.circle-economy.com/circular-jobs-initiative/monitor

Cities. United Nations Environment Programme. Accessed 16 August 2021. https://www.unep.org/explore-topics/resource-efficiency/what-we-do/cities

First Circular Economy Action Plan. European Commission. Accessed 16 August 2021. https://ec.europa.eu/environment/topics/circular-economy/first-circular-economy-action-plan_en

Linear Risks – how business as usual is a threat to businesses and investors. Circle Economy, 2018. Accessed 16 August 2021. https://www.circle-economy.com/news/linear-risks-how-business-as-usual-is-a-threat-to-companies-and-investors

The Key Elements of the Circular Economy. Circle Economy. Accessed 16 August 2021. https://www.circle-economy.com/circular-economy/key-elements

Bell, D., & Blanchflower, D. (2013). Underemployment in the UK revisited. *National Institute Economic Review, 224*(1), F8–F24. 10.1177/002795011322400110

Bocken, N., de Pauw, I., Bakker, C., & van der Grinten, B. (2015). Product design and business model strategies for a circular economy. *Journal Of Industrial And Production Engineering, 33*(5), 308–320. 10.1080/21681015.2016.1172124

Brein, H. G. (2020). How is a circular economy different from a linear economy? Accessed
16 August 2021. https://kenniskaarten.hetgroenebrein.nl/en/knowledge-map-circular-
economy/how-is-a-circular-economy-different-from-a-linear-economy/
Brynjolfsson, E. (2011). *Race against the machine: how the digital revolution is accelerating
innovation, driving productivity, and irreversibly transforming employment and the economy.*
Reprint, Digital Frontier Press.
Cambridge Econometrics, Trinomics and ICF. (2018). Impacts of circular economy policies on the
labour market. Reprint, Brussels: European Commission. Accessed 16 August 2021.
http://trinomics.eu/wp-content/uploads/2018/07/Impacts-of-circular-economy-on-policies-
on-the-labour-market.pdf
Circle Economy and New York Circular City Initiative. (2021). *Jobs and opportunities for New York
City in the circular economy.* Reprint, New York City: Freshfields. Accessed 16 August 2021.
https://www.circle-economy.com/resources/jobs-and-opportunities-for-new-york-city-in-the-
circular-economy.
Circle Economy and Zero Waste Scotland. (2020). *The future of work: Baseline employment analysis
and skills pathways for the circular economy In Scotland.* Reprint, Amsterdam: Circle Economy.
Accessed 16 August 2021. https://www.circle-economy.com/resources/the-future-of-work-
baseline-employment-analysis-and-skills-pathways-for-the-circular-economy-in-scotland
Circle Economy. (2020). *Baseline analysis of circular jobs in cornwall & the isles of scilly.* Reprint,
Amsterdam: Circle Economy. Accessed 16 August 2021. https://www.circle-economy.com/re
sources/baseline-analysis-of-circular-jobs-in-cornwall-the-isles-of-scilly
Circle Economy, European Environmental Bureau and Fair Trade Advocacy Office. (2020). *Avoiding
blindspots: Promoting circular and fair business models.* Reprint, Amsterdam: Circle Economy.
Accessed 16 August 2021. https://assets.website-files.com/5d26d80e8836af2d12ed1269/
5fc4d55d42ec471380dfb8e6_20201130%20-%20EEB%20-%20report%20web-%
20297x210mm.pdf
Circle Economy. (2018). *Circular jobs & skills in the amsterdam metropolitan area.* Reprint,
Amsterdam: Circle Economy. Accessed 16 August 2021. https://www.circle-economy.com/re
sources/circular-jobs-skills-in-the-amsterdam-metropolitan-area
Circle Economy. (2019). *Circular jobs in Belgium: A baseline analysis of employment in the circular
economy In Belgium.* Reprint, Amsterdam: Circle Economy. Accessed 16 August 2021.
https://www.circle-economy.com/resources/circular-jobs-in-belgium-a-baseline-analysis-of-
employment-in-the-circular-economy-in-belgium
Circle Economy. (2020). *Circular jobs: Understanding employment in the circular economy in the
Netherlands.* Reprint, Amsterdam: Circle Economy. Accessed 16 August 2021. https://www.cir
cle-economy.com/resources/circular-jobs-understanding-employment-in-the-circular-
economy-in-the-netherlands
Circle Economy. (2021a). *Jobs & skills in the circular economy – state of play and future pathways.*
Reprint, Amsterdam: Circle Economy. Accessed 16 August 2021. https://www.circle-economy.
com/resources/jobs-skills-in-the-circular-economy-state-of-play-and-future-pathways
Circle Economy. (2021b). *Key elements of the circular economy.* Reprint, Amsterdam: Circle
Economy. Accessed 16 August 2021. https://www.circle-economy.com/resources/the-key-
elements-of-the-circular-economy-framework
Circle Economy. (2021c). *The circularity gap report 2021.* Reprint, Amsterdam: Circle Economy.
https://www.circularity-gap.world/2021
Coleman, J. (2017). To find meaning in your work, change how you think about it. *Harvard Business
Review.* https://hbr.org/2017/12/to-find-meaning-in-your-work-change-how-you-think-about-it

Dohmen, G. Circular economy indicators: What do they measure? *Blog. GI-REC: Operationalizing Urban Metabolism At The City Level.* Accessed 16 August 2021. https://resourceefficientcities.org/2018/11/circular-economy-indicators-what-do-they-measure/

Egdell, V., & McQuaid, R. (2014). Supporting disadvantaged young people into work: insights from the capability approach. *Social Policy & Administration, 50*(1), 1–18. 10.1111/spol.12108

Enberg, H. (1992). *Industrial symbiosis in denmark.* Reprint. New York City: New York University, Leonard N. Stern School of Business.

Euro Cities. (2020). *Circular economy action plan: Speeding up the green transition of the EU's economy.* Accessed 16 August 2021. Accessed 16 August 2021. https://eurocities.eu/wp-content/uploads/2020/08/EUROCITIES_policy_statement_CEAP_FINAL.pdf

European Commission. (2020). A new circular economy action plan for a cleaner and more competitive Europe. Communication From The Commission To The European Parliament, The Council, The European Economic And Social Committee And The Committee Of The Regions. Reprint, Brussels: European Commission. Accessed 16 August 2021. https://eur-lex.europa.eu/resource.html?uri=cellar:9903b325-6388-11ea-b735-01aa75ed71a1.0017.02/DOC_1&format=PDF

Eurostat. *National accounts database.* Accessed 16 August 2021. http://ec.europa.eu/eurostat/web/national-accounts/data/

Ford, M. (2015). *Rise of the robots: Technology and the threat of mass unemployment.* Reprint, Basic Books.

Richard, R., & Schröder, P. (2016). Virtuous Circle: How the circular economy can create jobs and save lives in low and middle-income countries.

Green Alliance (2021). *Levelling up through circular economy jobs.* Reprint London: Author. Accessed 16 August 2021 https://green-alliance.org.uk/resources/Levelling_up_through_circular_economy_jobs.pdf

Gregson, N., Crang, M., Botticello, J., Calestani, M., & Krzywoszynska, A. (2021). Doing the 'dirty work' of the green economy: Resource recovery and migrant labour in the EU. *European Urban And Regional Studies, 23,* 541–555. 10.1177/0969776414554489

ILOStat. (2019). *The working poor or how a job is no guarantee of decent living conditions.* Spotlight On Work Statistics No. 6. Reprint, International Labour Organization. Accessed 16 August 2021. https://ilo.org/wcmsp5/groups/public/---dgreports/---stat/documents/publication/wcms_696387.pdf

International Labour Organization. (2021). *Decent work and the 2030 agenda for sustainable development.* International Labour Organization. Accessed 16 August 2021. https://www.ilo.org/global/topics/sdg-2030/lang---en/index.htm

International Labour Organization. (2021). *Skills for a greener future.* Reprint, Geneva: International Labour Organization. Accessed 16 August 2021. https://www.ilo.org/wcmsp5/groups/public/---ed_emp/---ifp_skills/documents/publication/wcms_709121.pdf

International Labour Organization. (2011). *Skills for employment policy brief: Greening the global economy – the skills challenge.* Reprint, Geneva: International Labour Organization. Accessed 16 August 2021. https://www.ilo.org/wcmsp5/groups/public/---ed_emp/---ifp_skills/documents/publication/wcms_164630.pdf

Jacobs, K. (2015). Job quality in an hourglass labour market. *HR Magazine.* Accessed 16 August 2021. https://www.hrmagazine.co.uk/content/features/job-quality-in-an-hourglass-labour-market

Kalleberg, A. (2011). *Good jobs, bad jobs: the rise of polarized and precarious employment systems in the United States, 1970S–2000S.* Reprint, Russell Sage Foundation.

Karabarbounis, L., & Neiman, B. (2014). The global decline of the labor share. *The Quarterly Journal Of Economics, 129*(1), 61–103. 10.1093/qje/qjt032

Kessler, S. (2018). *Gigged: The end of the job and the future of work*. Reprint, St. Martin's Press.

Laubinger, F., Lanzi, E., & Chateau, J. (2020). *Labour market consequences of a transition to a circular economy: A review paper*. Reprint, Paris: Labour market consequences of a transition to a CE: A review paper. Accessed 16 August 2021. https://www.oecd-ilibrary.org/environ ment/labour-market-consequences-of-a-transition-to-a-circular-economy_e57a300a-en;jsessionid=3Waeg_h64XOtC_tSFHD3m8Qd.ip-10-240-5-153

Lenzen, M., Moran, D., Kanemoto, K., & Geschke, A. (2013). Building eora: A global multi-regional input-output database at high country and sector resolution. *Economic Systems Research*, *25*(1), 20–49. 10.1080/09535314.2013.769938

Levy, F., & Murnane, R. (2004). *The new division of labor: How computers are creating the next job market*. Reprint, Princeton University Press.

Lovins, A., & Braungart, M. (2014). *A new dynamic – effective business in a circular economy*. Reprint, Ellen MacArthur Foundation Publishing.

Manolchev, C., Saundry, R., & Lewis, D. (2018). Breaking up the 'precariat': Personalisation, differentiation and deindividuation in precarious work groups. *Economic And Industrial Democracy*. 10.1177/0143831X18814625

Manolchev, C., Saundry, R., & Lewis, D. (2018). Breaking up the 'precariat': Personalisation, differentiation and deindividuation in precarious work groups. *Economic And Industrial Democracy*. 10.1177/0143831X18814625

Merli, R., Preziosi, M., & Acampora, A. (2018). How do scholars approach the circular economy? A systematic literature review. *Journal Of Cleaner Production*, *178*, 703–722. 10.1016/j. jclepro.2017.12.112

Ogunmakinde, O. (2019). A review of circular economy development models in China, Germany and Japan. *Recycling*, *4*(3), 27. 10.3390/recycling4030027

Rifkin, J. (1996). *The end of work: the decline of the global labor force and the dawn of the post-market era* (Reprint). Tarcher.

Rreuse. (2021). *Briefing on job creation potential in the re-use sector*. Reprint, Brussels: Rreuse. Accessed 16 August 2021. http://www.rreuse.org/wp-content/uploads/Final-briefing-on-reuse-jobs-website-2.pdf

Schaller, B., & Reitler, K. (2021). Skills for the green economy: Why investing in people is key. *The Adecco Group*. https://www.adeccogroup.com/future-of-work/latest-research/skills-for-the-green-economy/

Schröder, P., Lemille, A., & Desmond, P. (2020). Making the CE work for human development. *Resources, Conservation And Recycling*, *156*, 104686. 10.1016/j.resconrec.2020.104686

Schröder, P., Anggraeni, K., & Weber, U. (2019). The relevance of CE practices to the sustainable development goals. *Journal Of Industrial Ecology*, *23*(9), 77–95. 10.1111/jiec.12732

Schröder, P. (2020). *Promoting a just transition to an inclusive circular economy*. Reprint, London: Chatham House. Accessed 16 August 2021. https://www.chathamhouse.org/2020/04/promot ing-just-transition-inclusive-circular-economy

Somavia, J. (1999). Report of the director-general: decent work. In *International Labour Conference*. Reprint, International Labour Organization. Accessed 16 August 2021. https://www.ilo.org/public/english/standards/relm/ilc/ilc87/rep-i.htm

Sopjani, L., Arekrans, J., Laurenti, R., & Ritzén, S. (2020). Unlocking the linear lock-in: mapping research on barriers to transition. *Sustainability*, *12*(3), 1034. 10.3390/su12031034

Stahel, W. (2013). Policy for material efficiency – sustainable taxation as a departure from the throwaway society. *Philosophical Transactions Of The Royal Society A Mathematical Physical And Engineering Sciences*, *371*(1986), 20110567. doi:10.1098/rsta.2011.0567

Stahel, W. (2021). *The circular economy – a user's guide*. Reprint, Routledge.

Standing, G. (2011). *The precariat: The new dangerous class*. Reprint, Blomsbury.

United Nations Development Programme and United Nations Environment Programme. (2020). *A 1.5° world requires a circular and low carbon economy*. Reprint, New York City: United Nations Development Programme. Accessed 16 August 2021. https://www.oneplanetnetwork.org/sites/default/files/enhancing_ndcs_thru_circ_economy.pdf

United Nations Environment Programme. (2020). *Emissions gap report 2020*. Reprint, Nairobi: United Nations Environment Programme. Accessed 16 August 2021. https://www.unep.org/emissions-gap-report-2020

Webster, K. (2021). *The circular economy: A wealth of flows*. Reprint, Ellen MacArthur Foundation Publishing.

WIEGO. (2019). *Counting the world's informal workers: A global snapshot*. Reprint, Manchester: WIEGO Limited. Accessed 16 August 2021. https://www.wiego.org/sites/default/files/resources/files/WIEGO-Global-Statistics-Snapshot-Pamphlet-English-2019.pdf

Willeghems, G., & Bachus, K. (2018). *Employment impact of the transition to a circular economy: Literature study*. Reprint, CE Center: Steunpunt Circulaire Economie. Accessed 16 August 2021. https://ce-center.vlaanderen-circulair.be/nl/publicaties/publicatie-2/3-employment-impact-of-the-transition-to-a-circular-economy-literature-study

Wood, R., Stadler, K., Bulavskaya, T., Lutter, S., Gilijum, S., de Koning, A., . . . Tukker, A. (2015). Global sustainability accounting-developing EXIOBASE for multi-regional footprint analysis. *Sustainability, 7*(1), 138–163. 10.3390/su7010138

World Economic Forum. (2021). *Upskilling for shared prosperity*. Reprint, Cologny: World Economic Forum, Accessed 16 August 2021. http://www3.weforum.org/docs/WEF_Upskilling_for_Shared_Prosperity_2021.pdf

Zschirnt, E. (2016). Ethnic discrimination in hiring decisions: A meta-analysis of correspondence tests 1990–2015. *Journal Of Ethnic And Migration Studies*, 1–19. 10.1080/1369183X.2015.1133279

Fenna Blomsma, Geraldine Brennan

16 Resources, waste and a systemic approach to Circular Economy

Abstract: How can the benefits of a Circular Economy be captured? Assessing the potential for impact and uptake of circular practices is challenging without considering *why* certain problems occur, and *how* potential interventions will affect other parts of the industrial system. Therefore, we propose a novel method to circular oriented innovation that facilitates understanding the *problem space* – the current situation and the problem(s) that need to be addressed, as well as the *solution space* – how circular strategies can best be leveraged to address them. This method, titled Circularity Thinking, combines a systemic outlook with visual tools that can be used as a common conceptual 'red thread' to enable dialogues between different disciplines and diverse perspectives within organisations. Moreover, it can serve as the 'glue' between a range of tools and approaches that focus on specific elements of circular innovation such as product design, business modelling and impact assessment.

Keywords: Circularity Thinking, systems thinking, circular oriented innovation, circular configurations, method

Introduction

The core of the Circular Economy (CE) concept is to move away from linear 'take-make-use-dispose' practices and to replace these with cycling and cascading to enable the increased conservation, efficiency and productivity of resources. With the rise of the CE concept, a legitimate space for discussion is created where the appropriate and effective use of a range of waste and resource management practices can be examined. 'Circular strategies' that operationalise this concept include, but are not limited to: reducing, reusing, upgrading, remanufacturing, recycling, composting, cascading and industrial symbiosis. Although each of these individual strategies has existed previously, their grouping under the CE umbrella brings attention in a new way to their common capacity to improve resource conservation, efficiency and productivity (Blomsma & Brennan, 2017).

There seems to be broad agreement that operating in a more circular manner can generate value for businesses, reduce harm to the climate and the environment, and that it can even serve as a means to address social issues such as lack of employment and low investment in human capital (Material Economics, n.d.; Stahel, 2019) – provided it is implemented appropriately. However, a great many ideas exist about

https://doi.org/10.1515/9783110723373-019

what, exactly, operating a CE entails and how these benefits can be achieved. This is illustrated by the many circular frameworks that put forward different ideas with regard to what problems should be addressed, the mechanisms that solve these problems, what circular strategies are appropriate to use and which actors are best positioned to take action (Blomsma, 2018). See Figure 16.1 for an illustrative comparison of three well-known circular frameworks: (1) Waste Hierarchy, (2) Cradle-to-Cradle, (3) Performance Economy – but note that there are many more such unique perspectives. Figure 16.1 shows how even the same circular strategy – recycling – can take on very different roles and is envisioned to involve different processes as well as different outcomes, which complicates constructive dialogue on CE.

The existence of so many frameworks is little wonder; they are part of an ongoing societal discourse with regard to how to best manage waste and resources. After all, societies have not always expected the same from waste and resource management. When first applied on an industrial scale, during the second half of the eighteenth century, circular strategies primarily focused on keeping cities clean and increasing production through the use of by-products. Only gradually was this extended to include the entirety of the life cycle as it relates to both consumption and production and as a key lever for sustainable development (Blomsma & Brennan, 2017). In response, the type and role of different circular strategies changed and evolved over time, culminating in the current CE discourse.

Seen in the light of this historic narrative, no single author or organisation can be attributed with being the originator of the CE concept. Instead, the emergence of CE should be regarded as the outcome of a collective process to make sense of circular strategies and how they relate to each other. And as opposed to holding 'the' answer, the contributions of different circular frameworks are more helpfully regarded as visions that aid in imagining a different way of operating consumption and production systems.

Viewing CE in this way means that it is necessary to admit that we are still learning what it means to operate in a circular manner – and that we need to allow for discovery, emergence and experimentation. Through targeted learning, those practices that truly deliver on the CE promise can be identified and amplified – and those that do not dampened down. This is also in line with the acknowledgement that umbrella concepts, such as the CE, go through a phase of refinement and deepening on their way to becoming established (Hirsch & Levin, 1999). Two key implications of this for CE are (1) that the starting point of any circular oriented innovation (COI) effort (Brown et al., 2019) should be the systemic problems that it aims to address, and (2) that the outcomes of any proposed (sets of) circular strategies should be understood from the perspective of systemic impact.

All too often, however, both scholarly and practitioner contributions to CE tend to include neither. For example: many COI approaches focus on circular strategies largely separate from the problems they aim to address. This leads to 'circularity for circularity's sake:' to be seen engaging with circular strategies, but without regard for

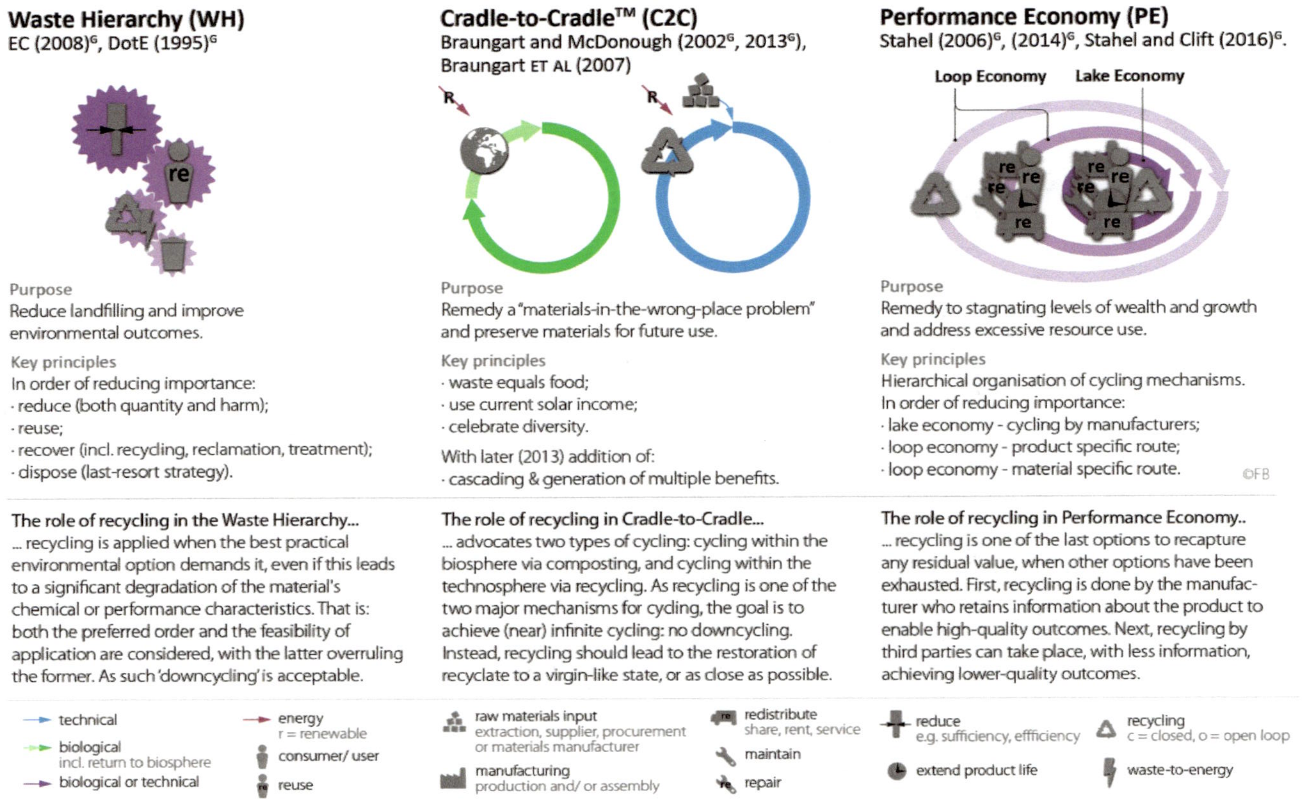

Waste Hierarchy (WH)
EC (2008)[G], DotE (1995)[G]

Purpose
Reduce landfilling and improve environmental outcomes.

Key principles
In order of reducing importance:
- reduce (both quantity and harm);
- reuse;
- recover (incl. recycling, reclamation, treatment);
- dispose (last-resort strategy).

The role of recycling in the Waste Hierarchy...
... recycling is applied when the best practical environmental option demands it, even if this leads to a significant degradation of the material's chemical or performance characteristics. That is: both the preferred order and the feasibility of application are considered, with the latter overruling the former. As such 'downcycling' is acceptable.

Cradle-to-Cradle™ (C2C)
Braungart and McDonough (2002[G], 2013[G]), Braungart ET AL (2007)

Purpose
Remedy a "materials-in-the-wrong-place problem" and preserve materials for future use.

Key principles
- waste equals food;
- use current solar income;
- celebrate diversity.

With later (2013) addition of:
- cascading & generation of multiple benefits.

The role of recycling in Cradle-to-Cradle...
... advocates two types of cycling: cycling within the biosphere via composting, and cycling within the technosphere via recycling. As recycling is one of the two major mechanisms for cycling, the goal is to achieve (near) infinite cycling: no downcycling. Instead, recycling should lead to the restoration of recytale to a virgin-like state, or as close as possible.

Performance Economy (PE)
Stahel (2006)[G], (2014)[G], Stahel and Clift (2016)[G].

Loop Economy   Lake Economy

Purpose
Remedy to stagnating levels of wealth and growth and address excessive resource use.

Key principles
Hierarchical organisation of cycling mechanisms. In order of reducing importance:
- lake economy - cycling by manufacturers;
- loop economy - product specific route;
- loop economy - material specific route.

The role of recycling in Performance Economy...
... recycling is one of the last options to recapture any residual value, when other options have been exhausted. First, recycling is done by the manufacturer who retains information about the product to enable high-quality outcomes. Next, recycling by third parties can take place, with less information, achieving lower-quality outcomes.

Legend:
- technical
- biological incl. return to biosphere
- biological or technical
- energy / r = renewable
- consumer/ user
- reuse
- raw materials input / extraction, supplier, procurement or materials manufacturer
- manufacturing / production and/ or assembly
- redistribute / share, rent, service
- maintain
- repair
- reduce / e.g. sufficiency, efficiency
- extend product life
- recycling / c = closed, o = open loop
- waste-to-energy

Figure 16.1: Comparison of three circular frameworks, focusing on the role and proposed outcomes of recycling within the frameworks.
Source: Authors.

systemic impact. Other COI approaches engage superficially with circular strategies – often driven by certain assumptions or preconceived ideas as to what 'should' cycle and 'how' to do this. But the potential for impact and uptake of circular practices in the real world can be easily misunderstood without considering *why* certain problems occur, *how* interventions impact other parts of the industrial system, and without taking into account path dependency and feasibility, and evolving possibilities.

To overcome these shortcomings, a systemic outlook on CE and COI is necessary. This means to acknowledge that both the problems that circular strategies address *and* the solutions they represent should be understood from the perspective of the entire life cycle – for multiple cycles. Therefore, we develop a COI approach that differs from those that came before, one that facilitates understanding the current situation and the problem(s) that need to be addressed, as well as how circular strategies can best be leveraged to address them. Our approach therefore allows for critically, reflexively and comprehensively examining both the problem *and* the solution space(s) associated with CE. With this, we contribute to a *mindset shift* that is grounded in systems thinking and we enable the implementation of CE to be truly transformative.

We start by outlining why a systems perspective is crucial to circularity, then we identify and bring together key concepts for understanding the physical reality of resources and waste. Next, we unpack how the roles and responsibilities of actors can be understood in enabling COI in circular configurations and demonstrate the insights that can be gained from taking a Circularity Thinking approach through its application to the case of Interface.

Why a systems perspective is crucial to circularity

Approaches for identifying different problems along production and consumption systems (e.g. Life Cycle Analysis(LCA)) and for mapping the journey of resources through the economy (e.g. Material Flow Analysis (MFA), and Sankey diagrams) are well-known. A range of COI tools covering product design, business modelling and impact assessment also exist. However, these and other COI approaches tend to not make an (explicit) connection between problem spaces and solution spaces (e.g. circular strategies). Even within well-known circular frameworks (see Figure 16.1), the different perspectives on what problems should be addressed and how this can best be done are implicit and only become apparent when deconstructed and compared with one another (Blomsma, 2018). Without a clear view on what's in focus, it is impossible to ask: "how can this whole system be optimised?"

Omitting such a life cycle–centred systems perspective from CE frameworks is problematic for a number of reasons. First, different types of waste occur at different life cycle stages – and they can be addressed in different ways. It is why so many different circular strategies fit under the CE umbrella and why each type of waste has a range of circular strategies associated with it. Whilst each circular strategy may be worthwhile, an exclusive focus on any individual one or even on any individual waste type is effectively still applying a linear mindset. Waste, after all, does not exist in isolation. Rather: **'waste begets waste.'**

To see this clearly, consider that a linear economy focuses on product throughput, which results in artificially cutting short a product's life before the end of its technical life through designed or perceived obsolescence. Components are not designed to last, to be upgraded, repaired or remanufactured, as there is no benefit to be gained from this under these circumstances – precluding any subsequent use. As a result, materials overflow landfills at the end-of-life and more waste is created during manufacturing – as increased throughput implies more co- and by-products. Put differently: *waste creates other waste*. To get to the **root cause** of why a specific waste is created, an understanding is needed of what is driving its creation – and the reason may lie in a different location in the system than where a problem is revealing itself. Only a systemic outlook can uncover this.

Moreover, different types of waste can have a different **impact type** and a different **impact magnitude** in relation to each other. Which problem is more important to address? For example, is it more important to prevent the loss of minute but crucial scarce rare earth metals used within an electronic device, or to ensure that the much larger volumes of plastics used in its casing return to a 'virgin like' state? And, is it more important to address the under-used capacity of cars (parked 93–97% of the time) or to prolong the use of their components? In some contexts, these answers may be evident; in others, it will only become clear after careful analysis. This is not to say that any of these problems can be left unsolved, but an understanding of the relative importance of issues may steer towards more salient impact. Also, interventions can be designed that allow for the addition of new circular strategies at a later time, thus addressing additional wastes – as opposed to creating system designs that preclude or complicate ('lock-out') the incorporation of these additional circular strategies. Rather, COI should enable continuous improvement. In this, it's not the purpose to do away with the entirety of the linear economy, but to examine how these processes can be adapted, redirected or repurposed.

Related to this is the need to consider the fact that circular strategies **interact** with each other. That is, **synergies, trade-offs** and **competition** exist between them. Synergies exist when circular strategies work together. For example, when renewables are used as substitutes for fossil-fuel based materials (prevention) that are also biocompatible and compostable at their end-of-life (closing loops). Trade-offs, on the other hand, are situations that involve optimising one circular strategy at the expense of another. For example, choosing durable and lightweight materials (prevention) such as composites can preclude recycling at the end-of-life. Moreover, competition between circular strategies can also exist – situations where circular strategies can be applied but with varying impacts, thus creating tension. For example, when recycled materials are not matched to appropriate applications. This occurs when using recycled materials for one application where the material is downcycled, alongside use of the same material for another application where it is truly 're'-cycled. Both are circular applications to a degree, but have very different impacts. Therefore, when faced with a choice between two competing circular strategies, the impact has to be reviewed from a system's perspective.

Other aspects of circular systems can also be better understood when taking a systemic outlook. For one, the **cost** of different strategies and different implementation scenarios of the proposed solutions will need to be weighed, including energy expenditure, labour investment, time, resilience and independence, exposure to risks, financial costs, etc. This extends to the 'total cost of ownership' (TCO) of resources – the total sum of all costs associated with the use of resources – which can only be fully understood when adopting a systems lens as these costs occur at different places in the system.

Importantly, the potential for the **creation of new wastes** and **circular rebound effects** (Zink & Geyer, 2017) needs to be addressed if circular solutions are to have an overall positive impact. That is: product-life extension is not always good, and likewise recycling shouldn't always be considered last; this will depend on the context and system boundary.

Moreover, a systemic outlook enables the exploration of the potential for change (e.g. which circular scenarios are more feasible than others). This involves interrogating what current capabilities can be leveraged, where barriers exist, and what enablers can be put in place. That is, when system maps are drawn the different levels of **agency** that exist for different actors or organisations can be better assessed, by examining their sphere of influence. Some actors may have a large influence over certain parts of the system; others less so. Naturally, large actors are more likely to wield a more powerful influence; but smaller actors may hold relational capital, too, based on a unique capability or contribution they are able to make. Moreover, **path dependency** exists. That is, the transition from one system to another will depend on the possibilities that exist for this in the current system.

Last, when practicing COI, it's vital to remember that CE is still an **evolving concept**. New applications of older ideas, such as product/service systems, can be seen, but also new ideas such as component/service systems and material/service systems have emerged (Aurisicchio et al., 2019; Blomsma et al., 2021). New technologies, in the form of new materials, new processes and new ways of sharing and accessing information (digitalisation, IOT, AI, etc.), combined with the establishment of new infrastructures, create new possibilities that are currently difficult to imagine. Furthermore, **industrial systems are complex adaptive systems** that cannot be steered by imposing a blueprint or a grand design given that there are so many different interacting components. It is impossible to understand all the intricate connections that exist between them and how they are continuously responding, adapting and changing.

Instead, **experimentation** will be a key part of exploring new possibilities and building momentum (Weissbrod & Bocken, 2017).

These considerations highlight that it is unwise to start COI with preconceived or *a priori* ideas about what 'should' be done without first developing a deeper understanding of the system – and they remind us why it is a necessity to place systems thinking at the heart of CE innovation and implementation. In fact, experience has shown that 'jumping to ideas' can be 'dangerous' and can lead to missed opportunities (Khoo, 2019). To understand how a systemic approach to circularity can be gained and how existing COI approaches can be linked to it, we first clarify and bring together key concepts for understanding the physical reality of waste and resources. Thereafter, we unpack how the roles and responsibilities of actors in enabling COI can be understood through the application of Circularity Thinking to the case of Interface.

'Resources,' or resource states

The starting point for our systemic approach to CE is the concept of *resources*: the physical 'stuff' considered worth preserving. But what, exactly, *is* a resource? This is not a straightforward question to answer since resources 'flow' from one place to the next in economies, taking on different forms along the way. Luckily, we can describe this journey in more detail using material entropy. *Material entropy* describes the degree to which materials are *ordered* – think of this as a scale where on the one end resources are distributed and diffuse, and on the other end they are concentrated and organised (Boulding, 1966). It is the act of organising that brings matter from a lesser to a more ordered state using technological and logistical means. Order is decreased through wear and tear, contamination and biological and chemical deconstruction.

Using the concept of material entropy, we can say that resources have different degrees of order relative to one another. Specifically, we can distinguish between three *resource states* or three different levels of order for the resources that are generally important in economies (Blomsma & Tennant, 2020). The first resource state is that of *particles,* where we speak of elements, substances, molecules or materials. This is where (bulk) materials are created through mining or harvesting raw materials followed by further treatment to purify, concentrate and improve their properties so that they can be used in industrial processes. Think of extracting oil to produce pellets for the injection moulding of plastic. Or the mining of ore to produce metals that are turned into rolls of sheet metal. Or the harvesting of cotton to produce bales of fabric.

Next, from these materials *parts* are created; this is where materials are given an intermediate level of organisation in the form of components, modules and (sub) assemblies. We say 'intermediate' because parts are more organised than materials, but they are usually not yet sufficiently organised to be useful on their own. That is, through a range of manufacturing processes such as extrusion, cutting, welding, moulding and gluing things like casings, gears and electrical motors, but also zippers and cut-to-size pieces of cloth – all the different elements that a product is made of – are created.

For end-users and customers to derive value from components, they have to be combined or assembled into *products* – the finished goods that are uniquely shaped to help people solve particular problems and accomplish certain tasks, delivered to the right location. Finally, this is where we get bikes, cars, kettles, clothes, shoes and the fast-moving consumer goods we all use on a daily basis.

So far, we have described the forward supply chain of 'take-make-use' processes. In a linear economy, these processes are followed by disposal in landfill, and in a recycling economy by incineration and downcycling. But with the help of material entropy, we can now also see how circular flows can increase the conservation, efficiency and productivity of resources. Through a resource states lens, we can understand the journey of resources in the economy in terms of supply chains working to first increase order in the particle state, then in the part state and finally in the product state. The aim for a CE, then, is to preserve this order in each of these levels of organisation as long as possible, feasible and desirable. Circular flows can take place in each of the three resource states, allowing for the (partial) reversal of order followed by its re-establishment. That means that we can distinguish between four cycling levels as outlined in the Circularity Compass, which uses the Resource State framework as its main organising principle, in Figure 16.2.

First, within the **product state**, opportunities can be sought to extend the useful life of products through optimising *performance cycles* in the use phase. A good understanding of your user is key to start with but also think of software updates, guidance on use of the product or services to optimise product settings. Useful life can also be extended through redistribution and as-is reuse that facilitate reuse by other users in *product cycles*. Crucially, products undergo minimal or only superficial changes when cycling in the product state (inspection, testing, cleaning) and no disassembly is involved.

Moreover, cycling can take place with the aim to preserve the **part state**. This involves (partial) disassembly into components and is implemented through repair, maintenance, refurbishment, remanufacturing, upgrading, component cascades, part cannibalisation, etc. Depending on the product design and contextual factors such as the difficulty of the intervention and the need for specialist tools or knowledge, these actions can be performed at the location of use, sometimes even by users themselves, or by trained technicians in dedicated facilities – involving different technical and logistical processes.

Last, but certainly not least, the concentration of particles can be preserved through cycling in the **particle state**. This entails the cycling of materials such as

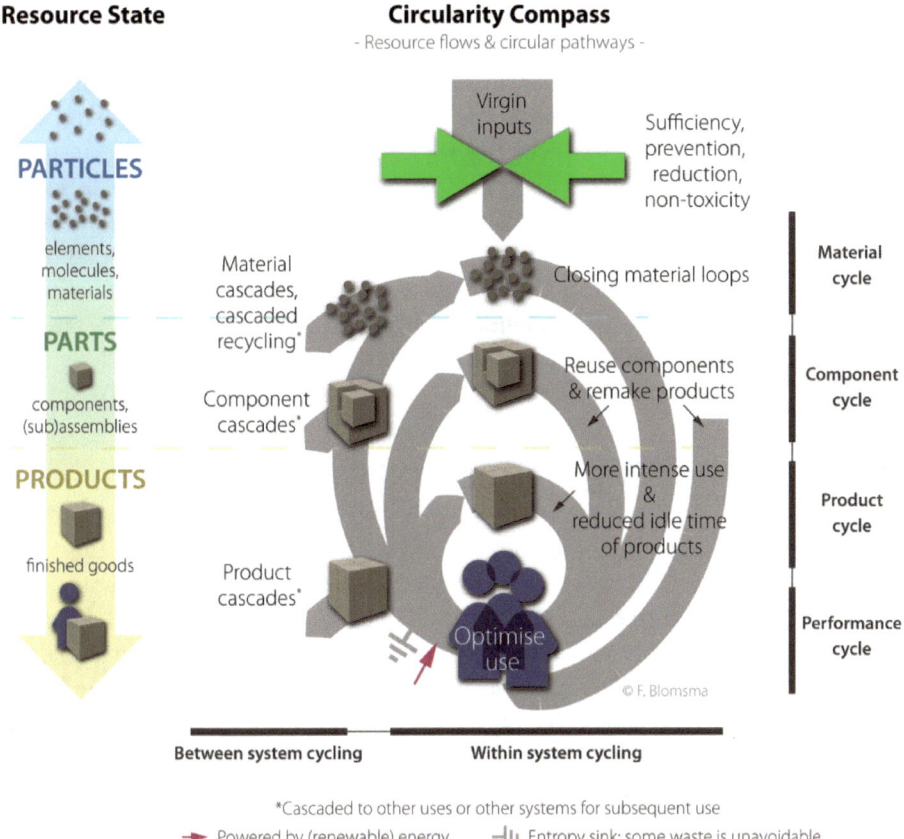

Figure 16.2: Depiction of levels of cycling in accordance with the three resource states. The resource states are shown on the left, adding the life cycle perspective creates the Circularity Compass.
Source: Authors.

through composting, recycling, downcycling and cascaded recycling/ substance cascades.[1] Here, parts and products lose their form or shape entirely, to allow for the recreation of new forms and shapes with cycled materials.

The resource states framework does not imply a hierarchy in the manner that the Waste Hierarchy does – prioritising the three Rs of 'Reduce, Reuse, Recycle' in descending order. The resource states approach does not ask "how best to preserve materials?" or even "how best to preserve products?" in the manner that frameworks such as Cradle-to-Cradle and Performance Economy do, respectively (see Figure 16.1).

---

1 We distinguish between recycling and downcycling or cascaded recycling to indicate that materials that are the outputs of these latter processes do not necessarily return to a 'virgin-like' state.

After all, in some contexts, material cycling is undesirable, and in others, product cycling does not make sense – making it risky to put in place heuristics and other rules of thumb that obscure the displacement of waste and hide new potential problems.

Instead, the question we put at the heart of circular economic thinking is to ask, for a specific context, "how can this whole system be optimised?" The resource states framework of Figure 16.2 offers one part of navigating this question by giving insight into 'what' it is that cycles within a system. Instead of prescribing 'how' to cycle, we contend that priorities for circular strategies can best be set depending on the waste that is present in a particular system. For this, it is necessary to understand what it means to 'waste' something so we can clearly see why specific circular strategies are applied and others are not.

'Waste,' or the Big 5 structural wastes

In the same way that it is not straightforward to define resources, it is also not so simple to define *waste*. How do we know when a resource is 'wasted'? Generally, it is agreed that waste is the loss and destruction of value – but this can take many forms. Some of these forms are clearly visible and identifiable, such as materials in a bin or a product with a clearly visible breakage. In contrast, other forms are inconspicuous, invisible and more difficult to point to. Think of products that are unused for a significant part of their life or products that are designed to fail after a certain amount of uses without outward signs or a clearly visible reason.[2] Such situations lead to more resources being needed and a higher level of material throughput than would strictly be necessary to fulfil human needs – and they can therefore also be seen as 'wasteful.'

So what do we mean by the loss and destruction of value and how this can be avoided? Preventing waste from being created usually means to 'preserve' or 'continue' something. There are three principal ways in which this can be achieved: (1) **addressing premature end-of-life through re-establishing performance**, (2) **addressing premature end-of-use through optimising functional life**, and (3) **addressing excess or harm through prevention** (Blomsma, 2018; Blomsma & Tennant, 2020), see Figure 16.3.

---

2 A practice also known as 'designed obsolescence.'

Addressing premature end-of-life through re-establishing performance

To **re-establish performance** is to 'renew' a resource. This describes the restoration or rejuvenation of a resource in a manner that signals the return to a pre-defined or evolving quality or functionality level. The occurrence of a limiting state that would signal the **end-of-life** (e.g. functional, stylistic or other constraint) is delayed, undone or reversed, and the original performance level is re-established or improved upon.

Re-establishing performance can be achieved through different means. Particles or materials can be renewed by recycling and, when allowing for nature to do part of the work, composting. Crucially, such strategies should keep materials in cycles of (near) equal quality. In the area of parts or components, strategies such as durability, repair, maintenance, refurbishment, remanufacturing and upgrading can be applied. These strategies may also be part of negating a limiting state for products. However, these operations usually involve (partial) disassembly, and therefore cannot be executed without involvement of the part state. However, sometimes it's not possible to re-establish the performance of a product, but some components may still be salvaged using these strategies. Therefore, while similar circular strategies can be used to address waste within both the part and product states, they are deployed to different ends.

Circular strategies that re-establish performance have in common that they prevent value from being lost unnecessarily as a result of neglecting the renewal of the resource. To be precise: **life cycles are regarded as 'wasted' when they are open-ended** – they simply 'stop.' The circular strategies that are used to address this type of waste turn would-be **end-of-life** into new or extended life that is usually not (significantly) different from the previous life.

Addressing premature end-of-use through optimising functional life

Second, to 'preserve' or 'continue' something can also be achieved through another pathway: **optimising the functional life** of a resource to maximally exhaust its value creation potential. Here, a resource can be fully 'consumed' or 'used up' as opposed to being **'wasted' by being only partially put to use** and thus reaching a **premature end-of-use**. This can be addressed by one or more transformations that utilise the fact that the end-of-use of a resource in one context does not have to be the end of value extraction from that resource altogether. After all, there may be other uses or contexts where the resource can continue to deliver value.

Circular strategies that leverage such transformations are, again, different for the three resource states. Particles or materials can be transformed by, for example,

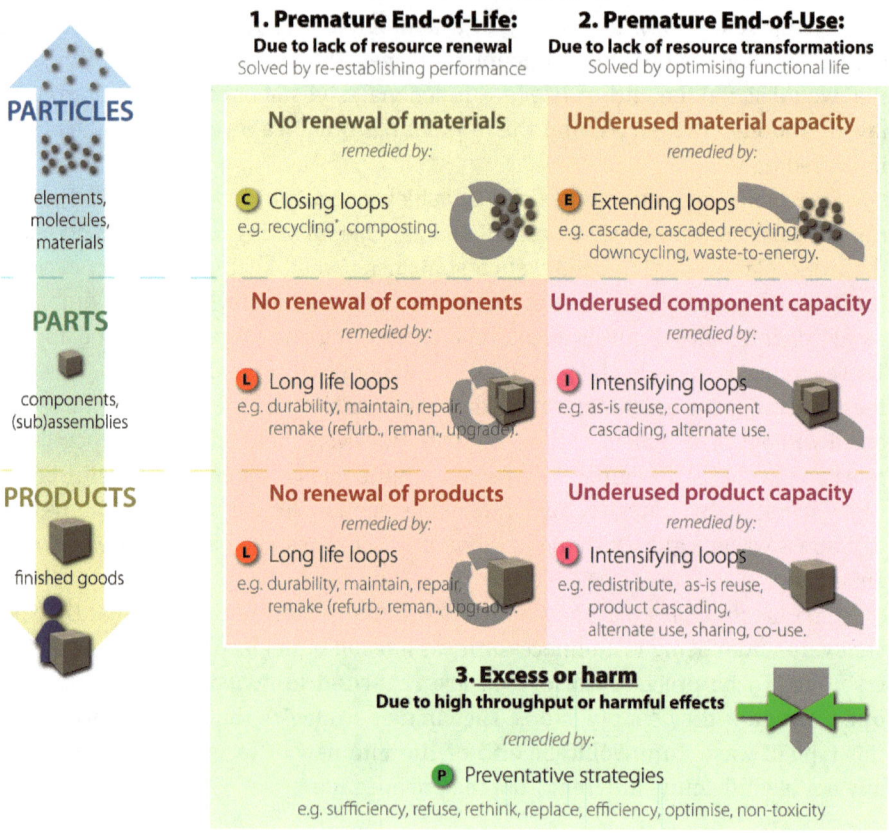

Figure 16.3: Overview of the five types of structural waste.
Source: Authors.

substance cascades (processing of materials into one or more substances with different properties), cascaded recycling (recycling into a lower grade material, e.g. 'downcycling') or waste-to-energy schemes (incineration with recapture of embedded energy). The value generation capacity of parts or components can be captured more fully by strategies such as as-is reuse, component cascading and alternate use. Products, similarly, can be transformed by redistribution to another use context, as-is reuse by a new user, product cascading, alternate use and sharing and co-use.

The common underpinning of circular strategies that optimise functional life share is to prevent value from being lost unnecessarily as a result of underused resource capacity. Crucially, circular strategies that prevent such premature **end-of-use** do not involve renewal. For this reason, they may involve changing or allowing for an evolution in the definition of 'quality' or 'utility' – what it means for a resource to deliver value or provide utility is redefined from one context to the next. It is through acknowledging and acting according to this changing definition that prevents a resource from being 'wasted.'

Addressing excess or harm through prevention

Lastly, the third category of strategies that prevent the creation of waste are those that forestall resource use, such as doing without ('sufficiency'/ 'doing without'), efficiency measures ('reduce'), functional replacement ('rethink'), or measures that prevent harmful effects, such as the choice for benign or non-toxic materials or processes. Collectively, we refer to strategies with the ability to prevent excess or harm as *prevention*. Here, again, we have to consider that preventative strategies can play a range of roles within the three resource states. For example, 'lightweighting' may offer a way to use less materials; parts or components could be designed to be energy efficient; and products could be constructed in such a way that they encourage sufficiency. Preventative strategies should not be seen as separate from circular strategies since they interact with each other. For example, while lightweighting beer bottles reduces the average refill rates due to increased fragility, the cumulative effect for resource efficiency will need to be assessed.

End-of-life, end-of-use and preventative strategies are not mutually exclusive; they may be combined. We expand on this later in this chapter when we introduce circular configurations.

Structural wastes and resource states

For reasons of simplicity and because parts and products are qualitatively different from particles, we can distinguish between five types of structural waste or the 'Big 5' Structural Wastes (Blomsma, 2018; Blomsma & Tennant, 2020). This term is adopted to make explicit that our interpretation of 'waste' goes beyond the colloquial meaning of the word – usually limited to something thrown out or discarded. Figure 16.3 depicts these 'Big 5' organised according to the resource states. The following five wastes can be identified, each linked to a set of circular strategies that can be used to address them:
- No renewal of materials – remedied through 'closing loop' strategies;
- No renewal of components/products – remedied through 'long life loop' strategies;

– Underused material capacity – remedied through 'extending loop' strategies;
– Underused component/product capacity – remedied through 'intensifying' strategies;
– Excess or harmful use of resources – remedied through 'prevention' strategies.

The 'Big 5' provides us with five problem spaces (areas in systems to investigate) and ask: "are these wastes present here?" Each part of the Circularity Compass can now be examined for the presence of these wastes (see Figure 16.4 for an illustration). Such a picture of the system can be used to optimise the whole as structural waste can be identified that occurs **'in between'** – in between production units, in between business units, in between companies and in between life cycle stages. Without this ability, it's easy to overlook structural waste, and that makes it such a big problem. Again, it's why a systemic perspective is so important to unlocking opportunities; waste has to be spotted before solutions can be developed that address it.

Therefore, to 'see' structural waste, we proceed with bringing together the Circularity Compass and the Big 5 to allow for a systems perspective of both the problems that are addressed (the problem space) and the solutions that are put in place (the solution space). This can be used as a starting point of COI by using it to map the current system state and to explore the different possibilities for new system designs, and the possibilities that exist to bring it about. For this, we next turn to *circular configurations* and explore both the flows of resources and the role of actors in 'making flows flow' (Baumann, 2012).

'Circular configurations': mapping the 'why,' 'what' and 'who' of CE

As outlined in the introduction to this chapter, there are two key steps to optimising whole systems: (1) understanding the problem space (the 'why'), and (2) developing an appropriate solution space (the 'what' and the 'who').

'Why': understanding the problem space

A system scan or system mapping is needed to be able to understand root causes of waste as well as different types and magnitudes of impact. The Circularity Compass and the Big 5 can be used to map the current or 'business-as-usual' situation for products, companies, sectors etc. That is, the Circularity Compass can be used as a template to identify the current flows. The Big 5 can then be used to interrogate this picture and ask "what wastes are present where in this system?" Usually, many different structural wastes can be identified. Moreover, the user is then in a position to assess the relative size and importance of these wastes: are they all equally important or are there particular structural wastes that deserve priority treatment?

The next step is to critically interrogate the mapping to uncover what drivers are behind a particular waste. That is, a waste may surface at one place in the system, but what is contributing to it being generated there? Does it stem from decisions or actions undertaken earlier or later in the system? And, will intervening at a particular place negate or reduce waste elsewhere? Through this reflective interrogation, the **'in between'** is made visible. When the system is sufficiently understood, the 'why' of COI can be clearly articulated – providing the anchor(s) for your circular approach, whether grounded in problems to be addressed, innovation opportunities currently uncaptured or a combination of both. The key or priority circular strategy(s) linked to this become(s) what we refer to as **anchor strategy(s)** from which to build a circular configuration – see Box 16.1.

Take for example the carpet tile manufacturer Interface – a company that has long since served as an archetypal example for both sustainable business as well as circular operation. Figure 16.4 (left-hand side) uses the approach described here to depict some of the key wastes that this company has addressed (e.g. Khoo, 2019, 2020; Interface, 2021). For example, similar to other manufacturing companies, the company's footprint is for a large part determined by the extraction and processing of raw materials, pointing to the need for reduction, alternative sources and replacement. Also, in the past, limited or no renewal of the material could take place, due to the choice of materials. In addition to this, installation as well as repair and maintenance resulted in waste (excess) and unattractive flooring due to mismatches in colour and patterning (no (high-quality) renewal of product). Finally, the absence of second-hand channels is worth highlighting, as it indicates the presence of underused product capacity. These latter issues further drive up material use.

If needed or wanted, greater detail and quantified impact assessment can further supplement such a mapping to create a more precise picture. That said, a first (indicative) picture can often be formed by collecting diverse perspectives *(from across an organisation or supply chain)* to create insights into different parts of the system. Other relevant supply chain issues such as conflict materials, unfair or unsafe labour practices, and low resilience to disruptions may also be included in such a mapping.

Exploring solution space(s)

Using the same frameworks, we can also identify effective solutions. Here, circular configurations play a major role. We speak of a *circular configuration* when two or more circular strategies are present and where interactions – synergies, trade-offs or competition – occur between them. Such configurations can be examined from the perspective of a) resource flows (the 'what') and b) actor relationships (the 'who').

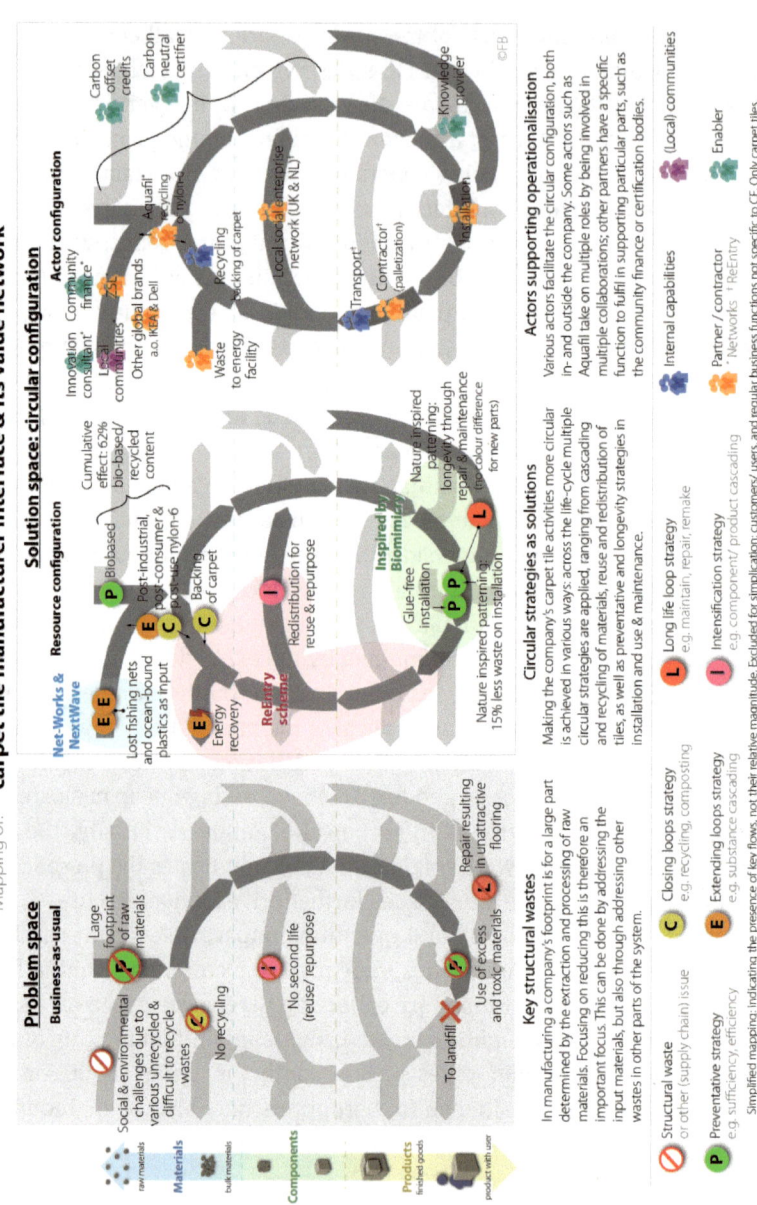

Figure 16.4: Depiction of the case of Interface, mapped using the proposed approach grounded in understanding the problem, before moving onto solutions.
Source: Authors.

'What': resource flows or 'resource configuration'

Once structural wastes have been identified, it's important to recognise that these are linked to a range of circular strategies each with a variety of implementation options. To create clarity, different scenarios can be explored that combine these different options in different ways, starting with the anchor strategies previously identified (see also Box 16.1).

Such scenarios should explore how much scope there is to preserve or continue using a resource. That is, does it make sense to (further) prolong resource use? If it's not obvious and if you notice a reliance on assumptions ('but this is how it's always been done'), try to get inspiration from elsewhere. For example, look for *best practice* in other (related) sectors, but also allow for exposure to *emergent practice* and new thinking to 'unfreeze' the current perceptions with regard to what can and cannot be done (Cummings et al., 2016). Also, is it **'within system' cycling** you are interested in (e.g. retaining resources within a single system), or should **'between system' cycling** also be a part of the solution (e.g. linking systems through the exchange of resources)?

Reflect on whether intervening at one particular place in the system may displace waste to elsewhere in the system, or create entirely new wastes or problems. For example, is your 'zero waste' initiative really zero waste, or does it merely mean that a supplier or customer is left with the waste instead? COI practitioners need to become attuned to how an intervention affects the overall production and consumption system. For example, reflect on what activities could become sources of circular rebound; does your product/service system contribute to the reduction of resource use, or does it increase it by drawing users away from more sustainable solutions, in the same manner that car-sharing services can compete with other transport options such as cycling (Bocken et al., 2020)?

Box 16.1: Prioritising in the face of complexity: anchor and supporting strategies

Anchor strategies are those circular strategies that are linked to the most salient issues or structural wastes identified in a system. They are the starting point and form the 'main dish' and focal point in COI. However, after identifying the major problem spaces and opportunities and clarifying anchor strategies the task may not yet be done; other structural wastes may still remain, new wastes may be created or the circular solution could be enhanced further through what we refer to as **supporting strategies**. For example, in a situation where under-used product capacity was addressed by implementing co-use and sharing models, further enhancement may be possible through long life strategies such as robust product design, the availability of high-quality spare parts, and predictive maintenance – facilitated by digital technologies and IoT networks. Such supporting strategies can be added later, or be an integral part for the solution to truly qualify as 'circular.' Distinguishing between anchor and supporting strategies allows for designing circular configurations in a way that a start can be made, whilst acknowledging and 'designing in' space for future continuous development.

Our case company Interface illustrates how circular configurations are becoming the norm rather than the exception. Figure 16.4 (middle) depicts the key circular resource flows this company has developed over the years to address the structural wastes they identified as key in their context. This figure shows several circular strategies,[3] among which are: closing loops (recycling) enabled by material choice and innovations in the production process; a preventative and long-life loop strategy, both enabled by nature-inspired patterning (through applying biomimicry); and a reuse scheme titled ReEntry that enables an intensification strategy through reuse and repurposing (alternate use outside of the office environment), as well as recycling for those products that can no longer be reused and – as a last resort – waste-to-energy. What's more, the company decided to address wider societal and environmental problems such as those associated with discarded fishing nets in coastal communities. Through the Net-Works initiative, Interface buys these discarded materials (waste as an input) and they are recycled into new yarn for carpet tiles by Aquafil. This moreover integrates socially restorative practices into their procurement activities. It's worth highlighting that not all flows are changed in the solution space; many forward processes remain similar to Interface's current business practices.

'Who': actor relationships or 'actor configuration'

Understanding how resources should flow to increase resource conservation, efficiency and productivity is but one part of the puzzle of circular configurations. Due to the specialised nature of many value chain actors, no single company can become circular all by itself (Kraaijenhagen et al., 2016). Therefore, we also need to look at 'who' or which actors are required to successfully implement a circular solution. Every resource configuration needs to be supported by an appropriate *actor configuration* – the sets of actors and their manner of interaction within a value chain that together orchestrate the circular resource flow(s). Of all the possible ways of organising value chains or networks, some will allow for more circularity, others less. And some configurations will allow for creating and capturing more or less value both overall (in total) and by different actors.

Some actor configurations supporting CE will represent only slight deviations from current linear operating procedures and existing supply-chain relationships – for example, the sales of renewable resources or secondary materials. Others may choose to develop or necessitate strategic partnerships, such as in our case company, Interface. For example, a close collaboration between Aquafil and Interface

---

3 Please note that this image is conceptual; it is not intended to depict the order of magnitude of the flows such as in a Sankey or MFA diagram.

was needed to develop a material and a manufacturing process using materials that could be recycled – larger suppliers were not willing to explore COI (Khoo, 2020). Similarly, the ReEntry scheme relies on a network of local partnerships to facilitate reuse. The Net-Works initiative, likewise, involves partnerships with local communities and the Zoological Society London to develop safe and effective recovery processes. A range of enabling actors can also be seen – organisations that do not directly participate in the orchestration of resources, but are nevertheless crucial in supporting the circular solutions. See Figure 16.4 (right-hand side) for an overview of the actor configuration for Interface; this mapping depicts which actors are involved in which circular strategies and it clarifies their role in enabling the overall resource configuration. In this, new relationships with existing partners are created, new partners are sought, and even new entities can be created to fill capacity gaps.

Yet other actor configurations may entail the development of ongoing or continuous relationships throughout the lifetime of the resources, such as in the case of car-as-a-service company Riversimple that has the ambition to upstream the service mindset into their supply chain through sourcing components-as-a-service and materials-as-a-service – necessitating the close exchange of information to allow for continuous improvement as well as profit sharing of key actors involved so that innovation processes across actors are aligned with the same goals (Blomsma et al., 2021).

Designing and implementing effective solution spaces

Designing and implementing circular configurations demands exploration of both the flows of resources and the role of actors in 'making flows flow,' which in turn requires a willingness to experiment and iterate based on learnings and insights gained over time.

Enabling the implementation of a truly transformative CE demands going beyond superficial engagement – and to acknowledge the complexity associated with it. CE offers significant opportunities, but we'd be remiss not to pay attention and take care of potential blindspots and unintended consequences. These range from **trade-offs** and **competition** between circular strategies to **circular rebound**, from **cycling risk** (e.g. harmful contaminants in recovered materials) to **circular washing**, and from local impacts to the role of the solution in creating healthy CE markets.

Conclusion: circularity thinking as a starting point for COI

We set out to contribute to a *mindset shift* that grounds CE in systems thinking. We have achieved this aim by introducing the Resource States framework, elaborated on this to create the Circularity Compass and introduced the 'Big 5' Structural

Wastes. Through the application of these tools to the case of Interface, we have also shown how a systems-based approach can provide insight into the problem space (the 'why' that drives the engagement with COI), and the solution space (the 'what' of resource configurations and the 'who' of actor configurations).

This approach, while grounded in research, is intentionally designed to be used and applied and be accessible to practitioners and industry. One of the key benefits of our approach is that it grounds the starting point of COI in *addressing structural waste* rather than in *implementing circular strategies*. This enables users to set priorities and identify possibilities for further development based on their respective contexts. Instead of offering a set of heuristics that is grounded in either the preservation of materials or products, we offer users a way of thinking and seeing in a systemic way, in order to gain agency and the ability to act systemically. We believe that this approach creates the conditions for discovery, emergence and learning, creating room for truly new systems and practices by mitigating the overwhelm associated with the 'tyranny of the possible' (Duncombe, 2007).

Our systems-based approach has four key implications for the CE discourse. First, we contend that – as long as it isn't co-opted to become an extension of the linear or recycling economy – the definition of the CE concept matters little; its definition needs to be sufficiently open and broad to encompass a wide range of practices and to allow for different actors to connect to it in their own way. However, it is on the next level down, on the level of circular strategies, that more precision and definitional clarity is important to support operationalisation. Precision at this level is a necessity to ensure that best and emergent practices can provide insight into new ways of doing things, that impacts can be fairly assessed and ultimately to push the boundaries of what is possible.

Second, our approach moves away from viewing circular strategies as a strict hierarchy and instead introduces a relational view that explicitly demands the reflexive consideration of synergies, trade-offs and competition between strategies. Circular strategies do not exist in isolation – in understanding CE and COI, it is important to acknowledge the complex relationships that exist between circular strategies that can be seen in reality.

Our third point relates to how circular business models (CBMs) are defined. In following the logic underpinning our approach, CBMs are interpreted to be applying circular configurations that rely on *multiple* circular strategies. It follows, therefore, that CBMs are unlikely to be defined in terms of a single circular strategy nor represent 'pure' types. Rather, CBM 'types' will have to acknowledge both anchor and supporting strategies.

Fourth and lastly, the preceding insights imply that it is not the single dyad of actor relationships that counts when it comes to implementing circularity, but the totality of the actor configuration. Therefore, it is not the business models of individual firms that are at the centre of COI, but the composition of business models of various actors and the interfaces that exist between them. The degree to which these are (mis)aligned with each other determines the capacity for circularity. In this sense, the field of CBMs and value chains/value networks can no longer be meaningfully separated.

These observations point to the necessity to interrogate resource and actor configurations simultaneously. Our approach, a systems-thinking approach for CE – hence titled *Circularity Thinking* – combines a systemic outlook with visual tools that can be used as a conceptual 'red thread' to enable dialogues between different disciplines and diverse perspectives within organisations. For example, design science can explore how product design can support circular configurations; sustainability science can develop impact assessment methods to link different circular configurations with sustainability impact; and management science can interrogate circular configurations from the perspective of shared value creation, business models and value chain management. Moreover, Circularity Thinking can serve as the 'glue' between a range of practical tools and approaches that focus on specific elements of circular innovation that exist in these areas.

In closing, COI is not about 'doing away' with the entirety of the linear economy. Instead, it's about repurposing those parts that have delivered great prosperity and progress and developing and further maturing circular processes so that they become as smart, effective and efficient as the forward part of the linear economy. While actors have the *freedom* to decide what CE means in their context, to ensure CE delivers on its transformational ambition, actors also have the *responsibility* to make sure that it addresses real problems and to not create new ones.

References

Aurisicchio, M., Zeeuw van der Laan, A., & Tennant, M. (2019). Material-service systems for sustainable resource management. *Proceedings of the Eco Design International Symposium*.

Baumann, H. (2012). *Using the life cycle approach for structuring organizational studies of product chains*.

Blomsma, F. (2018). Collective 'action recipes' in a circular economy – on waste and resource management frameworks and their role in collective change. *Journal of Cleaner Production*, *199*, 969–982. https://doi.org/10.1016/j.jclepro.2018.07.145

Blomsma, F., & Brennan, G. (2017). The emergence of circular economy: A new framing around prolonging resource productivity: The emergence of circular economy. *Journal of Industrial Ecology*, *21*(3), 603–614. https://doi.org/10.1111/jiec.12603

Blomsma, F., & Tennant, M. (2020). Circular economy: Preserving materials or products? introducing the resource states framework. *Resources, Conservation and Recycling, 156*, 104698. https://doi.org/10.1016/j.resconrec.2020.104698

Blomsma, F., Tennant, M., & Brennan, G. (2021). Exploring resource/service-systems – beyond product/service systems and towards configurations of circular strategies, business models, and actors. In: *Circular economy and sustainability* (Vol. 1). Elsevier.

Bocken, N., Jonca, A., Södergren, K., & Palm, J. (2020). Emergence of carsharing business models and sustainability impacts in swedish cities. *Sustainability*, *12*(4), 1594. https://doi.org/10.3390/su12041594

Boulding, K. (1966). *The economics of the coming spaceship earth.*

Braungart, M., & McDonough, W. (2002). *Cradle to cradle: remaking the way we make things* (1st ed). New York: North Point Press.

McDonough, W., & Braungart, M. (2013). *The upcycle: Beyond sustainability – designing for abundance.* New York, NY: Charles Melcher.

Braungart, M., McDonough, W., & Bollinger, A. (2007). Cradle-to-cradle design: Creating healthy emissions – A strategy for eco-effective product and system design. *Journal of Cleaner Production*, *15*, 1337–1348. https://doi.org/10.1016/j.jclepro.2006.08.003

Brown, P., Bocken, N., & Balkenende, R. (2019). Why do companies pursue collaborative circular oriented innovation? *Sustainability*, *11*(3), 635. https://doi.org/10.3390/su11030635

Cummings, S., Bridgman, T., & Brown, K. G. (2016). Unfreezing change as three steps: Rethinking Kurt Lewin's legacy for change management. *Human Relations*, *69*(1), 33–60. https://doi.org/10.1177/0018726715577707

DotE (Department of the Environment). (1995). *Making waste work: a strategy for sustainable waste management in England and Wales.*

Duncombe, S. (2007). *Dream: Re-imagining progressive politics in an age of fantasy.* New Press, Distributed by W.W. Norton. No further information available.

EC (European Council). (2008). *Directive 2008/98/EC of the European Parliament and of the Council of 19 November 2008 on waste and repealing certain directives (Waste framework).*

Hirsch, P. M., & Levin, D. Z. (1999). Umbrella advocates versus validity police: A life-cycle model. *Organization Science*, *10*(2), 199–212. https://doi.org/10.1287/orsc.10.2.199 Interface (2021). Corporate website. Accessed October '21.

Khoo, J. (2019). Interface: Net-Works—Lessons learnt turning nets into carpet. In M. Charter (Ed.), Designing for the Circular Economy. Routledge.

Khoo, J. (2020). *Interface's approach to circular economy.* ABIS Knowledge Into Action Forum.

Kraaijenhagen, C., Oppen, C. V., & Bocken, N. (2016). *Circular business: Collaborate and circulate.* Circular Collaboration.

Material Economics. (n.d.). *The circular economy – a powerful force for climate mitigation.* https://materialeconomics.com/material-economics-the-circular-economy.pdf?cms_fileid=340952bea9e68d9013461c92fbc23cae

Stahel, W. (2006). *The performance economy* (2nd ed.). Palgrave MacMillan.

Stahel, W. R. (2019). *The circular economy: A user's guide.* Taylor & Francis: Routledge.

Stahel, W. R., & Clift, R. (2016). Stocks and flows in the performance economy. In R. Clift & A. Druckman (Eds.), *Taking stock of industrial ecology.* Switzerland: Springer International Publishing. Pages 153–174.

Weissbrod, I., & Bocken, N. M. P. (2017). Developing sustainable business experimentation capability – A case study. *Journal of Cleaner Production*, *142*, 2663–2676. https://doi.org/10.1016/j.jclepro.2016.11.009

Zink, T., & Geyer, R. (2017). Circular economy rebound: Circular economy rebound. *Journal of Industrial Ecology*, *21*(3), 593–602. https://doi.org/10.1111/jiec.12545

Marta Ferri, Alison Stowell, Gail Whiteman

17 Plastic futures: mobilising circular economy contexts to address the plastic crisis

Abstract: The plastic crisis has demonstrated how single-use plastic waste is growing significantly and has a negative impact on natural ecosystems and human activities (UNEP 2018). Drawing upon Douglas' (1966, pp. 34–35) idea of pollution, single-use plastic waste is seen as a "matter out of place." In an attempt to put these materials back "in place," organisations invoke circular economy ideas. We use the International Alliance for Sustainable Business (IASB) case study to show how a business-driven global alliance attempts to organise circular solutions to help members face the plastic crisis. By paying attention to the process of 'contexting' (Asdal & Moser 2012), we argue that considering the interrelations between organisations' agendas (social values) and plastics material composition (material values) helps to understand how circular solutions get organised and what notions of responsibility these solutions enact – considering how and who invokes various contexts illuminates the processes that make certain solutions to the plastic crisis prevalent. Thus, the contexting becomes a political activity because the actors' interests and agenda inform its enactment.

Keywords: plastics, circular economy, circular discourses, contexting, waste studies

Introduction

In 1997, scientific researcher and sea captain Charles Moore encountered a massive 'plastic island' in the middle of the Pacific Ocean. He came back a few years later to get samples and demonstrated that the 'island' was primarily made of plastic debris floating in the ocean. Captain Moore and his crew named it the 'Great Pacific Garbage Patch' (Moore, 2011); this was one of the first examples of the widespread phenomenon called the 'plastic crisis.'

The plastic crisis has demonstrated how single-use plastic waste is growing significantly; the United Nations Environment Programme (UNEP) estimated that about 300 million tonnes of plastic waste were produced in 2015 (UNEP, 2018, p. 5), approximately 60 times more than in 1950. The significant increase of plastic waste carries numerous negative impacts, especially when leaking into the natural environment. Consequences include the disruption of natural ecosystems and human activities (e.g. tourism and fishery).

https://doi.org/10.1515/9783110723373-020

The plastic crisis showed how plastic waste that has been 'forgotten' (Hird, 2015), – i.e. escaped waste management networks – can leak, accumulate into the natural environment and become pollution. Drawing upon Douglas' (1966, pp. 34–35) idea of pollution, single-use plastic waste could be seen as materials 'out of place' when found in the natural environment. In the attempt to put these materials back 'in place,' some organisations have invoked ideas of the circular economy (Ellen MacArthur Foundation, 2013, 2015; European Commission, 2015a; Lacy, Long & Spindler, 2020). Circular solutions tend to focus on eliminating waste, maintaining materials' value and developing *sustainable* business models, i.e. being able to operate within 'planetary and social boundaries' (Schröder, Bengtsson, Cohen & Dewick, 2019; Whiteman, Walker & Perego, 2013).

There are several practitioner publications (e.g. Ellen MacArthur Foundation, 2013, 2015; European Commission, 2020) and academic studies regarding the circular economy and critiques of the uses of this concept (e.g. Blomsma & Brennan 2017; Calisto Friant, Vermeulen and Salomone, 2020; Corvellec, Bohm, Stowell & Valenzuela, 2020; Kirchherr, Reike & Hekkert, 2017). However, key gaps remain, such as understanding how circular solutions get organised by the actors, materials, objects and issues involved, how they interrelate and the possible impact on organisational practices (Corvellec, Stowell & Johansson, 2021).

Within this chapter, we address these gaps and aim to describe the attempt of a business-driven global alliance, here anonymised as the International Alliance for Sustainable Business (IASB), to organise circular solutions to help their members who are facing challenges related to the plastic crisis. In doing so, we look at the processes through which organisations' views on single-use plastics and ideas of the circular economy get mobilised. By paying attention to the relationships between organisations' interests (social values) and plastics material composition (material values), we aim to explain how certain circular economy initiatives become more prevalent rather than others.

To understand how organisations and materials interrelate, we draw upon Asdal and Moser's (2012) concept of 'contexting.' From this perspective, contexts are not 'fixed,' nor 'given,' but produced by the relationships between certain actors, materials, objects and issues. Because the context is an ongoing process, these authors suggest the term *contexting* encapsulates the dynamicity of these interrelations. We argue that the contexting activities are political, as organisations invoke specific circular economy contexts to support their agendas. In this regard, by promoting certain circular solutions, organisations enact particular notions of responsibility that help to justify their actions to tackle the plastic crisis.

By adopting a material semiotics perspective (Asdal & Moser, 2012; Callon, 1986; Latour, 1988, 1991), our findings contribute to the debates in the Waste Studies literature (Douglas, 1966; O'Brien 2008; Thompson, 1979) by recognising the material agency of materials and bringing greater attention to complex waste materials (Gille, 2010; Liboiron, 2016). Our contribution to these studies is to unpack the role of the material agency of single-use plastics within organising circular economy solutions to tackle the plastic crisis. Drawing upon the lessons learnt from the IASB case, our contribution is twofold. Theoretically, it is important to consider material agency within the process of organising, as specific materials may impact organisations' operations (e.g. plastics material composition could either support or disrupt recycling practices by being either recyclable or not). Practically, it is relevant to consider organisations' interests and agendas in order to understand the circular economy contexts that they invoke. In this sense, it will be possible to see if circular solutions are leading towards reproducing existing practices and how a business-driven alliance can organise their members towards the promotion of innovations in this area.

This chapter is structured as follows. First, we summarise and describe the relevant literature from Waste Theory, Science and Technology Studies and the Circular Economy to illustrate, respectively, the terms 'pollution, and 'contexting,' and different circular economy frameworks. Second, we move forward to the IASB case study, describing the Alliance's circular economy context and the contexting activities through the Sustainable Organisations Forum vignette. In the Forum, IASB interacts with other organisations, which invoke various circular contexts and views on single-use plastics. Third, we discuss our findings and theoretical and practical contributions that stress the relevance of considering material values alongside organisational values. Drawing upon Douglas' (1966) ideas of 'out of place,' we understand social values as organisations' agendas and interests, according to which other organisations and materials are judged to be 'out of place' or 'in place.'

Single-use plastics and the circular economy

Given that the focus of our chapter is to examine IASB's attempt to organise circular solutions, outlining the scholarly debates conceptualising single-use plastics as pollution, the activities undertaken to create this context (contexting), and the circular economy solutions to tackle the plastic crisis is an appropriate place to start.

Single-use plastics as pollution in the making

Douglas' (1966, p. 36) seminal text, where she suggests dirt (waste) is "matter out of place," draws upon the theological definitions of dirt and how polluting behaviour can defile the sacred (i.e. God). She proposes that dirt has yet to be assigned a cultural value in an ordered world, and when a value is given, dirt has the potential to disrupt and create chaos. Dirt is identified and pushed to one side until it starts to rot or decay and vanishes from social systems. In this way, dirt seems to become 'homeless,' i.e. a value that has momentarily been forgotten (Douglas, 1966; Hetherington,

2004; Stowell & Brigham, 2018; Thompson, 1979, pp. 79–80). For Douglas (1966, p. 2), "there is no such thing as absolute dirt: it exists in the eye of the beholder." In this way, dirt is culturally relative. It is symbolic, a herald of disorder that makes visible the social system. For an 'object, item or matter' to be categorised as dirt requires a social judgement to recognise what is 'in' and 'out of place' in specific social and cultural circumstances (Douglas, 1966; O'Brien 2008, p. 5).

Douglas' work is significant, as it provides insights into values associated with dirt and the meaning of dirt according to a particular society; it could be identified, for example, as different severities of polluting behaviour. However, it is relevant to notice that she discussed natural waste as dirt, such as food and human waste, throughout her work. However, Douglas did not investigate the impact of complex materials that modern-day society creates (Gille, 2010; Liboiron, 2016; Stowell, 2012), such as single-use plastics. Arguably the missing link to modern, complex waste could be attributed to the focus and time of her study, as waste was understood to be static and controllable (O'Brien 2008).

Saying that plastics are 'bad' ("out of place") or 'good' ("in place") is not enough, as the social judgement of plastics alone cannot grasp the influence these materials have on society (Liboiron, 2016). On the one hand, single-use plastics are often argued as a necessity. The retail and sanitary sectors, for example, rely on the use of disposable plastics respectively to preserve degradable goods and meet health and safety measures in terms of hygiene (Hisano, 2017; Kakadellis & Harris 2020; Plastics Europe, 2019). On the other hand, when becoming pollution, single-use plastic waste disrupts human activities (e.g. fishery and tourism) and natural ecosystems, as these materials' composition include toxic substances that leaked plastics can release (Almroth & Eggert, 2019). In this respect, single-use plastics cannot be ignored; they need to be seen. Some European organisations working within plastics have gotten used to plastics being 'taken care of,' first by China (before the plastics import ban in 2018) and then by South-East Asian countries (Plastics Europe, 2019; Qu et al., 2019). For decades, plastic waste was invisible to individuals, organisations and society; 'forgotten,' you may say. European organisations failed to see plastic waste as matter "out of place," through what Hird (2015) defines as rituals of 'forgetting,' – i.e. organisations and people forget about waste as soon as it goes beyond their sight. In the case of discarded plastics, the ritual of shipping abroad could be viewed as a mechanism for forgetting, resulting in the materials not being recognised as "out of place."

Post-consumption plastics rose to global attention after the media captured images of marine ecosystems being disrupted by plastic bags and bottles (e.g. Attenborough's Blue Planet episodes on the plastic crisis, 2017). Meanwhile, 'Garbage Patches,' like the one found by Captain Moore, have progressively started to disrupt human activities. Due to the disruption, plastic waste has increasingly been noticed by the public and has become judged as 'dirt,' and 'out of place.' Furthermore, the increasing amount of these materials in the natural environment, and the fact that plastics are largely an immortal matter that 'does not go away' (Gabrys Hawkins & Michael, 2013, p. 2) has led organisations to consider ways to put these materials 'in place,' like the Douglasian dirt.

Contexting plastics

What is perceived as 'waste' could be, to some extent, a matter of context; things are 'out of place' or 'in place' according to the actors that conceptualise them, their agendas and how they interrelate with one another. The relationships between certain actors, materials, objects and issues can be understood as 'context.' In social sciences, 'context' is often referred to as a particular setting where social interactions happen (Given, 2012). The context helps researchers describe a particular social phenomenon and explain the actors, relationships, rituals and values. In this regard, the context has become an explanatory tool for scholars to make sense of a social fact. This idea of context has led to critiques, such as those promoted within Science and Technology Studies (STS).

STS scholars, who adopt a material semiotics perspective (e.g. Asdal & Moser, 2012; Callon, 1986; Callon & Law 1982; Latour, 1988), criticised the previous use of 'context' as an explanatory tool. They argue that past research has made the context a 'fixed' and 'given' element in social sciences that ignores the role of the actors' performance in making a reality. For example, Callon (1986) discusses how actors and their performance enact a certain reality that can be transformed through an ongoing process. Instead of a fixed context, Callon suggested an ongoing process that makes sense of a particular reality and transforms constantly. This change in approach is significant because it shifts the focus on who the actors are, their interests and the dynamics of the interrelations between different actors, materials, objects and issues. Consequently, it becomes possible to examine how a particular context is enacted, rather than 'rhetorically' using the term to explain the social relationships.

Asdal and Moser (2012) built upon Callon's (1986) idea that contexts are particular and can change according to the actors' interests. They coined the concept of *contexting* – i.e. the activities between different contexts. In this way, they shifted the focus further, from the dynamics between the multiplicity of actors, materials, objects and issues that compose a certain context to the interrelations between different contexts. By paying attention to the performative dimension of the contexting activities, it will be possible to understand how certain relationships between actors, materials, objects and issues are more stable and successful than others. Successful contexts will be able to mobilise more actors and become predominant. From this perspective, specific contexts prevail over others because particular practices are enacted rather than others.

The conceptualisation of plastic waste as a matter 'out of place,' like pollution, depends on the context that actors invoke. Similarly, the way organisations attempt to tackle the plastic crisis is dependent on how they view plastics as 'out of place' and the social judgement related to that. For example, plastic bags are commonly judged as something 'bad' because many people in society perceive them as contributing to environmental degradation. Many European Governments have banned plastic bags in an attempt to tackle this issue, identifying these objects as the most likely to leak into the natural environment because they are commonly used as litter bags (European Commission, 2015b). However, it could be argued that judging plastic bags as 'bad' reflects a particular view – i.e. these materials are negatively judged because of what (often) happens when they become waste (i.e. they leak into the environment). The context that actors (i.e. governments, in this case) invoke to make sense of plastic bags focus on an agenda that considers the future of these materials, their future being 'out of place' as waste and contribution to the issue of environmental degradation. If we mobilise different contexts, actors with diverse interests, materials and issues, the conceptualisation of plastics might change. For instance, some actors may consider the issue of CO_2 emissions in manufacturing bags and argue that plastic bags are more 'environmentally friendly' to produce than canvas bags; thus, plastics become sustainable. Other actors may consider the issue of environmental degradation and argue that canvas bags are more 'environmentally friendly,' and in this instance, plastics become unsustainable.

We argue that paying attention to the performative dimensions of the plastics issue yields insights into how some contexts become predominant and others fail. This predominance has implications on how solutions are interrupted and developed in addressing the plastic crisis. Invoking a context rhetorically or practically is done for a reason. Depending on how we invoke a context, plastics can appear 'good,' 'bad,' 'in place,' or 'out of place' – considering how and who invokes various contexts illuminates the process that enacts certain solutions to the plastic crisis prevalent. Thus, the contexting becomes a political activity because it is informed by the actors' interests and agenda.

Circular solutions

Within industry and policy spheres, solutions to the plastic crisis focus on circular economy principles. For example, among business-driven organisations, a pervasive definition is by the Ellen MacArthur Foundation (2015, p. 2):

> A circular economy is restorative and regenerative by design and aims to keep products, components, and materials at their highest utility and value at all times, distinguishing between technical and biological cycles. This new economic model seeks to ultimately decouple global economic development from finite resource consumption.

The Foundation describes the circular economy within their guidance documentation and indicates that the biological and technical cycles are connected. By enhancing the flow of goods (biological cycle) and services (technical cycle), the circular economy is seen to enact circular business models that can rebuild capital, whether financial, manufactured, human, social or natural. These models are proposed to organisations in terms of rethinking their operations and policies.

The European Commission (2015a, 2020) launched a call for action in 2015 focusing on inspiring businesses to adopt a circular approach to their operations, while national governments were invited to support the industry efforts with adequate policies. Within the European Commission documentation, the circular economy is portrayed to progressively minimise waste by generating closed loops that maintain the value of materials through recycling and reusing practices (Kirchherr et al., 2017). Although the strategy is thought to be beneficial for both the natural environment and human societies, the main aim was to create competitiveness and market values for post-consumption materials so that these do not become waste (Lacy et al., 2020). In 2020, the Commission published a report that showed what actions were delivered within the European Union in terms of business strategies and policies to transition the region to a circular economy, stressing the technocentric and material-focused approach of the previous call.

Although organisations often invoke the narrative of the circular economy, this term has been identified as a contested paradigm (Calisto Friant et al., 2020; Corvellec et al., 2020). There are numerous definitions of this term (Kirchherr et al., 2017) and different frames these definitions refer to (Calisto Friant et al., 2020). For example, Kirchherr et al. (2017) made an inventory and analysis of the 114 different definitions of the circular economy, arguing that the circular economy is most frequently depicted as a combination of reducing, reuse and recycling activities, emphasising the focus on waste.

Blomsma and Brennan (2017) also draw attention to waste as they define the circular economy as an umbrella concept that brings together different waste and resource management strategies. Hence, the circular economy is enacted as a practice to manage post-consumption materials "to extend the productive life of resources [. . .] [. . .] to delay or prevent landfilling or permanent disuse" (Blomsma & Brennan, 2017, pp. 603–608) of materials (recognised as resources), such as recycling.

Corvellec et al. (2020, p. 97) stress the same focus on waste, outlining that the circular economy is often associated with the creation of "waste-free technical loops that resemble biological loops and make waste disappear at the same time as being restorative and regenerative by design." They identify this concept as an "empty signifier, which allows for a whole range of interpretations and approaches to be bundled together under the term 'circular economy'" (Corvellec et al. 2020, p. 97) and call for academics and practitioners to go beyond the coalescence of circular economy discourses that focus on technocentric practices to manage waste.

Hence, Calisto Friant et al. (2020) study shows that the circular economy is not always primarily conceptualised as technocentric strategies focusing on managing waste materials. They highlight the variety of circular economy frameworks since

1945 and identify the predominant discourse in different periods of time. Through Calisto Friant et al. (2020, p. 10) chronological analysis, they recognise discourses that refer to the 'circular economy' and 'circular society.'

According to these authors, the circular economy considers 'sceptical' and 'optimistic' technocentric discourses. The first ones are focused on population controls and resource efficiency with no mention of wealth distribution and social justice. Positive discourses, instead, lean toward a technocentric approach and are based on the assumptions that "capitalism is compatible with sustainability and technological innovation can [. . .] prevent ecological collapse" (Calisto Friant et al. 2020, p. 11) while progressively eliminating waste. According to these discourses, circular solutions are material-focused such as recycling, reusing, composting and any practice that, while attempting to avoid the creation of waste by closing resource loops, maintain the economic value of materials. Such discourses are predominant among the practitioners of business-driven organisations, as they focus on technological innovation and business models (Calisto Friant et al. 2020, p. 11).

'Circular society' discourses include approaches that "go beyond market-based solutions and economic considerations and see circularity as a holistic social transformation" (Calisto Friant et al. 2020, p. 8). Transformations are pursued through 'sceptical' and 'optimistic' attitudes; the first one by looking at ways to drastically transform society so as to achieve sustainability through reconnecting markets, communities and resources and boosting slow and local economies (economy of sufficiency); the second one "propose[s] a mix of behavioural and technological change, leading to a [. . .] sustainable future where scarcity and environmental overshoot has been dealt with by [. . .] social, economic, industrial and environmental innovations" (Calisto Friant et al. 2020, p. 11).

Calisto Friant et al. give us a significant analytical tool to understand IASB's circular economy agenda. The authors comment that discourses around the circular economy have been mostly developed by governments and the private sector with specific agendas often material-focused and technocentric. They argue that this has led to the failure of creating a systemic and holistic understanding of the implications of the circular economy, leaving this term to be a 'go-to concept' and to be easily discredited as greenwashing. For this reason, they contend that the circular economy is a contested paradigm, as it may have different meanings and be used according to different agendas. Hence, Calisto Friant et al. study represent a relevant analytical approach to understand the diverse circularity discourses encountered in the IASB case study and this global alliance's agenda. It also informs our attempt to show how IASB and their members invoke a context, their interests in choosing a context and the result of promoting particular circular solutions.

The next sections introduce the research methods and then move on to provide background information regarding the empirical context and the 'International Alliance for Sustainable Business (IASB) and plastic futures' case study.

Methods and analysis

The IASB case is drawn from a wider investigation into the interrelations between plastics and businesses in organising circular economy initiatives to tackle the plastic crisis, conducted by the first author, Marta. To preserve IASB's and the participants' anonymity, pseudonyms are used, and quotations from IASB documentation are paraphrased.

The IASB case study is based on six-month ethnographic research (Michrina & Richards, 1996; Yin, 2011) carried out between April and October 2019. During this period, Marta worked as an intern with IASB's Plastic Packaging Team (PPT). The PPT was a small unit made up of five employees (one Director, two Managers, one Associate and Marta, the intern). Data was collected from semi-structured and informal interviews, participant observation, shadowing PPT members, attending meetings and external events, internal and external documentation, and email correspondence. It is important to note that this research may present possible biases because of Marta's dual role as the researcher and intern. While working with the PPT, she produced research and documentation to navigate the relationships between the PPT and IASB, IASB's 'plastic' members and materials, therefore possibly influencing the circular solution outcomes.

The findings in this chapter emerged from a two-stage data analysis. Stage One focused on tracking the material semiotic relationships between plastic waste, IASB and their members; the goal was to identify relevant actors and relationships among the large amount of data collected. Stage Two looked at identifying significant threads to conceptualise the selected interrelations. In line with the material semiotic perspective adopted, the analysis was not linear; the two stages overlapped and entangled, showing what Law (2004, p. ii) describes as the 'mess in social science research.'

To illustrate how the relevant material semiotic relationships between organisations and plastics invoke circular solutions, we present the 'Sustainable Organisations Forum' vignette. This vignette refers to a significant event that Marta observed while shadowing James, the PPT Director, during a 3-day meeting organised by an external party named Green Organising. We use the term *vignette* to identify a story bracketed from Marta's fieldnotes that shows the dynamics between research participants (Hughes, 1998) – i.e. the organisations and materials relevant to the purpose of this chapter.

International Alliance for Sustainable Business (IASB) and plastic futures

IASB was a business-driven global alliance whose membership represents thousands of international organisations comprised of hundreds of multinational member corporations and national network partners acting on behalf of several thousand SMEs. These organisations varied in size and revenue, operating within various business sectors from agriculture, finance, mobility, construction, tyres to plastics. A Managing Director defined IASB as follows:

> a group of companies that creates solutions to sustainability to address the challenges they face [. . .] climate change, plastic waste etc. [. . .] IASB helps companies to understand how to progress, [. . .] to develop tools and interact [with] each other. (Interview with Managing Director Nadia, p. 1)

The Managing Director's words highlight how IASB supported collective action that was created through consensus and produces knowledge/business solutions, rather than a product, per se.

In consultation with its members' interests, IASB organised initiatives to address sustainability issues that affect them and arranged work activities accordingly. The Alliance's vision was to develop a system for the world population to thrive within the planetary and social boundaries by 2050. IASB's mission was to accelerate the transition to a sustainable world by promoting scalable and replicable business models through collaboration. At the time of this research, IASB had nine programs: consumer behaviour, development, finance, agriculture, forests, energy, construction, mobility and circular economy. Whilst IASB covered a wide range of programs, the focus of this chapter is on the Circular Economy Program (CEP) and, in particular, the project activities around plastics carried out by the PPT. The PPT was a team of five individuals set up by IASB senior management committee and CEO in collaboration with the CEP Director and team. The PPT was dedicated to developing circular solutions to the plastic crisis through the 'Plastic Project.' The Plastic Project was developed under the CEP.

Findings

This section explores the findings of our case study. The following two subsections explain how IASB enacted their circular economy agenda and how this global alliance's agenda informed initiatives to tackle the plastic crisis. The last subsection presents a vignette, the Sustainable Organisation Forum, that shows how paying attention to how material and social values relate, i.e. the contexting activities, aids our understanding of how circular solutions are proposed and the impact on organisational practices.

IASB and the circular economy

The CEP's mission was to create a platform to empower their members to create new practices and share experiences to transition towards a circular economy. In line with the Alliance's goal to develop a system for the world population to thrive within the planetary and social boundaries, the CEP team developed a definition of the circular economy based on the Ellen MacArthur Foundation's vision (2015):

> By promoting circular solutions, IASB can help the global economy to be more resilient, support civic society [. . .]. In our vision, the circular economy is a way to rethink the relationships between natural resources, materials, technology, consumers, and the industry to create a sustainable future where humans and the natural environment can thrive. (IASB's Organisations Guide to Circularity, p. 4)

Whilst still focusing on materials management, this definition highlights the need to re-organise 'relationships' between natural resources, materials, technology, consumers and business and making these changes will assist in transforming the global economy. According to the CEP's ambitions, the re-organisation of the economic activities would bring financial benefits by promoting innovative business models, the creation of new jobs, a safe environment and clean cities (Organisations Guide to Circularity, p. 4).

To help encourage their members to participate in this program, the team published the Organisations Guide to Circularity, which contained IASB's circular economy agenda. This publication aimed to address the difficulties that IASB members were experiencing with understanding and applying circular economy practices. Although the Guide was written mostly for a business audience, the actors who were mobilised to elaborate IASB's CE agenda were varied. They included IASB members (such as manufacturers, retailers, waste management companies and recyclers), consumers and policymakers. The activities of IASB and their member companies primarily utilised materials and resources such as steel, aluminium, plastic, cement, glass, wood, primary crops and cattle. These specific materials were the focus of mobilisation plans for the circular economy because they were viewed as being 'responsible' for a significant percentage of the global GHG emissions, water and land use, and also represented the predominant industry sectors that IASB members operated in retail, chemical, technology, automotive, cosmetic, metallurgic, construction and agricultural-farming.

> Steel, aluminium, plastic, cement, glass, wood, primary crops, and cattle are responsible for 20% of global GHG emissions, 95% of water use and 88% of land use. Adopting a circular

economy approach in these areas will tackle climate change and water and land use challenges. (IASB's Organisations Guide to Circularity, p. 9)

This extract accentuates IASB's need to consider materials when discussing circular solutions. Because of the negative impact these materials have on the planet, IASB had the opportunity to showcase what some of their members already did to tackle such challenges, demonstrating the role that organisations, under the Alliance guidance, could make in moving towards a circular economy.

The Guide did not call for actors to 'take responsibility.' However, it aimed to boost collaboration among businesses, policymakers and consumers while considering the planetary and social boundaries and the role of materials in understanding possible barriers and benefits to transitioning to a circular economy. IASB members were invited to showcase practical examples of circular solutions. These examples demonstrated that the circular economy could meet the need of a growing world population by enhancing the global economy with fewer resources available. At the same time, the Guide showed policymakers the value of circular solutions that deliver social and economic benefits (Organisations Guide to Circularity, p. 12). Despite mentioning the consumer and government, their roles were absent in this Guide. This implied that even if civic society, governments and businesses were invited to collaborate, industry was encouraged to lead the transition as illustrated through the practical examples of the business benefits of circular solutions.

A practical illustration of how to apply IASB's circular economy agenda is showcased in the Guide by a member organisation called Fly. Fly is a waste management company that focuses on plastic waste. Plastics were conceptualised according to a positive judgment that sees this material as 'good' but needed to be put 'in place.'

> Fly has rethought their plastic strategy to procure high-quality [Polyethylene terephthalate] PET flakes that have the same properties as virgin materials and that can be used to produce flowless recycled r-PET plastic bottles. (IASB's Organisations Guide to Circularity, p. 6)

Recycling is framed as the primary or critical way to make post-consumed PET bottles 'circular.' Furthermore, recycling PET into something with the 'same properties as virgin materials' and making something 'flowless' addressed organisational pressures to save raw materials and phase out plastic waste.

The reason for emphasising recycling as a circular solution is that the Guide and, therefore, IASB circular economy agenda, was written together with some of IASB's 'plastic' members – i.e. producers, retailers, waste management and recycling companies that appear to align with technocentric and material-focused views of circular solutions (Calisto Friant et al., 2020).

The data from this study indicated that the 'plastic' members used their sustainability reports to evidence their circular economy activities, and this influenced the contents of the Guide. Examples included Square, a large international plastic producer, who said it contributed to the circular economy

> with our Circular Chemical project, that aims at strengthening the already significant role of recycling by using chemically recycled plastic waste in manufacturing new objects. (Square, Sustainability Report 2019, p. 105)

Retailers like Blue and Star, which heavily relied on and worked with plastic packaging, referred to the circular economy as

> Redesigning products by . . . recycling and using post-consumer recycled materials in innovative ways. (Blue Sustainability Report 2018, p. 30)

> circularity by design, that includes reusing and recycling practices. To achieve this goal is necessary to understand the existing recycling infrastructure [. . .] and provide help to local entrepreneurs to build the infrastructure needed where there are gaps. (Star Sustainability Report 2018, p. 13)

Waste management and recycling companies (e.g. Fly) shared the same views on circular actions, seen as recycling practices that would enable a

> switch from a linear system to a circular loop approach. Industry and civil society will be able to contribute by encouraging waste recycling and recovery of secondary resources by participating to waste classification and collection. (Fly Sustainability Report 2017, pp. 2, 6)

Influenced by their "plastic" members IASB adopted ideas that related to "waste-free technical loops" (Corvellec et al., 2020) as a starting point and to show other corporations how to begin their adoption of circular economy practices.

IASB and the "plastic" circular economy initiative

After this conceptualisation of the circular economy agenda, IASB moved towards applying it to plastics. IASB was facing significant challenges in this sector, supporting their "plastic" members in creating scalable and replicable circular solutions to address the single-use plastic crisis. IASB's challenges reflected the most common issues that organisations faced. Challenges were related to:
- Loss of reputational capital triggered by being portrayed as the 'polluters' (e.g. the environmental activists Greenpeace 'Break Free from Plastics' campaign)
- Loss of revenue due to poor waste management, difficult operations between world regions given by different standards, waste management infrastructures and organisations' level of commitment to the circular economy

IASB's Senior Management Team saw the value of promoting an initiative to help their members solve the issues brought about by the plastic crisis. Members included plastics producers, retailers and consumer goods companies and recyclers. They set up the PPT to help IASB 'plastic' members deal with the plastic crisis.

James, the PPT Director, was responsible for designing the 'IASB plastic project.' To support this, he suggested that exploring what circular ideas were already 'out there' would be useful for designing the PPT's plastic initiative. In addition, this information would enable the plastic project work to be integrated within the CEP suite of circular economy solutions. The Sustainable Organisations Forum, an annual event organised by a not-for-profit, represented a significant opportunity to do so.

The 'Sustainable Organisations Forum'

The 'Sustainable Organisations Forum' was organised by the not-for-profit 'Green Organising' and ran for 3 days in a location in Europe. It gathered about 80 actors, such as practitioners, academics, NGOs and policymakers working on single-use plastics. The Forum vignette is an example of the process of invoking contexts depending on the actors, materials, objects and issues that interrelate. This vignette focuses on a particular activity called the 'Roundtable exercise,' which shows the 'contexting' activities and helps us understand who invokes certain circular solutions and how one of these solutions became more predominant over the others.

In the attempt to boost collaboration and problem-solving discussions, most of the Forum activities focused on organising solutions to the plastic crisis within simulated real-world scenarios. The Forum took place before the Covid-19 pandemic, and the 'Roundtable exercise' was run in one large room, hosting five roundtables that gathered up to ten people each. About 45 people participated in this exercise, which happened in parallel with other activities scheduled on that day. This task lasted a couple of hours and aimed at developing strategies to make single-use plastics 'circular.' All the roundtables had the same task:

> to save a country by finding a system that circulates materials. You have to focus on single-use plastics. Find a second life for these materials that do not end with landfilling. (Fieldnotes diary, Sustainable Organisations Forum, p. 39)

However, each table had a different, simulated real-life scenario designed by Green Organising. Because roundtables were composed of diverse actors with different agendas, the aim was to boost discussions towards finding common solutions to the plastic crisis. James, the PPT Director, participated in the group whose scenario was entitled 'a low-income island country' that relied on international aid, aquaculture and tourism. At the table sat representatives from plastic producer companies, environmental NGOs, a waste pickers' association, recycling companies, a policymaker and James. During the discussion, divergent contexts that invoked the opposite but competing ideas of plastics emerged. These ideas saw plastics as 'bad' or 'good,' but that needed to be put 'in place.'

The various participants at the table proposed eight solutions that presupposed a certain knowledge regarding the local community strengths, barriers and needs in that situation. According to Marta's analysis, these solutions could be grouped into two themes: "No plastics" and 'New life' to plastics; the former gathered solutions with a negative judgment of single-use plastics, while the latter gathered solutions that have a positive view of these materials. Criteria to group these solutions in such categories were according to the participants' discourses around plastics as 'bad' and 'good' that Marta collected in her fieldnotes whilst shadowing James.

The range of ideas, participants at the table (i.e. who is putting what solution forward), judgements toward plastics declared during the discussion of the exercise and notions of responsibility related to these themes are summarised in Table 17.1. The second column presents cherry-picked quotes from Marta's fieldnotes diary; these extracts are the most relevant to representing actors' judgment on plastics. Table 17.1 provides an overview of the analysis of the discussion observed during the Roundtable exercise.

Table 17.1: Circular solutions, theme, judgement towards plastics and notions of responsibility in the 'Roundtable exercise.'

Actors	Plastic solutions (fieldnotes)	Theme	Judgement towards plastics	Notions of responsibility
Waste Pickers Association	'Stop plastics by banning single-use plastics'	No plastics	Plastics are 'bad,' stop to plastics	Plastic organisations and consumers need to stop using single-use plastics
Environmental NGO 1	'Tax for tourists bringing plastics from abroad'	No plastics	Plastics are 'bad,' stop to (additional) plastics	Consumers need to stop introducing plastics
Policymaker	'Education for local people (about sorting and recycling) to avoid leakages'	New life to plastics	Plastics are 'good," education to avoid pollution	Consumers need to be educated on how to separate and sort post-consumption plastic
Environmental NGO 2	'Transformation of [disused] fishing nets into art'	New life to plastics	Plastics are 'good,' artistic reuse/recycling	Charities or private artists to give a new life to post-consumption plastics

Table 17.1 (continued)

Actors	Plastic solutions (fieldnotes)	Theme	Judgement towards plastics	Notions of responsibility
Recycling company 1	'Deposit system'	New life to plastics	Plastics are 'good,' reuse/recycling	Consumers need to place post-consumption plastics correctly; recycling companies required to set up recycling markets
Recycling company 2	'Making fuel out of plastics'	New life to plastics	Plastics are 'good,' recycling	Consumers need to place post-consumption plastics correctly; recycling companies required to set up recycling markets
James IASB, PPT'	'Collaboration between communities, government and the industry and set up recycling facilities'	New life to plastics	Plastics are 'good,' recycling	Consumers need to place post-consumption plastics correctly; policymakers and recycling companies required to set up recycling markets
Plastic Producer company 1 – 2 – 3 [they all agreed to the same solution]	'Plastics become of value (e.g. PET), leads to a market, (e.g. recycled PET)'	New life to plastics	Plastics are 'good' recycling markets	Consumers and recycling companies need to place post-consumption PET correctly

As Table 17.1 shows, the 'No plastics' themed solutions promoted by the Waste Pickers Association and Environmental NGO 1 considered the presence of plastic objects as 'bad.' Social values – i.e. organisations' agendas, seem to invoke circular ideas related to the reduction and elimination of single-use plastic waste through banning these materials. The 'No plastics' solutions are produced by the interrelations between these organisations' values, that would see plastics as future pollution, with plastics material value – i.e. the material composition that makes these objects accumulate and litter the natural environment. Notions of responsibility invoked by these organisations saw both consumers and plastic producers accountable for the plastic crisis.

Policymaker, Environmental NGO 2, Recycling company 1 and 2, Plastic Producer company 1, 2 and 3 and James (representing IASB) proposed the 'New life' themed solutions that went along with the idea that 'out of place' single-use plastic waste (i.e. misplaced in the natural environment) are 'bad' and they need to be 'in place' to be considered 'good.' Thus, there is hope for these materials to be redeemed, and the way to do that is by invoking technocentric circular economy ideas that focus on materials composition, such as recycling/reusing. Recyclable/reusable single-use plastics were considered "in place," therefore, "good." In this sense, by being recyclable/reusable or not, the material composition of plastics (material value) interrelate with organisations' agenda (social values) by supporting or disrupting recycling/reusing practices. The notion of responsibility produced by the 'New life' themed solutions, on the one hand, call for industry and policymakers to create a situation where single-use plastics can be recycled/reused. On the other hand, it perceives consumers as being mostly responsible for the fate of plastics (i.e. the population needs to get educated to sort waste materials properly and be creative with waste) and accountable for the plastic crisis.

Through the 2-h discussion, actors who promoted the 'New life' solutions managed to shift the attention from plastic *objects* to plastic *waste* as the matter 'out of place.' Strategies to make this shift happen included discourses around the possible value of post-consumption plastics and their role in improving the island nation's economy.

> [Plastic Producer company 1:] Plastics have value and could lead to [the creation of] a market [. . .] see that plant in Switzerland, for example, it used to be funded by the government as part of a government scheme to create jobs, and it managed to become independent and thrive because rPET is requested by companies. [. . .] PET is valuable and sustain a local economy. (Fieldnotes Diary, Sustainable Organisations Forum, p. 40)

Plastic Producer companies 2 and 3 immediately agreed to the idea of recycling plastics such as PET. The Policymaker, Recycling companies 1 and 2 and James followed up by bringing up examples of reusing/recycling plastics that would have improved the local economy through job creation.

The Policymaker sitting at the table stressed that because post-consumption plastics had value, i.e. represented a resource for recycling business and created jobs, incineration was seen as the last resource and only when relevant to stop leakages of plastic waste into the ocean. This contributed to making a significant connection that appeared to win over the 'No plastics' supporters; post-consumption plastics could be resources to improve the local economy. Therefore, giving market and social (creation of jobs) value to these post-consumption materials through reusing/recycling practices moved the focus on their *placement* rather than on the material composition. Discarded plastics, if placed in reusing/recycling networks through, e.g. 'deposit systems,' 'make fuel out of plastics' and 'set up recycling facilities' (Fieldnotes diary, Sustainable Organisations Forum, p. 40), would be kept away from the natural environment. This was passed as a possible solution to solve the plastic pollution problem in the short term.

Another aspect that played in favour of the 'New life' supporters was the language of the scenario that already addressed single-use plastic *waste* as the problem: "You have to focus on single-use plastics. *Find a second life for these materials* that do not end with landfilling" (Fieldnotes diary, Sustainable Organisations Forum, 39). In addition, although participants that proposed 'No plastics' solutions suggested a straightforward way to deal with the plastic crisis in that island nation, i.e. to reduce the consumption and disposal of these materials within the national borders, this was discarded as 'not relevant' as a few of the 'New life' supporters noted how ocean pollution is spread through currents and tides, and the 'No plastics' actors could not find a counter-argument.

The performance of the 'New life' solutions supporters demonstrated the existence of a certain political strategy carried out by a positive judgement of plastic objects, and that saw recyclable/reusable plastic waste as 'in place.' Moving the attention on plastic waste as a matter 'out of place' and arguing that ocean plastics are already in the water (thus, imported plastics are not the immediate issue), participants all agreed to the final solution:

> To aim to zero single-use plastics [waste], decreasing it over the years while implementing new initiatives such as reuse schemes, attracting artists and collaboration with other nations to develop recycling facilities. (Fieldnotes diary, Sustainable Organisations Forum, p. 40)

The 'New life' coalition seemed to accomplish their political agenda by stressing that the pervasiveness of plastic waste as ocean (e.g. discarded fishing nets) and land (e.g. PET bottles) pollution was the reason why single-use plastics were considered "out of place." Plastic waste disrupted the island-nation economy by "being there;" the problem was that these materials do not degrade but accumulate and litter. The presence of unwanted post-consumption plastics in the waters and on land was perceived as a problem to be solved as soon as possible, as the same scenario seemed to suggest that the country needed help: "*to save* a country by finding a system that circulates materials" (Fieldnotes diary, Sustainable Organisations Forum, p. 39). The emphasis on 'to save' implies that the island nation was facing an emergency. This means that solutions such as a deposit scheme, the creation of recycling infrastructures and "trash art" were likely to be considered, as they seemed to address the immediate problem.

Discussion

IASB's circular economy agenda is contextual, therefore, political

The IASB case has shown how paying attention to how and what actors invoke various contexts illuminate the process in which circular solutions are enacted.

In building their circular economy agenda, IASB invoked a context that aimed to go beyond a technocentric approach by including civic society and policymakers in the transition toward circularity, although they did not state clear roles for the civic society to cover. This addresses what Calisto Friant et al. (2020) describe as optimistic circular society discourses. However, the Organisations Guide to Circularity showcased material-focused solutions as examples of circularity, such as reusing/recycling practices, promoting a technocentric approach that Calisto Friant et al. considered in their definition of optimistic circular economy discourses. In this way, the Guide, as well as IASB's circularity agenda, contradicted itself. The CEP team attempted to clarify the ambivalence between the aspiration to build a more holistic understanding of circularity and the showcasing of material-focused solutions; they designed the Guide in a way that proposed practices related to 'waste-free technical loops' (Corvellec et al., 2020) as the starting point. In this way, CEP aimed to demonstrate how corporations began with circular economy practices and, possibly, give inspiration regarding achieving what Calisto Friant et al. (2020) defined as circular society.

The presence of conflictual elements in IASB' circularity agenda could be explained by the role that "plastic" members played in writing the Guide and developing such agenda. While collaborating with IASB on this task, these members mobilised a circular economy context that considered purely technocentric practices and mostly looked to their interests around plastics. Therefore, the Guide presented circular solutions that could be seen as a reproduction of existing practices, in line with "plastic" members' interests, such as reusing/recycling.

Hence, recycling/reusing solutions are identified as the circular practices able to put back 'in place' the single-use plastics considered 'out of place'; single-use plastics are 'good' when situated within the correct waste stream, while 'bad' when not recyclable/reusable.

For example, the recycling company Fly mentioned in the Guide decided to focus on PET bottles. The attention was not on the plastic bottle containing water itself, but on the future of that bottle. Fly's decision to focus on PET bottles and recycle these was in part given by this company conceptualising 'out of place' plastics. Because PET bottles were often in photos showing the disruption brought by ocean plastics, Fly made sense of these objects as possible 'future plastic pollution.' In this respect, materials become symbols of disorganisation when they are 'out of place' (Douglas, 1966) – i.e. when they are not placed correctly according to organisations' expectations and interests. A plastic bottle washed on a beach is pollution; the same object, placed within the correct waste management stream, is, instead, a secondary material; it will be recycled according to organisations' expectations.

The situation becomes more complex in the Sustainable Organisations Forum, as actors invoke different circular economy contexts to promote either 'No plastics' or 'New life' plastics-themed solutions. While 'No plastics' solutions address single-use plastic objects as 'bad,' 'New life' solutions advocate that these materials are 'good' but need to be put 'in place.' This ties in with Douglas' (1966) argument that there is no "absolute" dirt, but the enactment of dirt is culturally and socially contextual. The Roundtable exercise shows how policymakers, certain environmental NGOs, industry and business-driven organisations such as IASB exercise their political agenda toward moving 'No plastics' actors (e.g. waste pickers associations and other environmental NGOs) to support 'New life' solutions. The way they do that is showing how post-consumption plastics are dangerous when 'out of place' (i.e. in the environment) and stress the urgency to put these materials 'in place' through existing practices seen as the "quickest" solutions to tackle the plastics emergency. The contexting activity is political and enacted through the interrelations of social values (organisations' agendas) and material values, i.e. plastics material composition that might support or disrupt organisations' solutions to the plastic crisis.

Responsibility and the Circular Economy

When looking at how contexts are mobilised, it is relevant to consider how this enacts a certain understanding of waste and the related notion of responsibility. Douglas' (1966) work is significant to highlight the social judgements that actors invoke when mobilising particular contexts; these views also reflect a certain conceptualisation of "in place" and "out of place" plastics. Asdal and Moser's (2012) argument enables the connection between social judgments, part of organisational social values, and material values (i.e. single-use plastics material agency). It is possible to understand how a particular notion of plastic waste and responsibility is enacted by paying attention to how these values interrelate within the contexting activities. Contexts are political, used to support organisations' agendas, and so is the conceptualisation of waste and the related notion of responsibility. However, as Hird (2015, p. 10) comments, "It is difficult to take responsibility for forgotten actions"; therefore, actors tend to invoke contexts that justify their actions and pass the responsibility to other agents. For example, invoking a context that gives responsibility to consumers (e.g. recycling is the solution; therefore, people should buy recyclable plastics) or plastic producers (e.g. reusing plastics is the solution; therefore, companies should design and produce reusable plastics) also changes the notion of responsibility that is performed. The political loading of the contexting operation creates a specific strategy that organisations carried out to gain consensus, attract other possible agents and prevail on the other proposed solutions to the plastic crisis like the Roundtable exercise showed.

The notions of responsibility enacted by invoking circular economy contexts are driven by organisations' perspectives and how waste is enacted according to these. On the one hand, organisations that focus on plastics as 'bad,' e.g. the 'No plastics' supporters in the Forum vignette, look at plastic producers and recyclers as the 'polluters,' 'waste creators.' Most of the responsibility is given to these actors, without considering the implications of eliminating plastics overall and focusing on the figure of the 'guilty business' that produced waste likely to leak into the environment and become pollution, such as plastic bags and PET bottles. An example is Greenpeace's 'Break Free From Plastic' campaign that has targeted plastic manufacturing companies (Greenpeace, 2019). On the other hand, responses that look at plastics as 'good,' e.g. the 'New life' supporters, focus the responsibility on consumers. The emphasis on reproducing existing practices that look at materials values, such as recycling/reuse solutions, seem to stand on the creation of the figure of the 'guilty consumer' who uses a plastic bag or drinks from a PET bottle, objects likely to leak when discarded and pollute.

The final circular economy solution produced in the Roundtable exercise is based on the key figure of the 'guilty consumer' and the pervasiveness of plastic waste. Although including the role of businesses (creation of recycling/reuse networks) and the government (policies towards decreasing the generation of plastic waste through the years), it sees civic society as responsible for sorting waste materials correctly (otherwise they could not be recycled/reused), engaging in deposit schemes, being creative with post-consumed single-use plastics. This notion of responsibility and the negative conceptualisation of plastic waste when not recycled/reused are performed by the interrelations between organisational social and material values. For example, in performing their social values, the Roundtable participants had to confront the plastic waste materiality and consider these materials as possible *pollution to come* if not channelled within the right networks. The final solution reflected the political process of contexting that saw predominant the technocentric circular economy agenda because able to deal with the *emergency* represented by the plastic crisis. In this fashion, circular economy solutions seem to consider both materials and organisations when applied to real-life scenarios.

Within the Forum vignette, the presence of different circular economy definitions challenges the application of circular solutions in real-life scenarios and reinforces the notion of the circular economy as a contested paradigm (Calisto Friant et al., 2020), an 'umbrella concept' (Blomsma & Brennan, 2017) that could be adopted in different situations without an actual, unique, meaning; an "empty signifier" (Corvellec et al., 2020). Hence, in the Roundtable exercise, participants relied on invoking a circular economy identified with existing practices (recycling/reusing).

Contributions

The IASB study shows that social and material dimensions (Wright et al., 2013) matter, and these interrelations give insights into how complex waste materialities, such as plastics, are re-conceptualised as "in place." This is helpful to understand how alliances such as IASB and their members develop judgements depending on how material and social values perform. The IASB case shows that considering how the contexting activity impacts material and social values leads to relevant lessons.

The theoretical lesson contributes to Waste Studies (Douglas, 1966; Gille, 2010; Liboiron, 2016; O'Brien 2008; Thompson, 1979) by considering the material agency of single-use plastics through a material semiotic approach (Asdal & Moser, 2012; Callon, 1986; Latour, 1988, 1991). Drawing upon Asdal and Moser's (2012) discussion around contexts and contexting highlights the importance of considering materials in the process of organising. Materials like single-use plastics have not been paid attention to as much as they should have, as they tend to be "forgotten" by organisations. However, by looking at the contexting activity, it is possible to see how plastics can influence circular solutions by being 'out of place' or 'in place' and possibly performing as a disruptor or supporter of circular economy practices. The Roundtable exercise illustrated this when the proposals of banning imported single-use plastics did not get mobilised in the final solution because plastics pervasiveness and disruptive behaviour led actors to consider solutions that seemed to address the threat immediately. Thus, the contexting activity shows us how contexts that refer to technocentric approaches and a positive judgement of plastics objects prevailed.

The practical implications are for business-driven alliances, such as IASB, to identify their members' agendas and interests and consider how that influences their definition of the circular economy and related initiatives. As the IASB's Organisations Guide to Circularity showed, when it comes to plastics, business organisations tend to adopt a positive conceptualisation of plastics and see these materials as 'in place' when recycled/reused. This might reproduce existing practices that do not necessarily aim for innovative solutions. The recognition that members might mobilise practices that tend to preserve their approach to single-use plastics could help alliances manage these members' influence on their conceptualisation of future plastics and the circular economy. In this sense, considering the role of the contexting activities in formulating ideas and designing circular solutions could drive business-driven alliances in their mission to help members transition to a circular economy.

Conclusion

The IASB case has shown how important lessons can be learned by paying attention to the contexting activity (Asdal & Moser, 2012). Theoretically, the role of complex waste materials and their agency expands the discussion in Waste Studies (Douglas, 1966; Gille, 2010; Liboiron, 2016; O'Brien 2008; Thompson, 1979); practically, what and whom to consider in organising a circular economy for single-use plastics matters as it determines the type of circular economy solutions enacted. Explicit consideration of 'contexting' helps us to see how material and social values interact, i.e. how organisations and materials composition interrelate (Asdal & Moser, 2012; Callon, 1986; Callon & John, 1982; Latour, 1988) when attempting to organise circular solutions, and how complex waste materials, such as plastics, may affect these solutions. Furthermore, because organisations invoke contexts that are political (i.e. in support of actors' agendas), specific notions of responsibility are produced, and particular circular solutions are enacted over other options. These solutions could reproduce existing practices rather than promote innovation. By considering who, what and how contexts are invoked, global alliances, such as IASB, can identify existing patterns and manage the influence of their members on projects and initiatives. This approach would help global alliances in their mission to support members while promoting innovation and sustainability.

References

Almroth, C. B., & Eggert, H. (2019). Marine plastic pollution: sources, impacts, and policy issues. *Review of Environmental Economics and Policy*, *13*(2), 317–326. https://doi-org.ezproxy.lancs. ac.uk/10.1093/reep/rez012

Asdal, K., & Moser, I. (2012). Experiments in context and contexting. *Science, Technology & Human Values*, *37*(4), 291–306. https://doi-org.ezproxy.lancs.ac.uk/10.1177/0162243912449749

Attenborough, D. (2017). The dangers of plastics in our Ocean. *Blue planet II*, BBC One, https://www.bbc.co.uk/programmes/p05q49hq (accessed May 25th, 2020)

Blomsma, F., & Brennean, G. (2017). The emergence of circular economy. A new framing around prolonging resource productivity. *Journal of Industrial Ecology*, *21*(3), 603–614. https://online library-wiley-com.ezproxy.lancs.ac.uk/doi/epdf/10.1111/jiec.12603

Calisto Friant, M., Vermeulen, W. J. V., & Salomone, R. (2020). A typology of circular economy discourses: Navigating the diverse visions of a contested paradigm. *Resources, Conservation & Recycling*, *161*, 104917. https://doi.org/10.1016/j.resconrec.2020.104917

Callon, M. (1986). Some elements of a sociology of translations. Domestication of the scallops and the fishermen of St. Brieuc Bay. *Sociological Review*, *32*, 196–223. (accessed September 2nd, 2021) https://search-ebscohost-com.ezproxy.lancs.ac.uk/login.aspx?direct=true&db= sih&AN=127596037&site=ehost-live

Callon, M., & Law, J. (1982). On interests and their transformation: Enrolment and counter-enrolment. *Social Studies of Science*, *12*(4), 615–625. https://www.jstor.org/stable/284830

Corvellec, H., Bohm, S., Stowell A. & Valenzuela, F. (2020). Introduction to the special issue on the contested realities of the circular economy. *Culture and Organization*, *26*(2), 97–102. https:// doi-org.ezproxy.lancs.ac.uk/10.1080/14759551.2020.1717733

Corvellec, H., Stowell, A. F., & Johansson, N. (2021). Critiques of the circular economy. *Journal of Industrial Ecology*. early view. https://doi.org/10.1111/jiec.13187

Douglas, M. (1966). *Purity and Danger: An analysis of concepts of pollutions and taboo*. London: Routledge.

Ellen MacArthur Foundation. (2013). *Towards the Circular Economy. Economic and business rationale for an accelerated transition*. https://emf.thirdlight.com/link/x8ay372a3r11-k6775n /@/preview/1?o (accessed September 16th, 2021)

Ellen MacArthur Foundation. (2015). *Towards a circular economy: Business rationale for an accelerated transition*. https://emf.thirdlight.com/link/ip2fh05h21it-6nvypm/@/preview/1?o (accessed September 27th, 2019)

European Commission. (2015a). *Closing the loop – An EU action plan for the circular economy*. http://eur-lex.europa.eu/resource.html?uri=cellar:8a8ef5e8-99a0-11e5-b3b7-01aa75ed71a1.0012.02/DOC_1&format=PDF (accessed March 30th, 2020)

European Commission. (2015b). Directive (EU) 2015/720 of the European Parliament and of the Council of April 29th, 2015 Amending Directive 94/62/EC on Packaging and Packaging Waste to Reduce Non-Consumption of Lightweight Carrier Bags. https://eur-lex.europa.eu/legal-content/EN/TXT/?uri=CELEX:32015L0720 (accessed May 04th, 2021)

European Commission. (2020). *Circular Economy: Implementation of the circular economy action plan*. https://ec.europa.eu/environment/strategy/circular-economy-action-plan_en (accessed August 08th, 2021)

Gabrys, J., Hawkins., G. & Michael. M. (2013). *Accumulation. The material politics of plastics* (pp. 1–14). London and New York: Routledge.

Geyer, R., Jambeck J., R., & Law, K., L. (2017). Production, use, and fate of all plastics ever made. *Science Advances, 3*, 31700782. https://www.science.org/doi/10.1126/sciadv.1700782

Gille, Z. (2010). Actor networks, modes of production, and waste regimes: reassembling the macro-social. *Environment & Planning A*, 2010-05, *42*(5), 1049–1064. https://doi-org.ezproxy.lancs.ac.uk/10.1068/a42122

Given, L. M. (2012). Social context. In: *The SAGE encyclopedia of qualitative research methods*. Thousand Oaks: Sage Publications. https://dx.doi.org/10.4135/9781412963909

Greenpeace. (2019). *We're going after Nestlé. Here's why*. https://www.greenpeace.org/international/story/21712/were-going-after-nestle-heres-why/ (accessed August 17th, 2021)

Hetherington, K. (2004). Secondhandedness: Consumption, disposal, and absent presence. *Environment and Planning. D, Society & Space, 22*, 157–173. https://doi-org.ezproxy.lancs.ac.uk/10.1068/d315t

Hird, M. J. (2015). Waste, environmental politics and dis/engaged publics. *Theory, Culture and Society, 34*(2–3), 187–209. https://doi.org/10.1177/0263276414565717

Hisano, A. 2017. Cellophane, the new visuality, and the creation of self-service food retailing, *Harvard business school general management unit working paper no. 17–106*. http://dx.doi.org/10.2139/ssrn.2973544.

Hughes, R. (1998). Considering the vignette technique and its application to a study of drug injecting and HIV risk and safer behaviour. *Sociology of Health & Illness, 20*(3), 381–400. https://doi-org.ezproxy.lancs.ac.uk/10.1111/1467-9566.00107

Kakadellis, S., & Harris Z. M. (2020). Don't scarp the waste: The need for broader system boundaries in bioplastic food packaging life-cycle assessment. A critical review. *Journal of Cleaner Production, 274*(2020), 122831, 1–13. https://doi.org/10.1016/j.jclepro.2020.122831

Kirchherr, J., Reike, D & Hekkert, M. (2017). Conceptualising the circular economy: An analysis of 114 definitions. *Resources, Conservation and Recycling, 127*, 221–232. https://doi.org/10.1016/j.resconrec.2017.09.005

Lacy, P., Long, J. & Spindler, W. (2020). *The circular economy handbook: Realising the circular advantage*. London: Palgrave Macmillan.

Latour, B. (1988). *The pasteurisation of France*. Cambridge: Harvard University Press.

Latour, B. (1991). Technology is society made durable. In: *A Sociology of Monsters: Essays on Power, Technology and Domination*. London and New York: Routledge. 103–131.

Law, J. (2004). *After method: Mess in social science research*. London: Routledge.

Liboiron, M. (2016). Redefining pollution and action: The matter of plastics. *Journal of Material Culture, 21*(1), 87–110. https://doi.org/10.1177/1359183515622966

Minchirina, B. P., & Richards, C. (1996). *Person to person. Fieldwork, dialogue, and the hermeneutic method*. New York: State University of New York Press.

Moore, C. (2011). *Plastic ocean: How a sea captain's chance discovery launched a determined quest to save the oceans*. New York: Penguin Random.

O'Brien, M. (2008). *A crisis of waste? Understanding the rubbish society*. Oxford: Routledge.

Plastics Europe. (2019). *Plastics – The facts 2019*. https://www.plasticseurope.org/en/resources/publications/1804-plastics-facts-2019 (accessed October 7th, 2021)

Qu, S., Guo, Y, Chen, W-Q, Liu, J, Ganga, L. . . . Xu, M. (2019). Implications of China's foreign waste ban on the global circular economy. *Resources, Conservation & Recycling*, 144, 252–255. https://doi.org/10.1016/j.resconrec.2019.01.004

Schröder, P., Bengtsson, M., Cohen, M. & Dewick, P. (2019). Degrowth within: Aligning circular economy and strong sustainability narratives. *Resources, Conservation and Recycling*, 146, 190–191. https://doi.org/10.1016/j.resconrec.2019.03.038

Stowell, A. F., & Brigham, M. P. (2018). Extractivism, value and waste: Organizational mining of e-waste in the United Kingdom. *Etnografia E Ricerca Qualitativa*, (1), 75–95. 2018

Stowell, A. F. (2012). *Organising e-waste*, PhD. Lancaster University.

Thompson, M. (1979). *Rubbish theory: The creation and destruction of value*. Oxford and New York: Oxford University Press.

United Nations Environment Programme. (2018). Single-use plastics. *A roadmap for sustainability*.

Whiteman, G., Walker, B. & Perego, P. (2013). Planetary boundaries: Ecological foundations for corporate sustainability. *Journal of Management Studies*, 50(2), 307–336. https://doi-org.ez proxy.lancs.ac.uk/10.1111/j.1467-6486.2012.01073.x

Wright, C., Nyberg, D., De Cock, C. & Whiteman G. (2013). Future imaginings: Organising in response to climate change. *Organisation*, 20(5), 647–658. https://doi-org.ezproxy.lancs.ac.uk/10.1177/1350508413489821

Yin, K. R. (2011). *Qualitative research from start to finish*. New York: Guilford Press.

Kim Poldner, Domenico Dentoni

18 Aesthetic engagement: material practices of organising towards regenerative futures

Abstract: We propose aesthetic engagement as a valuable construct for organisation studies to advance its contribution to organising for sustainability. Aesthetic engagement is defined as a set of material practices that re-engage humans and systems to trigger and accelerate transitions towards regenerative futures. We adopt an aesthetic, practice-based approach to study the emerging field of circular fashion, zooming in on six research projects evolving around bio-based textile design. Our results show that matter needs to matter more in sustainable organising in three key material practices: (1) re-presenting alternative systems, (2) re-imagining affective materialities and (3) re-claiming embodied ethical agency. Matter that reflects new 'imagined' realities – whether in artefacts, bodies or socio-material spaces – could greatly support stakeholder engagement and collective identity-building towards transitioning to regenerative futures.

Keywords: aesthetic engagement, circular fashion, practice-based research, materiality, regenerative futures

Vignette 1: mycelium shoe

A slightly tense bioprocess engineer awaits us in her lab. 'I can't show you much yet, the material grows really slowly.' It's the end of a long day of filming on campus. Scientists and students have explained their research projects: some of them in shy English, others more confident, convinced of the value they have created with their innovations. We are tired but have been asked to drive to a different location to film the last scientist. When we arrive, the late afternoon sun shines into the glasshouse and stimulates us to quickly take some contextual shots. We must put on lab coats and wear protective glasses to be allowed in her lab. 'Don't worry,' we say: 'do your thing with the mushroom and we'll film you – it's about the process, not the final product.' Afterwards, we position ourselves in the middle of an open space and let her do the talking: camera runs.

Four weeks later, she whatsapps me: 'I can't do it Kim; I don't have enough material and will have to withdraw from the exhibition.' I try to invigorate her as I have done often over the past few months. Recently graduated, she only just started her business and is making a steep learning curve in entrepreneuring. Things are going fast for her, too fast. With the requests she gets for media interviews, to collaborate

https://doi.org/10.1515/9783110723373-021

and to participate in contests, there is barely any time to spend in her lab, to develop the material that sparked her story. 'Leave it to me' the shoe designer says when I mention her concern to him during a meeting that morning. 'I'll visit her lab tomorrow and will talk to her.'

One day before the exhibition opens, she sends me a photo that speaks for itself. The first vegan shoe with mycelium straps, ever. 'WOWWWW!' I write 'Congratulations!' She is happy and relieved: 'He encouraged me to give it to him – even though I thought it was not a perfect sample. And then he made this.' The power of the imaginary.

Figure 18.1: Mycelium shoe.
Photo: Kim Poldner, Iris Houthoff.
Note: The bioprocess engineer being filmed in her lab. The first mycelium shoe.

Introduction

What if the future of fashion could be regenerative? What if shoes could be made from a bio-based material such as mycelium? A material that is biodegradable nurturing the soil after use instead of polluting the planet through incineration or waste generation? From an environmental standpoint, fashion is responsible for 10% of global CO_2 emissions and 40% of clothing produced is unsold and discarded as waste (Ellen McArthur Foundation, 2017). From a social standpoint, a disaster like the Rana Plaza factory collapse in 2013 that killed 1134 industry workers has led to a fashion revolution mobilising millions of people to become aware who made their clothes. Unfortunately, textile companies are still caught in conventional supply chains and models of business as usual (Niinimaki, 2018). As a response to the status of the industry, sustainable fashion has been 'in fashion' since the turn of the century, gradually leading to moving sustainability into the high street fast fashion landscape (Niinimaki, 2018). Several studies have focused on how sustainability adds value to a fashion company (Hill & Lee, 2015; Lueg et al., 2015). Still, many problems persist in the business of 'doing fashion' and thus more recently, circular thinking has been adopted as a promising approach to reconsider, redesign and restructure textile supply chains (Fischer & Pascucci, 2017).

We adopt a definition of circular fashion as clothes, shoes or accessories that are designed, sourced, produced and provided with the intention to be used and to be circulated conscientiously and efficiently in society for as long as possible in their most beneficial form, and at the end, will return safely to the environment when they are of no use anymore (Niinimaki, 2020). For over 200 years, our economy has been based on degenerative design, taking Earth's precious materials, making them into stuff we want, using them for a while and then trashing them. This one-way system that runs against Earth's cyclical processes of life destroys the living systems on which we, as humans, fundamentally depend (Raworth, 2017). To make our society regenerative by design is the greatest design challenge for twenty-first-century professionals; this chapter unravels how fashion and textile designers enact three key material practices of organising towards regenerative futures: (1) re-presenting alternative systems, (2) re-imagining affective materialities and (3) re-claiming embodied ethical agency.

In particular, we zoom in on a handful of pioneers that have been experimenting with innovative technologies of designing fabrics, i.e., by growing them from fungi, developing them from food waste or dyeing them with bacteria. Starting to theorise from our deep embeddedness in the circular fashion context, we introduce the concept of aesthetic engagement as a set of practices that re-engage humans and systems to trigger and accelerate transitions towards regenerative futures. Specifically, we propose aesthetics as a tool to invigorate not just the 'what,' namely the spotlight on the practices of aesthetic engagement, but also the 'how' as a form of aesthetic inquiry to decipher practices and artefacts.

We aim to add value to the discourse in two ways. First, the main theoretical contributions of this study are introducing the construct of aesthetic engagement, describing the material practices of changemakers in circular fashion, showing how their practices fit with the construct, and identifying ways in which aesthetic engagement can contribute to regenerative futures. Second, methodically we invite scholars who are impassioned about studying organisation in the context of sustainability, to consider an aesthetic practice-based approach (Nicolini, 2017) in grappling with novel phenomena. The chapter is structured as follows. First, we unpack the key concept of aesthetic engagement and explain its relation to aesthetics, sustainability and organisation. Second, we elaborate on the research setting and data analysis. Third, we disclose the findings after which we turn to discussion and conclusion. Throughout the chapter, we illustrate our argument with vignettes and visuals to enhance the storytelling capacity and agency of the materials at the centre of this study.

Theory

The following section first provides insight into the relation between aesthetics and the organisation as discussed in the literature, thereby disclosing the current disconnect with the sustainability and CE discourse. After that, the concept of aesthetic engagement is contextualised and proposed to establish a conceptual bridge between the study of aesthetics in organisational lives and the necessary transitions towards more regenerative futures.

Aesthetics, sustainability and organisation

The dominant paradigm in business is that of instrumental rationality, which is valuable as it is geared to the external goods of efficiency and productiveness. Unfortunately, it also often underestimates the reality of practices, which are linked to the internal good of beauty in community caring, compassion and sustainability issues. Until now, a fully fledging strand of literature (Stigliani & Ravasi, 2018) focuses on the role of design and aesthetics in multiple facets of organisational life (Stigliani & Ravasi, 2012). More recently, studies have illustrated how material objects and bodies carry agentic potential, in interaction with human agency, in triggering change at multiple scales of organisations (Pascucci et al., 2021), institutions (De Vaujany et al., 2019) and society (Barinaga, 2017). What is persistently missing, though, is a deeper analysis of the role of aesthetics in the broader landscape of sustainability challenges (Ferraro et al., 2015; Waddock et al., 2015); and, more specifically, an analysis of how human-material entanglements can trigger and support regeneration in socio-ecological systems.

We build on the notion of 'aesthetic rationality' as being complementary to instrumental rationality (Shrivastava et al., 2017) to forge organisations equipped to navigate and cope with paradoxes inherently emerging for those seeking to support transitions towards more sustainable systems. As such, the chapter adopts an 'aesthetic awareness' approach (Strati, 1999) by interpreting organisational life as a multi-sensory experience (Shrivastava et al., 2017). Beyond words and rational processes of deliberation, organisations and its members influence each other through emotions generated and shaped by sounds, colours, shapes and bodies. The aesthetic turn has "highlighted shortcomings of causal theories of organizing" (Strati, 1999, p. 13) and focuses on corporeality (Linstead & Höpfl, 2000), sensory experience (Taylor & Hanson, 2005, p. 1212) and the beautiful (Strati, 1999). Within the sustainability discourse, the literature has mainly focused on cognitive understanding of external spaces (land, air, water and people) and not on the internal spaces of the human mind and emotions (Shrivastava, 2011, pp. 1–2). Rational knowledge creation has been favoured over intuitive and emotional development. Conversely, the cultural (Kagan, 2013) and aesthetic (Harper, 2017) sustainability literature involves the inclusion of the arts as a form of aesthetics into "informing the transition to more sustainable practices" (Kagan, 2013, p. 16). This literature often associates the concept of aesthetics with ecology, thus linking ecological aesthetics with cultural sustainability (Nassauer, 2004). Aesthetic sustainability humanises the concept of sustainability by shining light on how we as humans emotionally connect with the objects that surround us in everyday life.

We build upon this notion of aesthetics as a form of organisational artefacts that can shift the paradigm (Waddock et al., 2015) and contribute to processes of change beyond organisational boundaries (Waddell et al., 2015). From this sustainability perspective, while aesthetics are recognised to be the key to trigger transformational change, little investigation has taken place on the practices that spur emotional values that appeal to, nourish and engage the users. A notable exception is represented by the study done by Akemu et al. (2015), which highlights the role of artefacts as 'boundary objects' that allow previously disconnected agents in a system to collectively re-imagine the future of an industry through a vivid 'visual feel' (e.g. an alternative phone that shifts people's values inherent to the original product). In this chapter, we extend the notion introduced by Akemu et al. (2015) to comprehend multiple forms of aesthetic engagement beyond the visual, as we discuss in the following sub-section.

Aesthetic engagement

Initially coined by Berleant (1991), aesthetic engagement involves active participation in the appreciative process and can be perceived as the opposite of the traditional claim of aesthetic disinterestedness. Berleant suggests an opposing theory of aesthetic perception based on the notion of engagement. We distil three elements from the literature on aesthetic engagement. First, applying the concept to a range of different art forms led to the acknowledgement that the arts evoke experiences with their own claims to reality (Berleant, 1991). Second, aesthetic engagement emphasises the holistic contextual character of aesthetic appreciation. It involves active participation in the appreciative process, sometimes by overt physical action, but always by creative perceptual involvement (Berleant, 2013). Third, aesthetic engagement acknowledges aesthetics' etymological origins by highlighting the primacy of sensible experience (Berleant, 1991). Aesthetic engagement is commonly used in philosophy and the art world but, to the authors' knowledge, not yet applied in organisation studies.

We borrow and adapt the notion of aesthetic engagement to the context of organisation studies to advance its contribution to organising for sustainability. Our definition is that aesthetic engagement means *engaging* humans, organisations and systems through the aesthetic materiality of artefacts (Bruggeman, 2018). We see that contemporary critical fashion practices can help de-territorialise fashion's dominant structure to affirm ways of seeing the world differently and to craft alternative systems (Bruggeman, 2018). This study aims to unveil these practices of aesthetic engagement. We argue that a designer's degree of aesthetic engagement may affect his or her commitment to, and practice of, sustainability. A study of circular fashion pioneers is especially useful for exploring an aesthetic engagement approach as their *designerly* way of knowledge creation 'naturally' takes the interaction between materiality, sustainability and organisation into account.

Vignette 2: 3D-printed dress

Imagine yourself ten years from now. You wake up, brush your teeth, have breakfast and get ready for work. On your way to work, you drop your garbage at a waste collection hub. At the end of the day, you pass by the hub again to pick up the cellulose 3D filament that has been created out of your waste. You go home and do a body scan in front of the mirror to measure your exact sizes. Then you quickly design your dress online, and soon enough your 3D-printed outfit is ready to wear, off to the party!

Research setting and data creation

Using a practice-based approach (Nicolini, 2013), we zoom in on six projects that were part of a large research collaboration between a life science university and a university of the arts (see Table 18.1 for the overview of the six projects). One of the projects is called Living Waste, which is captured in a graphic narrative representing

Figure 18.2: Obsolete catwalk.
Photos: Sven Menschel.
Note: Entrance. Ludi Naturae dress Iris van Herpen. Apparatus catwalk installation.

a new system of cellulose extraction from waste which can then be used to 3D print your own dress (Vignette 2, see Figure 18.2). The objective of the projects was to develop innovations that can disrupt the fashion industry towards more regenerative futures. First, we explicate our practice-based, aesthetic approach to inquiry. Second, we narrate about our process of data creation, and third, we elaborate on how we went about the analysis.

Table 18.1: Overview of the projects.

Project	Topic	Time	Exemplary prototypes from the research projects
Living Systems	Biomimicry and sustainable fashion principles	January–February 2018	A range of drawings depicting biomimicry innovations A set of black/white microscopic photographs showing the wax structure of the juniper tree
Living Leather	Bio-based alternatives for leather	January–February 2018	A range of small different fruit leather samples, such as leather made from mango and orange peel
Living Skin	Sustainable skin-like materials	November–December 2017	An iphone case from mycelium The first small samples of mycelium 'leather'
Living Colours	New ways of textile dyeing	March–April 2018	Degradation of the purple waste ink colour to obtain green
Living Waste	Extracting raw materials from waste streams	September–October 2017	A graphic narrative representing a new system of cellulose extraction from waste, which can then be used to 3D print your own dress (Vignette 2)
Fresco ensemble	A combination of Living Leather, Living Skin and Living Colours	March–September 2018	Four items designed by a total of six designers, thereby employing four innovations leading to a dress, scarf, shoes and watchband

A practice-based approach

To help us study and analyse the emerging phenomenon of circular fashion, we depart from a practice-based approach both in theory and method (Gherardi, 2012; Nicolini, 2013; Schatzki, 2006). This section will explain our ontological and epistemological foundations and how we employ a practice-based perspective in this empirical study (Bourdieu, 1990). We give depth to our zooming in/zooming out practices by adopting four strategies of studying practices: situational, genealogic, configurational and conflict-sensitive orientations (Nicolini, 2017). The situational refers to "the analysis of the concerted accomplishment of orderly scenes of action." The genealogic captures "the study of how individual practices emerge and disappear," the configurational examines "how concerted accomplishments hang together to form constellations and what consequences descend from this" and the conflict-sensitive inquires "into the co-evolution, conflict and interference of two or more practices" (Nicolini, 2017, p. 25). We rely on Nicolini who argues that these strategies, which build on the different traditions that go under the umbrella term of *practice-based*, allow practice theory to present a view of the social that is richer, thicker and more convincing than that offered by competing paradigms.

Aesthetic inquiry

The understanding of a creative, aesthetic approach to sustainability requires a new epistemological stance since it is associated with transformation in worldviews and paradigmatic bases (Kagan, 2013). This epistemology is based on Morin's understanding that "complex knowledge contains its own reflexivity" (Morin, 1992, p. 13). This reflexivity concerns the individual researcher's subjectivity, which must be acknowledged and made visible (Kagan, 2013). Phenomenological, social-constructionist inquiry seeks to describe phenomena beginning with how one experiences them. When studying aesthetics, *how* researchers question is at least as important, perhaps even more important, than *what* they know. The qualities of presence scholars bring to their inquiry influence both what they and 'the others' can understand and how they act (Ivanaj et al., 2014). In this study, the authors sought a deep understanding of sustainability and creativity in organisations that requires experiencing it in an embodied way (Shrivastava, 2011, p. 7). Aesthetic inquiry helps to apply this new form of knowing by examining creative aesthetic artefacts and analysing them to render plausible inter-subjective judgements (Shrivastava & Ivanova, 2015). Aesthetic inquiry is not just a form of qualitative research since there is an essential element of 'subjectivity' in the sensorial and emotional experiencing of the artefacts (Poldner et al., 2016). After our elaboration of the methods we use, we now turn to our research setting.

Vignette 3: obsolete catwalk

A hot afternoon in June. Industrial premises located next to the river Rhine. A big gate marks the entrance of an old milk factory on a barren piece of land. A sense of squatter camp dooms when you see several unusual looking people wading through the high weeds, appearing to be busy searching for something. The new luxury?

You've bought your ticket and push aside the vibrant plastic strips to enter cool darkness. Steel staircase constructions, a large video screen and friendly volunteers await you inside. 'Welcome,' they say. To a new fashion world where new values are given shape. You enter the first space: clean and light. Captured by a dress, you try to figure out what material it is made of. A video provides wordless explanation in an interplay of 3D technology, beauty and materiality. Your hands are itchy, but it's prohibited to touch the fabric. Far too exquisite.

Through a dark alley with cavities separated by black curtains, you continue to explore the multiplicities of new luxury. A black Osklen ensemble made of pirarucu fish skin. Four H&M-conscious luxury dresses seem to be borrowed from a black and white movie. A G-star creation consisting of dozens of denim leaves patched together in an avant-garde ballroom dress. The Most Sustainable Jeans Ever, they state. So how unsustainable were they before? Finally, you find your way to a grid contrived of cubes in different sizes. Here you can touch things! Orange fiber. A recycled fish skin jacket. A pinatex coat. Algae-dyed threads. A solar-powered backpack. You indulge in the tactile experience of rubbing all these innovations between your fingertips. The last space hosts a catwalk with small stools to the side. Every full hour the show starts and you're lucky: it's 3:55 pm. In suspense, you sit down and wait; you're the only visitor in the room. Then suddenly, smoke is blown on stage. Nothing else happens and you stay seated. Bewildered. A new reality as a crack of dawn.

Data creation

Five of the research projects brought together scientists and students with designers in the field of textile, fashion and product design to intensively collaborate during periods of two months for each project between September 2017 and July 2018. In each project, they jointly developed new material and conceptual solutions to sustainably innovate fashion supply chains and wrote a comprehensive research report about their project. The results of the five projects (artefacts, reports, new technologies) were translated to expo materials and videos that were displayed during the international exhibition State of Fashion that showcased a circular vision of the future of fashion from 1 June until 22 July 2018 in Arnhem, The Netherlands (Vignette 3 describes one element of that exhibition, see Figure 18.3). Data collection took place specifically during the 6-week process in April and May 2018 and consisted of creating five short videos that visualised the research projects for a large audience. Five designers involved in a total of three of the research projects were invited to craft a circular ensemble for the executive president of the life sciences university. The entire process of developing this ensemble, including the research that needed to be done, the media attention it brought and the 'making of . . . ' video that was created, was included as the sixth project. Even though the videos have been transcribed and quotes have been sourced from these transcriptions, this is only the textual side of the data. We argue that this chapter can't be viewed separately from the six videos that serve as the core data for this study. In fact, we suggest that it is not possible to properly understand our main construct of aesthetic engagement without embedding (parts of) these six videos into the final publication of this chapter.

(Teara, n.d.)

Figure 18.3: Lichen dye – Growing colours on clothes (drawing by a student). Source: Authors.

The first author of this study was involved in the entire collaboration as a project leader – positioning herself "amid the scene of action" (Nicolini, 2017, p. 29). The action research approach and total immersion of the first author enabled in-depth insights into the processes and practices underlying the development of the circular fashion discourse. Throughout the process of conducting the "5 + 1" research projects and preparing for and then participating in the three events, the first author also supervised five MSc thesis projects and traineeships of her students within organisations that are key stakeholders in the landscape that was studied. The second author served as the critical outsider supporting the inductive process of early theorisation. While as co-authors we looked at and discussed the data together, we realised that the core data – the videos that were produced – presented coherent narratives, while there were many paradoxes at play. Perceiving paradoxes as bringing together different logics and ways of looking at things, we acknowledged that we had to include other sources of data.

Table 18.2 shows an overview of the frontstage/backstage data that we collected, divided in core data and contextual data. Core data focuses on the six videos (frontstage) and fieldnotes, emails and whatsapp messages related to the process of 'the making of the videos' (backstage). Contextual data comprises a collection of videos, photos, reports and media publications centred on three key events: the Fashion Colloquium, State of Fashion opening event and Textile Exchange conference – all three taking place and overlapping on 30 and 31 May and 1 June 2018. The first author was actively engaged in these events as a speaker, moderator and member of the scientific committee of the Fashion colloquium.

The frontstage/backstage perspective enabled us to unravel the discrepancies arising between the polished 'frontstage' performances – such as the five videos – and the 'behind the scenes' conversations and struggles. In addition to the primary

Table 18.2: Frontstage/backstage data.

	'Outside' data	Source	'Inside' data	Source
Core data	Videos of the five research projects	http://innovatie.artez.nl/2018/06/the-future-of-living-materials-state-of-fashion-t-m-22-juli/	The process of 'the making of . . .' these videos	Whatsapp contact, emails and phone calls – collected in Dropbox folder
	Reports of the five research projects	Dropbox folder	Coaching sessions with the research teams	Notes collected in Dropbox folder
Contextual data for 3 events	*Event 1: Fashion Colloquium (31 May–1 June)*			
	Website Fashion Colloquium	http://fashioncolloquium.artez.nl/	Reviewer academic & creative practice contributions	Documents and notes collected in Dropbox folder
	Live recordings days 1 and 2	httdp://innovatie.artez.nl/	Conversations with organisers, researchers, students and companies	Documents and notes collected in Dropbox folder

Event 2: State of Fashion opening (31 May, exhibition ran until 22 July)

Bricolage of business cards, folders, name tags and conference paraphernalia	 Photo: Authors.	
Photos & video opening StoF	https://www.facebook.com/pg/stateoffashion2018/photos/?tab=album&album_id=204444102255096 https://vimeo.com/274128740	Conversations with curator and producers of the StoF expo	Fieldnotes – collected in Dropbox folder
Website StoF	https://stateoffashion.org/nl/	Conversations with curator and producers of the StoF expo	Fieldnotes – collected in Dropbox folder

Event 3: Textile Exchange Conference (30 May)

Reports & photos of Textile Exchange meeting	https://certifications.controlunion.com/en/news-media/events/towards-circular-fashion https://textileexchange.org/event/towards-circular-fashion/	Conversations with textile companies, researchers, organisers and other stakeholders	Fieldnotes – collected in Dropbox folder

data collected, we also looked at the discourses that were crafted in media attention that was generated by the State of Fashion exhibition. Table 18.3 shows a non-comprehensive overview of this source of data.

Table 18.3: Media attention State of Fashion.

Name	Media type	Date	Source	Type of data	Time/ min
Modemuze	National platform	February 2018	https://www.modemuze.nl/ agenda/state-fashion-2018	Article and photos	
NRC	National newspaper	May 25	https://www.nrc.nl/nieuws/ 2018/05/25/dit-zijn-de-beste-nederlandse-modefestivals-van-deze-zomer-a1603974	Article and photos	00:24
WUR	Website	May 30	https://www.wur.nl/en/news-wur/Show/Sustainable-cloth ing-made-from-fungi.htm	Article and photos	–
Omroep Gelderland	Regional newspaper	May 31	https://www.omroepgelder land.nl/nieuws/2316632/Inter nationale-mode-event-State-of-Fashion-van-start	Article and video	00:56
NPO 1	National radio	May 31	https://www.nporadio1.nl/we tenschap-techniek/9875-zo-maak-je-mode-duurzaam	Interview with Jose Teunissen & Anneke Smelik	7:09
NPO 1	National radio	May 29	httpaps://www.nporadio1.nl/ langs-de-lijn-en-omstreken/on derwerpen/457669-state-of-fashion-met-mode-de-wereld-mooier-maken	Interview Bert van Son (Mud Jeans) and Liselore Frowijn	14:38
AD	National newspaper	May 31	https://www.ad.nl/video/pro duction/state-of-fashion~ vp39105	Video	02:44
FD	National newspaper	May 31	https://fd.nl/economie-polit iek/1256583/in-beeld-state-of-fashion-2018-in-oude-melkfabriek	Article & photo	–
Omroep Gelderland	Regional newspaper	June 1	httpls://www.youtube.com/ watch?v=7FhAfwYKgBs	Video	01:34
Duurzaam Nieuws	Website	June 5	https://www.duurzaamnieuws. nl/circulaire-mode-leer-maken-uit-fruitafval/	Article and photos	–

Table 18.3 (continued)

Name	Media type	Date	Source	Type of data	Time/ min
Elsevier	Magazine	June 8	https://www.elsevierweekblad. nl/stijl/achtergrond/ 2018/06/arnhem-modestad- kleren-zonder-schade- 136526w/	Article and photos	–
AD	Website	June 19	https://www.nrc.nl/nieuws/ 2018/07/02/paddestoelenleer- wordt-de-nieuwe-luxe- a1608562	Video	02:22
Eenvandaag	National TV	June 28	https://eenvandaag.avrotros. nl/item/een-t-shirt-van-algen- of-liever-een-jas-van-een- vrachtwagenband/	Video	04:18
Blog Fresco	Website	July 3	https://www.nrc.nl/nieuws/ 2018/07/02/paddestoelenleer- wordt-de-nieuwe-luxe- a1608562	Article	–

The data reflected in these two tables provided us with plenty of material to build a coherent story for this study.

Analysis

We were so captivated by the phenomenon as a potential empirical source of finding new insights in organising for regenerative futures that we started with making general sense of the data by watching and reading all data sources. Table 18.4 shows the three-staged analysis process, revealing how the discovery surfaced (Stage 1) that aesthetic, sensorial, performative enactments can have a powerful impact on systemic change towards sustainability.

In Stage 2, we took Nicolini's orientations as our conceptual lens to a second-order coding of the data (Gioia, 2013). We conducted an aesthetic inquiry, which unravelled the three key practices of aesthetic engagement. In this stage, we selected boundary objects that best exemplified the key practices. We realised that we wanted to show real 'products' and not just the small prototypes that were developed during the research projects. Our choice was limited, as only a few 'products' have been created over the course of this study. The mycelium shoe, for example, did not end up in one of the tables, as it was already accompanying Vignette 1. Table 18.5 provides an

Table 18.4: Methodological overview of analytical steps.

	What	How	Assumptions	Questions asked	Outcome
First stage of analysis	Understanding circular fashion	Project leader (first author) Full immersion in the data (second author)	Circular fashion is a unique phenomenon that can shed light on the engagement of business and organising for sustainability	In what way is circular fashion different from ethical fashion? How can breakthrough textile/supply chain innovations spur achievement of sustainability? What is the discovery?	Tables 18.1 and 18.2 – data sources The discovery is the construct of aesthetic engagement as a way of re-engaging humans, artefacts, systems and organisations in alternative ways.
Second stage of analysis	Identifying three key practices for aesthetic engagement	Empirically derived from the data through the conceptual lens of orientations (Nicolini, 2017)	Paradoxes inhibit the impact of aesthetic enactments on sustainability	What are the main practices of aesthetic engagement? When are they applied? Which practices can be analysed with Nicolini's orientations?	Three key practices of aesthetic engagement.
Third stage of analysis	Choice of boundary objects	Empirically derived from the data by means of aesthetic inquiry	Artefacts can serve as 'boundary objects' that allow previously disconnected agents in a system to collectively re-imagine the future of an industry (Akemu et. al, 2015)	Which artefact is exemplary for each key practice? Which sensorial, affective aspects of these artefacts are crucial in aesthetic engagement?	Three boundary objects that help to comprehend multiple forms of aesthetic engagement beyond the visual

overview of answers given to the different questions belonging to Nicolini's orientations (2017) and the resulting practices.

To avoid exceeding page numbers, we refer to the tables, as they provide detailed insight into the process of coding and analysis. Tables 18.6, 18.7 and 18.8 in the Appendix show the sub-practices and quotes from the data related to the three key practices.

Findings: practices of aesthetic engagement

Aesthetic engagement means re-engaging humans, systems and organisations through the materiality of artefacts. Our data disclose that the objective of moving the fashion industry towards regenerative futures requires redefining ways in which we engage with materiality, agency and systems. Materials play an active role in (1) re-presenting alternative systems, (2) re-imagining affective materialities and (3) re-claiming embodied ethical agency. Through these bundles of practices, circular fashion emerges from the relationships between humans who do research, craft artefacts, write texts and speak at events, and the materials they engage with.

1) Re-presenting alternative systems

We observe that designers depart from a first sub-practice of *problematising* the current fashion system and developing the objective to redesign supply chains. From this macro issue, they then zoom in on one material, such as leather, to substitute its designerly properties as well as technical properties. From the second sub-practice of *substituting*, they then move on to the third sub-practice of *collaborating* (meso-level) with scientists, as designers don't have the technical know-how needed to further develop these materials. These three sub-practices can be summarised by 're-presenting alternative systems;' the enactment of depicting a substitute for how the fashion system currently is organised in a linear way. Table 18.6 shows the coding for the three sub-practices and illustrative quotes.

The boundary object that is exemplary here is the ensemble that was developed for Louise Fresco, president of the executive board of Wageningen University & Research. The dress was made of rest ink printed on deadstock (unsold) silk; the scarf was dyed with bacteria on peace silk (vegan silk that does not harm the silkworms). The core of both items is the desire of the designers to come up with a more sustainable alternative for current toxic textile dyeing practices. They had developed the techniques of dyeing textiles, respectively, with rest ink and bacteria and through the collaboration with students and scientists within the project Living Colours, they could further refine their innovations. Due to the development of the chemical industry, textiles can be dyed and printed in all colours of the rainbow. Colour has

Table 18.5: Orientations (Nicolini, 2017) and practices.

Orientations	Questions	Answers (core data)	Answers (contextual data)	Practices of aesthetic engagement
Situational: Practice is experienced from within a particular scene of action, where several practices intersect and are knotted together (nexus)	Which practices am I observing? What are the practices that are circulating in and through this scene of action? Which practices are relevant?	Experimenting Collaborating Writing Presenting Explaining Designing Discussing	Collaborating Mingling Presenting Asking Discussing Eating & drinking Exchanging	Aesthetic engagement means re-engaging humans, artefacts, systems and organisations in alternative ways
Genealogic: Practice emerges from or is constituted by the association of three elements: meaning, skills and tools	How are these elements associated? By whom and under what conditions do they become a practice?	Focus is on building the narrative of the value that is created through interdisciplinary collaborations between scientists and designers	Main narrators of the circular fashion narrative are the actors that organised the events as they decide who were keynote speakers/ performers	Three core practices 1. Re-presenting alternative systems
Configurational: Practices form constellations and broader configurations	How do social phenomena transpire amid and through constellations, bundles and regions of practice? What other material practices and localised performances can we distinguish?	Circular fashion arises through bundles of practices created by humans who do research, craft artefacts, write texts and speak at events		2. Re-imagining affective materialities 3. Re-claiming embodied ethical agency

Table 18.5 (continued)

Orientations	Questions	Answers (core data)	Answers (contextual data)	Practices of aesthetic engagement
Conflict-sensitive: Inquiry into the co-evolution, conflict and interference of two or more practices	Are these practices aligned among themselves? Are they good at the purpose they were set up to serve? How does a practice gain superiority over a competing one?	Conflict arises when novel practices arise, but are not understood or properly enacted (i.e. organising a fashion expo that is different from others); collaborating is still often more like competing		

become cheaper and more democratic, but designers develop new colour palettes using micro-organisms or rest ink. The rest ink is a residual flow of digitally printing on textiles: a dark purple solution that remains out of ink from the eight standard colour cartridges of industrial printers. Currently this residue goes to chemical waste, but it is a valuable raw material with various interesting properties and potential applications. As designer Aliki van der Kruijs explains:

> In my own textile printing, I discovered that there is a residue, a waste stream, generated by the inkjet printers and I am researching this purplish colour to see if I can bring it back into a stream and to give it value.

She proposes an *Afterseason*, integrating material harvested from the industrial waste stream of the previous collection: *"A range of colours not chosen on prognosis or intuitive base but as a result from the 'previous' season."* The Fresco ensemble is the exemplary boundary object here as it represents 'an alternative system.' The goal of the project was to bring several innovators together to create an entire circular outfit for Fresco: to simply show that it is possible to develop *real* clothes out of these material innovations. When we proposed the project to Fresco, the one requirement she had was that she should feel comfortable in the outfit and that it should not fall apart during the day, start smelling or otherwise be uncomfortable to the skin. Fresco wrote a newspaper article about her (and first author's) journey of developing the outfit in which she explains the choices we had to make and the trade-offs we had to consider coming to the final range of products. The video of 'the making of . . .' shows how the outfit was created and which failures we encountered during the process. The ensemble thus becomes a boundary object for re-presenting alternative systems; systems in which scientists work closely together with designers on textile innovations, systems in which chemical dyeing practices are being replaced by more sustainable alternatives and systems in which different material innovations are brought together to craft an entire circular outfit.

2) Re-imagining affective materialities

The collaboration between scientists and designers (meso-level) is centred around developing the new materials (micro-level) to answer the required system changes (macro-issues). The focus on these new materials supports the second key practice of aesthetic engagement: re-imagining affective materialities. Innovators focus on matter in three sub-practices (often employing all three ways): (1) starting from, (2) co-creating with and (3) returning to matter. Table 18.7 shows the coding for these three sub-practices and illustrative quotes. The data – and especially the social science questions asked within each research project – shows that the creation of novel bio-based materials strongly affects stakeholders, not just the designers that develop them but also consumers who are curious and at the same time experience it as very scary to think of having to wear fungi leather on their skin.

First of all, in search of closing the loop in textiles, protagonists experiment with new sources of matter (sub-practice 1), for example, departing from waste. The project 'Living Waste' looked at distilling cellulose from industrial and agricultural waste streams. As the scientist involved in this project explained: "For instance, they looked at rice husks and they extracted cellulose out of that in an enzymatic way, a biological way or a chemical way." The cellulose could then be upcycled into new textile products. Designer Aliki van der Kruijs started in a similar fashion with the waste from inkjet printers within the project Living Colours. In the projects Living Leather and Living Skin, we also see how our protagonists depart from a certain matter; in both projects, new 'leather/skin'-like materials were developed from substrates such as fruit peel or mycelium (a mushroom/fungi type).

The chosen boundary object for the practice of re-imagining affective materialities is a shoe and purse created by Emma van der Leest, made of kombucha, a fermented type of green tea drink. The designer explains and shows in the video how she creates this 'living skin' material, starting from matter, which she frames as the mother culture:

> Bacterial cellulose is made by bacteria and yeast in a fermentation process and it basically grows you a 'skin' on the top in your container. If you dry it down, you get something like a leather material. So this is a culture of bacterial and yeast and we call it the mother culture.

Technology plays a crucial role in the practice of starting from matter, as we see in the contextual data with renowned designer Iris van Herpen who always departs from the question: *"How can I adjust my way of designing to how the material is being developed?"* Her couture piece showcased at the exhibition (see Figure 18.2) pushed the limits of what has been done with 3D printing in terms of the fabrication of garments. For the design of the piece, van Herpen worked with scientists from TU Delft,[1] who stated:[2]

> During the printing process, ultraviolet light was used to cure the structures, making them set . . . Before this, no one had succeeded in effectively combining plastic with different properties with textile and we're proud of what we've achieved . . . Normally, you would start with a design With a project like this, you take a new production method and see what's possible. It's venturing into uncharted territory. (3d printed 2021)

Protagonists start with matter, but then need to work together with this matter to develop the desired material (sub-practice 2). Emma van der Leest explicates in the video what her process of co-creating with matter looks like:

> The mother culture spins little threads of pure cellulose in this fermentation process and over time it will give you a material, which eventually can be used as a leather material in different products.

In the case of Living Colours, the bacteria need to be treated in a specific way for them to 'dye' textiles. As designer Laura Luchtman explains in the video:

> what we do is that we grow the bacteria in a petri dish (. . .) and then we put the liquid medium in a petri dish (. . .), add the textile to it and then add a little bit of the bacteria to it. (. . .) And then we just grow the bacteria on the textiles. The bacteria are live dyeing the textiles, they are growing on top of the textiles and dyeing the textiles at the same time.

Here we observe the capacity of the matter to grow by itself, but it needs to be supported in doing so. In the project Living Leather, protagonists take this one step further by really manipulating the matter to become material as one MSc student plant biotechnology and molecular breeding explains in the video:

> The first step is to take the fruits and juice them to remove as much of the water content as possible, so we tried many different types of fruit like banana, grape, mango, orange. We even tried some vegetables like turnip. The next step is then to blend the fruits to a nice pulp that is easier to work with. You then mix in the adhesive agents so then you have a nice paste that you can easily spread, which is the next step. And then you finally put it in an oven and then finally you can take it out and you can give it a coating with a material that would give it some water resistance and you have a product at the end of it. We managed to get something that you could possibly use.

Co-creating with matter requires, in some cases, more agency from the protagonist and in other cases, the matter shows more agency by growing by itself with the protagonist only slightly intervening.

---

1 https://www.tudelft.nl/en/.
2 https://www.tudelft.nl/en/2018/tu-delft/tu-delft-scientists-work-on-3d-printed-dress-designed-by-iris-van-herpen/.

Acknowledging "the embodied 'intelligence' of living botanical matter" enables a third sub-practice of re-imagining affective materialities, as materials can return to matter at the end of the use phase. We observe this practice with bioprocess engineer Iris Houthoff and her mycelium shoes (Vignette 1 and Figure 18.1) that could be turned into living materials again when composted. Designer Aniela Hoitink suggests:

> We don't need to abandon fast fashion, but we can embrace the common practice of wearing garments only five times and then trashing them by making our clothes out of materials that can be composted.

In the project Living Leather, students developed 'fruit leather:' an alternative to leather made from orange peels that could be composted after using it. In the case of Emma van der Leest as well as with most of the other protagonists, we see how they enact all three sub-practices: both starting from matter (waste or a certain material such as kombucha mother culture), co-creating with and returning to matter (compost).

3) Re-claiming embodied ethical agency

Building on the first two key practices of aesthetic engagement, the third key practice explores the role of the body and embodiment. We found that an in-depth understanding of 'embodiment' is a necessary precondition when moving towards alternative systems and imagining new realities. Material 'matter' immediately engages bodily 'matter' and bodies are required to engage in different ways, as active co-creators of regenerative fashion futures. Fashion is often perceived as a 'second skin' and an expression of identity. Personalising clothes is essential in fashion that revolves around expressing unique identities. Shaping one's second skin – materials that literally 'move on the body' – can be perceived as the ultimate form of personalisation. For example, drawing on biomimicry, the project Living Systems studied how micro-organisms such as lichen can directly grow colour on T-shirts while users are wearing it on their bodies (Figure 18.4).
Another example from that same project is explicated by plant scientist Anja Kuipers:

> Another example of biomimicry is the juniper tree, which produces a waxy layer on the leaves, which captures particulate matter from air. It could be an idea to impregnate T-shirts with that same type of wax, so that people wearing those T-shirts can also walk around and capture particulate matter from air.

Figure 18.5 depicts microscopic images of the leaves of the juniper tree that were created by the students in the research project. Fashion is about the intimate relationship between body and materiality, and we see how some fashion practitioners focus on *exploring these material encounters* (sub-practice 1). Instead of focusing on the creation of a visual aesthetics on the surface, they explore the relations and interactions between body and garment – as material itself. The collaboration between Aniela Hoitink

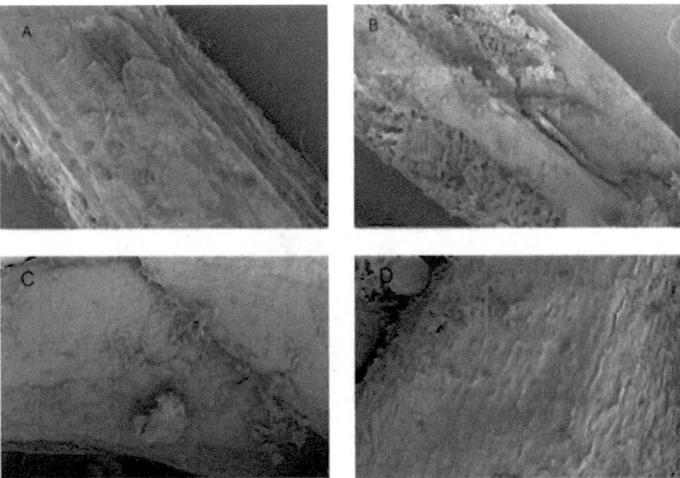

Figure 18.4: Overview pictures of the wax structure of the leaves (created by students).
Source: Authors.
Note: **A:** Leaf of the Chinese Juniper 'Aurea' young leaf, 1200x magnification. **B:** leaf of the Chinese Juniper 'Aurea' old leaf 250x magnification. **C:** Leaf of the Chinese juniper 'Blaauw' young, 500x magnification. **D:** Leaf of the Chinese Juniper 'Blaauw' Old. A FEI Magellan 400 microscope was used to take the pictures.

and Karin Vlug within the project Living Skin explores the boundaries of the body in engaging matter (see Table 18.8). As the designers explain in the video:

> So we start by growing mycelium, which is the root of mushrooms in order to create a new material, and then we use this material directly on a 3D mold. And so this is where our collaboration started. (Aniela Hoitink)

> For me as a fashion designer, I am developing the mold system and we're researching how these molds should be designed for Aniela to be able to grow her material on. Because it won't be traditional pattern making, so I am basically researching this new way of design, so it's more a design approach. (Karin Vlug)

Embodiment can be understood as the lived experience of human beings rather than reducing the body to the commodified objects to be looked at, which is common practice in fashion. A focus on embodiment in the field of fashion is always an ethical act, as many bodies do not have a voice in the system but are exploited as dehumanised and objectified instruments to produce fashion. As designer duo Vin + Omi, who exhibited at State of Fashion (contextual data), stated: "*When you buy at Primark, you make yourself complicit to injustice.*" Our data discloses that protagonists increasingly recognise fashion's human dimensions – embodiment, subjectivity, agency and human values – which lead to involving all human beings as active agents. These dimensions are fundamental for a more ethical engagement with each other and with material resources. For example, the exhibited project 'Conscious

Figure 18.5: Conscious contemporary tailoring by Fondazione Zegna.
Photo: Eva Broekema.

contemporary tailoring' (Figure 18.6) by Fondazione Zegna gave agency to all workers involved, focusing on their own personal development and interpersonal relationships instead of being primarily occupied with economic value and profit. As designer Bethany Williams explains (Conscious 2018):

> First off, the women in prison here in the UK created these really beautiful pieces and joint which were then sent to the women in San Patrignano as gifts with notes. And now the women in San Patrignano are responding with textiles, with letters to the women in the prison here in the UK. (. . .) The human interaction is as important as the outcome.

In other initiatives, we observe how agency and a strong emotional bond between subject and object, consumer and producer is crucial in shaping alternative systems and re-imagining affective materialities. Empowering people to make ethical and aesthetical choices (sub-practice 2) to limit overconsumption is yet another element of reclaiming the practice of embodied ethical agency. While the body is, in the current fashion system, rather passive, in this new system and new way of engagement, it becomes an active agent (sub-practice 3) in more ethical practices of purchasing, wearing and disposing of garments. Karin's study of crafting moulds that more accurately reflect real body shapes and Aniela's mycelium 'pads' that can be 'draped' around the body reflect the desire of designers to re-claim ethico-aesthetic agency. What we observe is that the third practice leads back to the first key practice of aesthetic engagement in a continuous loop. As Aniela says when speaking about mycelium and how it can be moulded around the body:

> It is also a new take on how you can, should in my opinion of course, start looking for new answers for our current supply chain.

To be able to return to matter involves the agency of the user, who needs to consciously take the decision to put the garment in the compost.

Discussion

Our discussion connects two different perspectives of contemplating on our empirical results. First, we deliberate on the specific role of artefacts in aesthetic engagement; second, we discuss the role of material practices in supporting regenerative futures.

The role of artefacts as boundary objects in organising sustainable transitions

Our discovery on the role of aesthetic engagement practices in the context of circular fashion relates to the broader conversation on the role of artefacts as boundary objects in catalysing action towards sustainability. Based on shared interpretations of an artefact (Zott & Huy, 2007), boundary objects play pivotal roles in triggering the development of collective identities (Wry et al., 2011) and supporting institutional change (Zietsma & Lawrence, 2010). In relation to the attempt of purposively disrupting an industry towards sustainability, for example, in the context of the cell phone sector (Akemu et al., 2016), "members of the effectual network self-select and pre-commit resources to the social enterprise not because they expect immediate calculative benefits from the artefact, but because the artefact embodies and symbolizes their beliefs and values" (Akemu et al., 2016, p. 872). Therefore, these boundary objects, as Nicolini et al. (2012, p. 614) observed, "have a 'deep emotional holding power' and are potent enough to mediate interactions and trigger commitment of resources and expression of values by these dispersed actors" (Akemu et al., 2016, 874). By eliciting emotions, boundary objects play a critical role also in crossing disciplinary boundaries to facilitate knowledge co-development and learning in emerging sustainable communities of practice (Benn et al., 2013).

Our findings from the phenomenon of circular fashion add a critical "*how*" perspective to this on-going conversation of the role of boundary objects for catalysing actors in supporting institutional change and addressing sustainability. Where are boundary objects coming from? And where are they going after their function as boundary objects has been fulfilled? Our discovery from the circular fashion industry is: from the matter. The boundary objects, per se, do not trigger nor support regenerative futures; instead, it is the entanglement between material and human agency that does so. Seeking the matter, moulding the matter and regenerating its use over time embeds the boundary object into a recursive process which shapes regeneration across scales. Through this recursive process around the material artefact, individuals, organisations and systems may engage, disengage and re-engage over time, depending on how the artefacts stimulate their senses. Therefore, relative to Benn et al. (2013) and Akemu et al. (2016), our discovery on the practices of aesthetic engagement opens new research questions around the role of artefacts as boundary objects towards addressing sustainability. Interesting questions for deeper theorisation in this domain may include, for example: how do practices of aesthetic engagement shape processes of stakeholder engagement, collective identity-building or institutional change towards addressing sustainability issues?

The role of material practices towards regenerative futures

The promise of the CE as an industrial economy aiming at enhanced sustainability through restorative intent and design is often inhibited by institutional barriers posed by the current linear economy of make, use and waste (Ghisellini et al., 2016). The report 'The Pulse of the Fashion Industry' (2017) was critiqued due to its focus on profit and growth, and the general anxiety is that circularity will become yet another tool to "feed the Ego of capitalist fashion" (Bruggeman, 2018, p. 68). With our study, we offer an attempt to distil practices derived from the empirical phenomenon of circular fashion to navigate this anxiety. We argue that aesthetic, practice-based approaches can help organisation studies to create radical ideas and alternative approaches on how to transition towards sustainability (Nicolini, 2017). We urge organisation scholars to take aesthetics into account when grappling with emerging sustainability challenges.

We expand earlier work that has been done on aesthetics and sustainability in organisation (Shrivastava, 2011; Shrivastava et al., 2017) by including affective materiality and embodied agency. We could argue that moving beyond the perceived dominance of the triple P (profit, people and planet) to include the cultural (Kagan, 2013) and aesthetic dimension of sustainability (Harper, 2017) no longer suffices. Instead, we should embrace matter as mattering the most. The recent cultural and philosophical theory of 'new materialism' can support giving special attention to matter and to the actual materiality of matter (Bruggeman, 2018). New materialism proposes a renewed focus on materiality that incorporates bodily matter as the "embodied or enfleshed subject" (Braidotti, 2011, 15). Matter, both bodily matter and material objects, has agential qualities (Barad, 2003): matter and meaning are mutually articulated in that "materials play an active role in relating to the body, in becoming garments and in creating meaning" (Bruggeman, 2018, 51). This philosophical perspective of "a new materialist aesthetics" is especially relevant in relation to the use of new bio-based materials or bacteria to develop or grow textiles or fashion objects (Bruggeman, 2018). For example, mycelium-based fabric can be seen as a new material 'category' (Sudjic, 2008), positioned between conventional wovens made from natural fibers (e.g. cotton and linen) or synthetic fibers and non-wovens (e.g. felt or leather-like materials). Yet, in some cases, these new materials don't have a direct relationship to fashion's human dimension and thus it is important to further explore the future of these kind of living materials for a more sustainable fashion system and society. As Harper (2017) suggests: we need a design strategy devoted to nurturing, maintaining and caring for things. Just like we need to craft more socio-material spaces such as State of Fashion that enable conversing, performing and enacting new practices.

The three key practices we distinguished can lead to a more comprehensive understanding of the emancipatory potential of organising (Calás et al., 2009; Rindova et al., 2009), societal change (Spinosa et al., 1997) and the effects on subjectivities in the making (Poldner et al., 2019). Markets do not exist, per se, but only through the many practices of assembling that bring them into being (Callon, 2007). To understand the human engagement with matter, it is important to further explore the affective and sensorial dimensions of sustainable development. In addition, the relationship between new living materials and the bodily matter of the wearer – the embodied subject – requires further research in the context of achieving sustainability.

Practical implication and limitations

The changing role of designers and the practices they start to enact challenges fashion schools to re-design their curricula. The Amsterdam Fashion and Management Institute has recently launched a new MA in Circular Fashion Business, while the Willem de Kooning Academy has started including courses on material design, following the practices of graduates such as Emma van der Leest and the Fruitleather designers. The ArtEZ MA in Fashion Design has recently been restructured by Pascal Gatzen, idealising 'back to basics' in a curriculum that focuses on practices of teaching students growing, spinning and weaving hemp. We already witness that those alliances built by industry stakeholders that collaborate around the creation of an artefact often display best practices in achieving their determined objectives. Our study has confirmed that matter that reflects new 'imagined' realities – whether in artefacts, bodies or socio-material spaces – could greatly support stakeholder engagement and collective identity-building towards transitioning to sustainable futures. A limitation is that we studied a single industry in a European context, which makes it worthwhile to reflect on how practices of aesthetic engagement would work in other contexts and different industries.

Conclusion

This chapter introduced the concept of aesthetic engagement as a way of re-engaging humans, artefacts, systems and organisations to accelerate transitions towards regenerative futures. More broadly, we see aesthetic engagement as holding potential to steer and accelerate sustainable transitions in less conventional – yet equally powerful – ways than technical, rational and discursive ways. Hence, our methodological invitation to organisation scholars is to consider aesthetic, practice-based approaches (Nicolini, 2017) as alternative methodologies when studying sustainability issues. Our findings show that matter needs to matter more in three practices: re-presenting alternative systems, re-imagining affective materialities and re-claiming embodied ethical agency. The first key practice is the enactment of depicting a substitute for how the fashion system currently is organised in a linear way by means of developing a new material 'category' (Sudjic, 2008). The collaboration between scientists and designers (meso-level) is centred around developing the new materials (micro-level) to answer the required system changes (macro-issues). The focus on these new materials supports the second key practice of aesthetic engagement: re-imagining affective materialities. Building on the first two key practices of aesthetic engagement, the third key practice explores the role of the body and embodiment. We found that an in-depth understanding of 'embodiment' is a necessary precondition when moving towards alternative systems and imagining new realities. Material 'matter' immediately engages bodily 'matter' and bodies are required to engage in different ways, as active co-creators of regenerative fashion futures.

Appendix

Table 18.6: Key practice 1 – re-presenting alternative systems.

First order constructs	Illustrative quotes	Boundary object	Exemplary second-order quote
Problematising the current system	"Since leather is quite a polluting process and it's harmful to the environment due to it creating between 30 to 40 billion liters of wastewater per year, globally." "Traditional pigments have quite a limited range of colours."	Circular ensemble designed for Louise Fresco, President of the Board of Wageningen University Designed by Aliki van der Kruijs 	"I am suggesting a new system, which I call the after season, so the idea is that you work with ink that you harvest from the previous printed collection. So there will also be an after season stream, generating this colour, and maybe this will become the new black, who knows?"
Substituting a specific material	"The aim of the project was to create an alternative material that would act as a substitute for leather." "If you find the right pigments that have the properties that you want and the colours that you want, we could technically substitute these very toxic and synthetic chemicals for natural compounds. . ."		

Table 18.6 (continued)

First order constructs	Illustrative quotes	Boundary object	Exemplary second-order quote
Collaborating with other disciplines	"If you want to innovate, you need to collaborate. It sounds like science fiction, but it's science fact as it is happening." "The first time I was working with designers, they could see a hat, they could think of new utilities, new ways – how to implement this material onto different type of products."		

Table 18.7: Key practice 2 – Re-imagining affective materialities.

First-order constructs	Illustrative quotes	Boundary object	Exemplary second-order quote
Starting from matter	"Cellulose is the most abundant organic compound found in plants, algae, fungi and bacteria. So therefore, we thought it is a good idea to use this cellulose into a fashion product."	Prototypes made from kombucha, designed by Emma van der Leest Photo: Sven Menschel 	"I know that in five years from now we will walk with a pair of shoes made from bacterial leather, or maybe a leather that has been grown in a lab."

Table 18.7 (continued)

First-order constructs	Illustrative quotes	Boundary object	Exemplary second-order quote
Co-creating with matter	"The mother culture spins little threads of pure cellulose in this fermentation process and over time it will give you a material, which eventually can be used as a leather material in different products."		
Returning to matter	"These materials provide nutrients for our soil, which in some countries is depleting. It's also a regenerative system in that way."		

Table 18.8: Key practice 3 – Re-claiming embodied ethical agency.

First-order constructs	Illustrative quotes	Boundary object	Exemplary second-order quote
Exploring material encounters	"Mylium is a bio-based material made from mushrooms that is flexible and feels soft *to the skin*. It's lightweight and quite strong."	The work of Aniela Hoitink and Karin Vlug, exhibited at State of Fashion Photo: Sven Menschel 	"The properties of mylium could find different applications such as handbags and shoes. The material looks very nice so it can find different accessories applications. The first time I was working with designers, they could see a hat, they could see new utilities how to implement this material onto different types of products."

Table 18.8 (continued)

First-order constructs	Illustrative quotes	Boundary object	Exemplary second-order quote
Empowering people to make ethical and aesthetical choices	"Many people are purchasing items on the go, based on aesthetics, and then you just dispose them. But if you think of disposing it you can just simply put it in the compost and grow other plants if you like."		
Discovering the body as active agent	"Living Skin focuses on the behavior of new materials such as mycelium, kombucha and algae when worn on the skin of the human body, and on the personalisation of these materials." "It's a new take on how we should start looking for new answers for our current supply chain."		

References

Akemu, O., Whiteman, G., & Kennedy, S. (2016). Social enterprise emergence from social movement activism: The Fairphone case. *Journal of Management Studies, 53*(5), 846–877.

Barad, K. (2003). Posthumanist performativity: Toward an understanding of how matter comes to matter. *Signs: Journal of Women in Culture and Society, 28*(3), 801–831.

Barinaga, E. (2017). Tinkering with space: The organizational practices of a nascent social venture. *Organization Studies, 38*(7), 937–958.

Benn, S., Edwards, M., & Angus-Leppan, T. (2013). Organizational learning and the sustainability community of practice: The role of boundary objects. *Organization & Environment, 26*(2), 184–202.

Berleant, A. (2013). What is aesthetic engagement? *Contemporary Aesthetics*. Online available: https://contempaesthetics.org/newvolume/pages/article.php?articleID=684

Berleant, A. (1991). *Art and engagement*. Philadelphia: Temple University Press.

Bourdieu, P. (1990). *The logic of practice*. Redwood City, CA: Stanford University Press.

Braidotti, R. (2011). *Nomadic subjects. Embodiment and sexual difference in contemporary feminist theory*. New York: Columbia University.

Bruggeman, D. (2018). *Dissolving the ego of fashion*. Arnhem: ArtEZ Press.

Callon, M. (2007). What does it mean to say that economics is performative? In D. Mackenzie, F. Muniesa & L. Siu (Eds.), *Do economists make markets? On the performativity of economics* (pp. 311–357). Princeton, NJ: Princeton University Press.

Calás, M., Smircich, L., & Bourne, K. (2009). Extending the boundaries: Reframing 'Entrepreneurship as social change' through feminist perspectives. *The Academy of Management Review, 34*(3), 552–569.

Conscious Contemporary Tailoring. (2018). Claire Swift and Bethny Williams in Conversation. *Youtube*. https://www.youtube.com/watch?v=-vy8QFo_Ffs&ab_channel=LondonCollegeofFashion

De Vaujany, F. X., Adrot, A., Boxenbaum, E., & Leca, B. (2019). Introduction: How can materiality inform institutional analysis? In *Materiality in institutions* (pp. 1–31). Cham: Palgrave Macmillan.

Ellen McArthur Foundation. (2017). *A new textiles economy: Redesigning fashion's future*. Report available at: https://www.ellenmacarthurfoundation.org/publications/a-new-textiles-economy-redesigning-fashions-future

Ferraro, F., Etzion, D., & Gehman, J. (2015). Tackling grand challenges pragmatically: Robust action revisited. *Organization Studies, 36*(3), 363–390.

Fischer, A., & Pascucci, S. (2017). Institutional incentives in circular economy transition: The case of material use in the Dutch textile industry. *Journal of Cleaner Production, 115*, 17–32.

Gherardi, S. (2012). *How to conduct a practice-based study: Problems and methods*. Edward Elgar Publishing Cheltenham, UK.

Ghisellini, P., Cialani, C., & Ulgiati, S. (2016). A review on circular economy: The expected transition to a balanced interplay of environmental and economic systems. *Journal of Cleaner Production, 114*, 11–32.

Gioia, D. A., Corley, K. G., & Hamilton, A. L. (2013). Seeking qualitative rigor in inductive research: Notes on the Gioia methodology. *Organizational Research Methods, 16*(1), 15–31.

Harper, K. (2017). *Aesthetic sustainability: Product design and sustainable usage*. Taylor & Francis Ltd Oxfordshire, UK.

Hill, J., & Lee, H. (2015). Sustainable brand extensions of fast fashion retailers. *Journal of Fashion Marketing and Management, 19*(2), 205–222.

Ivanaj, V., Poldner, K., & Shrivastava, P. (2014). Hand, heart, head: Aesthetic practice pedagogy for deep sustainability learning. *Journal of Corporate Citizenship, 54*, 23–46.

Kagan, S. (2013). *Art and sustainability: Connecting patterns for a culture of complexity* (Second amended edition 2013 2013). Bielefeld: Transcript Verlag.

Linstead, S., & Höpfl, H. (2000). *The aesthetics of organization*. London: Sage.

Lueg, R., Pedersen, M., & Clemmensen, S. (2015). The role of corporate sustainability in a low-cost business model – A case study in the Scandinavian fashion industry. *Business Strategy and the Environment, 4*(5), 344–359.

Morin, E. (1992). *Method: Towards a study of humankind*. The Nature of Nature (vol. 1). New York: Peter Lang Publishing Inc.

Nassauer, J. I. (2004). Monitoring the success of metropolitan wetland restorations: Cultural sustainability and ecological function. *Wetland, 24*(4), 756–765.

Nicolini, D. (2013). *Practice theory, work and organization: An introduction*. Oxford: Oxford University Press.

Nicolini, D. (2017). Practice theory as a package of theory, method and vocabulary: Affordances and limitations. In M. Jonas, B. Littig & A. Wroblewski (Eds.), *Methodological reflections and practice-oriented theories* (pp. 19–34). Switzerland: Springer.

Niinimäki, K. (2018). Sustainable fashion in a circular economy. In K. Niinimäki (Ed.), *Sustainable fashion in a circular economy* (pp. 12–41). Finland: Aalto ARTS Books.

Niinimäki, K., Peters, G., Dahlbo, H., Perry, P., Rissanen, T., & Gwilt, A. (2020). The environmental price of fast fashion. *Nature Reviews Earth & Environment*, *1*(4), 189–200.

Pascucci, S., Dentoni, D., Clements, J., Poldner, K., & Gartner, W. B. (2021). Forging forms of authority through the sociomateriality of food in partial organizations. *Organization Studies*, *42*(2), 301–326.

Poldner, K., Steyaert, C., & Branzei, O. (2019). Fashioning ethical subjectivity: The embodied ethics of entrepreneurial self-formation. *Organization*, *26*(2), 151–174.

Poldner, K., Dentoni, D., & Ivanova, O. (2016). Aesthetic mediation of sustainability, creativity, and organization. *Journal of Cleaner Production*, *140*, 1936–1947.

Raworth, K. (2017). *Doughnut economics: Seven ways to think like a twenty-first century economist*. UK: Penguin Random House.

Rindova, V., Barry, D., & Ketchen, D. (2009). Entrepreneuring as emancipation. *Academy of Management Review*, *34*(3), 477–491.

Schatzki, T. R. (2006). On organizations as they happen. *Organization Studies*, *27*(12), 1863–1873.

Shrivastava, P., Schumacher, G., Wasieleski, D., & Tasic, M. (2017). Aesthetic rationality in organizations: Toward developing a sensitivity for sustainability. *The Journal of Applied Behavioural Science*, *53*(3), 369–411.

Shrivastava, P., & Ivanova, O. (2015). Inequality, corporate legitimacy and the occupy wall street movement. *Human Relations*, *68*(7), 1209–1231.

Shrivastava, P. (2011). Enterprise sustainability 2.0: Aesthetics of sustainability. In A. Hoffman & T. Bansal (Eds.), *The Oxford handbook of business and the natural environment* (pp. 630–638). Pratim: Oxford University Press.

Spinosa, C., Flores, F., & Dreyfus, H. (1997). *Disclosing new worlds: Entrepreneurship, democratic action and the cultivation of solidarity*. Cambridge: MIT Press.

Stigliani, I., & Ravasi, D. (2012). Organizing thoughts and connecting brains: Material practices and the transition from individual to group-level prospective sensemaking. *Academy of Management Journal*, *55*(5), 1232–1259.

Stigliani, I., & Ravasi, D. (2018). The shaping of form: Exploring designers' use of aesthetic knowledge. In *Organization studies*. In Press.

Strati, A. (1999). *Organization and aesthetics*. London: Sage.

Sudjic, D. (2008). The language of things. In *Design, luxury, fashion, art. How we are seduced by the objects around us*. London: Pinguin Group.

Taylor, S., & Hanson, H. (2005). Finding form: Looking at the field of organizational aesthetics. *Journal of Management Studies*, *42*(6), 1211–1232.

Waddell, S., Waddock, S., Cornell, S., Dentoni, D., McLachlan, M., & Meszoely, G. (2014). Large systems change. *Journal of Corporate Citizenship*, *53*, 1–13.

Waddock, S., Meszoely, G. M., Waddell, S., & Dentoni, D. (2015). The complexity of wicked problems in large scale change. *Journal of Organizational Change Management*, *28*(6), 993–1012.

Wry, T., Lounsbury, M., & Glynn, M. A. (2011). Legitimating nascent collective identities: Coordinating cultural entrepreneurship. *Organization Science*, *22*(2), 449–463.

Zietsma, C., & Lawrence, T. B. (2010). Institutional work in the transformation of an organizational field: The interplay of boundary work and practice work. *Administrative Science Quarterly*, *55* (2), 189–221.

Zott, C., & Huy, Q. N. (2007). How entrepreneurs use symbolic management to acquire resources. *Administrative Science Quarterly*, *52*(1), 70–105.

[3]D Printed Dress from Iris Van Herpen Pushes Boundaries of Fashion – 3dprint. (2021). *The voice of 3D printing / Additive manufacturing*. 3DPrint.com, October 15, 2021. https://3dprint.com/201774/3d-printed-dress-iris-van-herpen/

Part III: **Industrial vignettes: exploring industry transition**

Fiona Charnley
Introduction

The Circular Economy (CE) offers a simple, yet compelling framework based on a set of principles that decouple resources from economic growth through innovation and entrepreneurship. Analysis has indicated that scaling up circularity provides a multi-billion-dollar economic opportunity (Ellen MacArthur Foundation 2017a' 2017b), driving up resource productivity, driving down material costs, improving resource security and reducing negative externalities and their human and environmental costs (Scottish Government, 2016, World Economic Forum, 2020). In the wake of COVID-19, recent reports propose that moving to a more CE can provide an essential element of the global recovery plan, delivering increased clean growth, net jobs, higher resilience and regenerating natural capital (Committee on Climate Change, 2020, WRAP, 2020).

To drive a CE at scale across value chains requires a systematic and systemic approach, including the future design of materials, components, products, services and infrastructure (EMF 2012-2014, WBCSD, 2020); business models that promote access and performance over ownership (Stahel 2019); closed-loop and reverse logistics (Kortmann 2016); whole system enablers and innovation, including the use of data, emerging technologies and behaviour change (Schroder, 2020).

The role and responsibility of business in achieving this vision is fundamental. Despite the varying definitions, theories, perspectives and schools of thoughts presented in Parts I and II of the CE Handbook, businesses of all sizes, across sectors and across all parts of the world are recognising the opportunity presented by a CE. Numerous examples of CE in practise exist and, in this section of the CE Handbook, we demonstrate the wealth of insight, knowledge and best practice that can be gleaned from those pioneering businesses.

The role of business and business models in a circular economy

The circulation and cascading of products, components and materials at their highest value, a key principle of the CE, can be achieved in many ways. It may require small or more complex shifts in the way a company configures its value chain and can open opportunities for innovative and new forms of collaboration. It may also alter the relationship with the consumer, shifting to the role of user or 'prosumer'

Acknowledgements: The editors would like to thank the 30 businesses who have shared their honest, detailed and inspirational stores with us and the researchers who crafted the resulting vignettes: Georgie Hopkins, Jamie Wheaton, Ruth Cherrington, Isabelle Housni, Lucy Chamberlin and Oke Okorie.

https://doi.org/10.1515/9783110723373-022

and highlighting the crucial role of reverse supply chain management in closing the implementation gap. Circular Business Models (CBMs) are commonly referred to as the way in which businesses capture and create new value from applying the principles of a CE. According to Nußholz (2017):

> A circular business model is how a company creates, captures, and delivers value with the value creation logic designed to improve resource efficiency through contributing to extending useful life of products and parts (e.g. through long-life design, repair and remanufacturing) and closing material loops.

CBMs have been classified by Bocken et al. (2021) into three different strategies: (1) narrowing resource flows (increasing resource efficiency), (2) slowing resource loops (enabling longer life through service models, resale, reuse and remanufacture) and (3) closing loops (capturing value from waste through recovery and recycling). When looking towards implementation, the founders of Circularity Capital, a private venture capital fund, have drawn upon over 20 years of combined experience to organise different types of CBMs into five categories (Figure III.1).

Differences are seen in the application of these models and, as with all innovative business models, the test of any value proposition is determined by the market. This throws up challenges around potential customer demand, first-mover advantage (or disadvantage) and accounting for the value of environmental and social benefits in a world that currently prioritises economic calculations over all others.

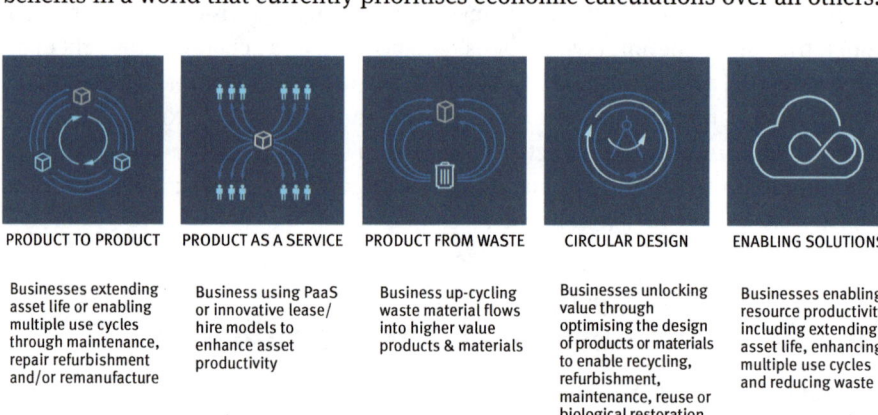

PRODUCT TO PRODUCT	PRODUCT AS A SERVICE	PRODUCT FROM WASTE	CIRCULAR DESIGN	ENABLING SOLUTIONS
Businesses extending asset life or enabling multiple use cycles through maintenance, repair refurbishment and/or remanufacture	Business using PaaS or innovative lease/ hire models to enhance asset productivity	Business up-cycling waste material flows into higher value products & materials	Businesses unlocking value through optimising the design of products or materials to enable recycling, refurbishment, maintenance, reuse or biological restoration	Businesses enabling resource productivity, including extending asset life, enhancing multiple use cycles and reducing waste

Figure III.1: Circular business models.
Source: Circularity Capital (2021).

Across the different business models associated with the CE there are some recurring patterns regarding how companies can position themselves to capture and distribute circular value. Three common examples are:

Transformers: Typically larger organisations, these businesses control a high share of the value chain, with the ability to capture value by innovating and improving their own operations. This internalization typically leads to further value capture through distribution into secondary markets. An example would be a manufacturing company that implements a take-back scheme for its products (or some components), building a remanufacturing capability to circle these products back into the marketplace, often through new channels.

Enablers: Mostly innovative new or spin-off companies that provide the missing capability to other companies, in effect filling existing gaps in value chains. Such businesses more frequently apply innovative technological solutions, often working with CE Transformers to become part of their value distribution. An example would be a company that develops access to a new market from an existing waste stream, such as surplus food.

Communities: These groups of companies use collaboration to develop CE practices across their value chain through addressing material leakages in existing chains. Often through a form of collective internalization, these firms improve resource efficiency symbiotically. Examples can be seen in take-back bottle syndicates, design standards agreements and co-creating services.

Stories of implementation

This section of the CE Handbook is composed of a series of 30 unique and inspirational vignettes, relaying the stories of businesses at different stages in the transition towards a CE. They include internationally diverse businesses of different sizes from start-ups of two or three people to multi-national organisations. They cover a multitude of sectors from fast-moving consumer goods, cosmetics and consumer electronics to furniture, mobility and office printing. The three different types of business previously mentioned (transformer, enabler and community) are all represented, as are the five different types of CBMs depicted in Figure III.1, operating in both business to business and business to consumer environments. Some of the common factors that enable success as well as the barriers to progress are as follows:

Leadership and organisational mindset: Strong leadership has long been recognised as an enabling condition for business success, and successful circular organisations are no different. All of the vignettes share accounts of passionate, driven and supportive leaders who set out a clear direction and ambition for the business, yet also recognise the wider role they have to play within the wider system, including the urgency and enormity of the challenge. In many of the vignettes, the inspiration has been driven by the CEO and/or founder who are restless in pursuit of environmental innovation and many for whom their business has been a lifelong commitment. It is not only leadership that is seen to be a key ingredient for success but also the creation of an organisational culture and mindset, often consisting of an appreciation for continuous experimentation, lifelong learning, a team of entrepreneurs (and intrapreneurs) and encouraging all staff to continuously push, question and challenge the business. The transition towards a CE requires contribution from all functions and roles within a business and ensuring that all of the team are aligned with the organisational vision and have a good understanding of the benefits of a circular approach is essential.

Strategic approach: Many of the vignettes provide detail of the strategic approach that has been adopted, including the use of guidelines, pillars and frameworks to support coordinated and consistent implementation. Factors include identification of the areas of the business with the highest impact (environmentally and socially), areas of value leakage, appreciating the full lifecycle of materials, components and products and the design of materials, components and products in parallel to the development of the CBM.

Technical enablers: Multiple examples have involved the development of technology either as a central offering as in the cases of Riversimple and Winnow or as part of the infrastructure to enable circulation of materials such as BAM. Blockchain, machine learning and artificial intelligence are some of the digital technologies that have been employed alongside others such as new manufacturing, remanufacturing and recycling technologies. Measurement was another common theme across multiple vignettes and often cited as a challenge to know what to measure and what metrics and indicators to use to inform decision-making and to evidence progress.

Whole system appreciation: The transition towards a CE requires a system-level change, and many of the vignettes recognise that their business or product is one 'cog' in a complex system of parts that need to move together. Many examples reflect the important role of partnerships across the supply chain not only between manufacturers and suppliers but also between public and private bodies and between industrial organisations, research and educational institutes and government, charities, NGOs, regional authorities and consumers. Several vignettes detail the need for networks to share physical resources and also for the sharing of ideas, design, technology, labour, skills and learning. Pioneers in this space are generous with their time, knowledge, expertise and experiences in a drive to ensure that the whole system works and transitions together.

Barriers to success: As evidenced throughout this section, the CE is a complex concept requiring multi-stakeholder collaboration across a broad spectrum of sectors, regions and scales of business and the wider community. There are many challenges involved, and implementing CBMs can feel both daunting and economically risky. Each business needs to work collaboratively with others to develop both the understanding and application of the concept and find innovative methods to capture value (Stumpf, 2021). Many existing 'linear' business models are currently not equipped with the core know-how, support or resources to capture value from the CE (Ludeke-Freund et al, 2019), research in the area is still nascent (Tura et al, 2019), and there is a knowledge gap when it comes to understanding and competences relating to CE. Upskilling is therefore needed throughout communities and organizations, with a particular focus on programmes relating to return systems, remanufacture and repair, and developing a common understanding between companies, product developers and customers.

CBMs can also face barriers with regards to 'business as usual' assumptions around profit and product development since making things to last longer or for multiple use cycles can clash with trends for fast-moving low-cost goods. Both technological and societal obsolescence can impact adoption of CE, particularly where customers are encouraged into short life cycles of use, lacking both care or attachment to the product and an understanding of durability (Ludeke-Freund et al, 2019). Conversely, where CBMs are presented as a route to profitability and business growth (e.g. creating new revenue streams out of 'waste'), social perspectives can be omitted, whilst environmental and ecological benefits are sometimes assumed as outcomes, yet this is not always the case. Increased efficiency measures, for instance, can lead to greater demand which then outstrips the gains made through efficiency, whilst so-called best practice can also result in greater environmental degradation and increased energy use if a systems perspective is not taken and careful life cycle analyses are not conducted (Whalen & Whalen, 2020). The World Economic Forum has grouped some of these barriers into five categories, as shown in Figure III.2.

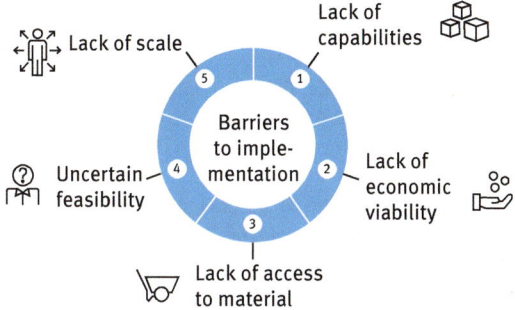

Figure III.2: Challenges in implementing the circular economy (The World Economic Forum, 2020). Source: World Economic Forum.

The eclectic collection of vignettes that follows this introduction demonstrates the diversity of approaches and perspectives present across the CE, yet all are aligned in their passion for innovation, entrepreneurship and challenging 'business as usual.' All provide honest stories of the challenges businesses face, some over many years, to design new materials, products, business models, mindsets, behaviours and relationships in their pursuit of an alternative future in which finite resources are decoupled from economic, social and environmental prosperity. The vignettes provide a rich resource of insights, knowledge, expertise and experiences for individuals who want to learn about CE in practise.

References

Bocken, N., Weissbrod, I., & Antikainen, M. (2021). Business experimentation for sustainability: Emerging perspectives. *Journal of Cleaner Production*, *281*, 124904. https://doi.org/10.1016/j.jclepro.2020.124904

Circularity Capital (2021). *Circular business models*. Available at: https://circularitycapital.com/our-approach

Committee on Climate change. (2020). *Reducing UK emissions 2020 progress report to parliament.*

Ellen MacArthur Foundation. (2012–2014). *Towards the circular economy vol. 1-3.* Available from: http://www.ellenmacarthurfoundation.org

Ellen MacArthur Foundation. (2017a). *The new plastics economy: Rethinking the future of plastics & catalysing action.*

Ellen MacArthur Foundation. (2017b). *A new textiles economy: Redesigning fashion's future.*

Jensen, P., Laursen, L., & Møllerhaase, L. (2021). Barriers to product longevity: A review of business, product development and user perspectives. *Journal of Cleaner Production*, *313*. https://doi.org/10.1016/j.jclepro.2021.127951

Kortmann, S., and Piller, F. (2016). Open business models and closed- loop value chains: Redefining the firm-consumer relationship. *California Management Review*, *58*, 3.

Ludeke-Freund, F., Gold, S., & Bocken, N. (2019). A review and typology of circular economy business model patterns. *Journal of Industrial Ecology*, *23*(Issue), 1.

Nußholz, J. (2017). Circular business models: Defining a concept and framing an emerging research field. *Sustainability*, (9), 1810. 10.3390/su9101810

Scottish Government. (2016). *Making things last: A circular economy strategy for Scotland.*

Schroder, P. (2020). *Promoting a Just Transition to an Inclusive CE.* Chatham House research paper: https://www.chathamhouse.org/sites/default/files/2020-04-01-inclusive-circular-economy-schroder.pdf

Stahel, W. (2019). *Circular Economy A user's guide.* Cowes: EMF.

Stumpf, L., Schöggl, J., & Baumgartner, R. (2021). Climbing up the circularity ladder? – A mixed-methods analysis of circular economy in business practice. *Journal of Cleaner Production*, *316*. https://doi.org/10.1016/j.jclepro.2021.128158

Tura, N., Hanski, J., Ahola, T., Ståhle, M., Piiparinen, M., & Valkokari, P. (2019). Unlocking circular business: A framework of barriers and drivers. *Journal of Cleaner Production*, *212*. https://doi.org/10.1016/j.jclepro.2018.11.202

WBCSD. (2020). *Circular economy practitioner guide.* Available at: https://www.ceguide.org

Whalen, C., & Whalen, K. (2020). Circular economy business models: A critical examination. *Journal of Economic Issues*, *54*(3). https://doi-org.uoelibrary.idm.oclc.org/10.1080/00213624.2020.1778404

WRAP. (2020). *How moving to a CE can help the UK to build back better.*

World Economic Forum: Circular Economy for Net-Zero Industry Transition. https://ceclimate.weforum.org/

World Economic Forum. (2020). *CE and material value chains,* Available at: https://www.weforum.org/projects/circular-economy

Georgie Hopkins
19 BAM bamboo clothing
Sector focus: Fashion and textiles

Rationale

The sustainability challenges of the fashion industry are well-documented, with average global annual consumption of textiles having doubled in the last 20 years, from 7 to 13 kg per person, estimated to represent global trade of $1.3 trillion per year. Textile manufacturing is deemed to be one of the most polluting sectors, using significant energy and water, with a high carbon footprint and producing large volumes of waste. Heightened by the Covid pandemic, many consumers are becoming increasingly aware of these sustainability issues; however, the challenges of low utilisation rates and 'fast fashion' remain. Change within the industry requires a system-level approach, often working across extended global supply chains with varying degrees of transparency.

Leadership

Launched by David Gordon in 2006, BAM's foundation sits firmly on environmental sustainability, primarily producing clothing from bamboo (a fast-growing, regenerative material) and organic cotton. Driven by David's commitment in this space, in 2018 the company set itself an ambitious goal of becoming impact positive by 2030. In David's words: "If 'going green' is hiking to the top of a hill, Impact Positive is scaling Everest, but it's where every business needs to be headed." With a strong company culture, this commitment is a focus of the entire team, with work progressing across six key pillars: carbon, people, land, waste, chemicals and water. Through the adoption of this framework, each initiative and intervention can be identified according to the pillar(s) it addresses and as such, BAM is now successfully incorporating circularity within its business model.

Approach

BAM currently operates with an existing linear business model, focusing on seasonal sales targets and volume throughput; however, the challenge to remain financially viable while shifting to a circular regenerative model is at the forefront of the

https://doi.org/10.1515/9783110723373-023

company strategy. In the words of MD Ryan Shannon, working out how to "keep the lights on while rewiring the house" is critical.

BAM started the journey with measurement, working with dedicated professionals to understand every element of their business activity and quantify the full impact of their clothing through life cycle analysis. This involved gathering and collating novel data from four tiers of supply chain all the way to customer use, washing and end-of use disposal methods. Utilising this knowledge to identify areas of the business with the highest impact, BAM has applied circular design principles to rethink production of selected clothing categories, following the logic that "If we don't know how we'll dispose of it, we shouldn't make it."

The company demonstrates an innovative approach in this regard, appreciating that novel methods being explored may not always work, but continuing with a mindset of learning. An example of this is BAM's first fully recyclable product: 73 Zero jeans, which use 90% less irrigation water, no pesticides and are designed to be fully recyclable, with the exception of rivets and a button that can be unscrewed. The product development took three years to bring together materials, durability, recyclability and traceability to meet standards as set out by the EMF Jeans Redesign project. This development has outlined key challenges, given that some of the technology and infrastructure needed for circularity in clothing production does not yet exist. Equally, BAM have experienced constraints of scale, given the relative size of the business against the industry as a whole. One notable area is zips for which, at the time of publication, no recyclable option exists in the supply chain.

Based on this success, BAM has also launched performance waterproof jackets, as seen in figures 19.1 and 19.2, that are constructed using 98% recycled materials and again, designed to be recyclable. The test here was sourcing one material that could be used to make the ten different components needed in each jacket, a challenge they met, with the exception of Velcro cuff tags and aluminium toggles. Beyond this, a challenge recognised by the company is the system-level change required, with BAM being just one cog in the process. From farmers to yarn manufacturers, consumers to recycling centres, there is a need for both understanding and action across the value chain to drive effective change. BAM is working with partners to raise awareness, especially with customers through campaigns such as 'dare to wear longer' and clothing return, partnering with charity Sharewear.

A key enabler of this circular movement has been the training and education extended across all disciplines of the BAM team, developing knowledge and understanding of circular economy principles and business implementation. Moving forward, BAM is working on further material developments, with a particular focus on technological capabilities to recycle bamboo viscose yarn. Alongside this, new business models are being explored to operate within the inner loops of a circular economy model, extending product life through repair and resale schemes.

On reflection

BAM is a great demonstration of the journey to circularity; continuously innovating and improving, using learnings from one activity to stimulate another via a multi-faceted approach. Having set bold ambitions, the company could be seen to be stealing a march on competition, although equally, is there a degree of first-mover disadvantage given the investment required for such innovation?

Applying a critical lens to both upstream and downstream processes, questions still remain: On the supply side, what level of client demand is needed for manufacturers to develop yarn fit for circularity? For farmers and manufacturers alike, are the skills and investment needed to implement business model change accessible? On the demand side, will consumers change behaviour to purchase less clothing and be committed to extending the life of each product? Can the BAM business model adapt to capture equivalent (or greater) economic value from alternative value streams, such as repair and resale? Will material recycling become widely available and will consumers adopt return processes to take advantage of them?

Figure 19.1: BAM 73 Zero Jacket.
Source: BAM Clothing

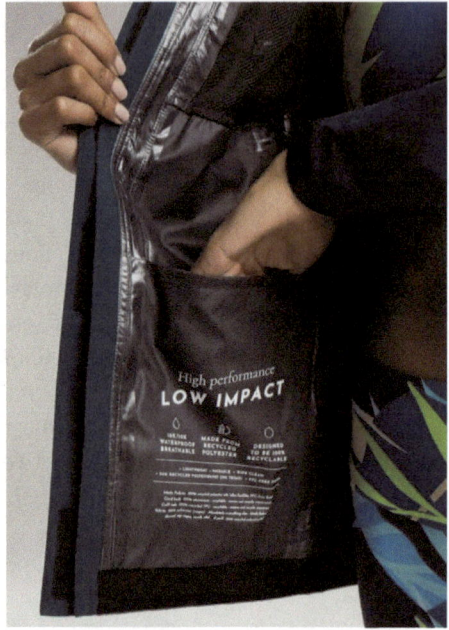

Figure 19.2: BAM 73 Zero Jacket Case Instructions.
Source: BAM Clothing.

Source

http://www.bambooclothing.co.uk

Georgie Hopkins
20 Winnow Solutions Ltd
Sector focus: Agriculture and food systems

Rationale

Global food waste is a documented problem, with an estimated 931 million tonnes of food going to waste each year, representing 17% of global production. Beyond contributing to food scarcity, reports estimate that up to 10% of global greenhouse gas emissions are associated with this waste. A key focus of transitioning to a circular economy involves moving to a regenerative agricultural system that builds natural capital, improving soils and biodiversity. Alongside this, designing out waste through effective value capture and redistribution is an essential part of building a sustainable food system. Tackling this challenge needs a multifaceted approach, with companies such as Winnow focusing on the food service sector specifically. Hospitality and restaurant systems account for around 26% of food waste, generating a wide range of negative environmental impacts and disposing of edible food that could be put to socially productive and more valuable alternative uses.

Leadership

Founded in 2013 by Marc Zornes and Kevin Duffy, Winnow is creating a movement to inspire chefs to see that food is too valuable to waste. Initially established in a single restaurant kitchen, the team recognised the problem of food waste and created a solution to capture the value leakage, based on the premise that 'what gets measured gets managed.' As a start-up employing an innovative business model, the company has attracted investment and grown exponentially, operating in over 45 countries, with a clear vision of using technology to empower chefs to cook and use food smarter. Winnow became a certified B Corporation in 2017 and has enabled clients to save the equivalent of 36 million meals each year.

Approach

The business has developed technology to enable commercial kitchens all over the world to record and analyse exactly what is put in the bin. Employing Winnow Vision gives food service establishments true visibility of all food waste, enabling them to become more efficient and gaining the highest value from food, saving money and

https://doi.org/10.1515/9783110723373-024

reducing environmental impact simultaneously. Facilitating a move from the 'pen-and-paper' model, the system employs artificial intelligence technology, smart scales and cameras to document food as it is discarded, with machine-based learning accurately recognising different types of food waste. Through use, the system gets smarter, providing the additional benefit of reducing human error and saving time in the kitchen, with each food waste 'entry' being estimated to take just three seconds.

Winnow reports that employing their technology can help reduce food waste by up to half within the first year and reduce food costs by between 3–8%, providing a positive return on investment within the first 12 months in around 95% of cases. As an example, IKEA have been working with Winnow since 2017, implementing the system in 23 UK&IE stores and have seen a reduction in food waste of 50%, equating to cost savings in excess of £1.4 million in 2018, the equivalent of over 1.2 million meals.

A challenge of introducing such a system has been users' concern over surveillance and assigning blame for waste; however, through communicating with staff in a clear and positive way, businesses have overcome these barriers and found increased engagement and team action on waste. Supported by Winnow Hub, easy-to-use daily reports help kitchen staff identify top food waste streams, adjusting menus and production accordingly. In the words of a member of the IKEA Wembley Kitchen Team: "Using the Winnow system, you can quickly see where you have issues or problems. It starts the conversation about the waste we have and why we have it. Nobody wants to throw away food away needlessly."

Looking to the future, the Covid pandemic has forced many consumers and food service businesses alike to embrace significant digitalisation, employing new front-of-house technology and digital touch points. Through this period of change, additional opportunity exists to implement novel back-of-house technology, with Winnow's systems offering a tangible method for food services to capture value leakage, meeting the demand for sustainability to be at the heart of the recovery. As with any novel technology, market exposure increases use and acceptance, with Winnow's expansion likely to continue, already partnering with large organisations such as Compass Group, IKEA, IHG, Accor Hotels, Hilton and Costa Cruises.

On reflection

Winnow's innovative value capture mechanism can be characterised as a product service business model, allowing restaurants to internalise costs of food waste and potentially increase revenues via re-selling what would otherwise have been lost. The technology is currently targeted to businesses operating over ten sites with cumulative annual budget of £1 million. Looking at the wider market, with households accounting for 61% of food waste, what adaptations may be possible to employ the technology effectively at a scale appropriate for smaller food services and households?

Additionally, while waste is reduced using the Winnow system, what additional support or network developments could Winnow offer to forge relationships with industry partners to capture further value from the remaining waste? Is there opportunity to employ waste separation at source, which would help maintain a higher residual value in the waste stream? What other application areas could this technology be applied to?

Sources

http://www.winnowsolutions.com
https://circularitycapital.com

Oke Okorie
21 Riversimple
Sector focus: Mobility

Rationale

Sustainability has become embedded in the expectations of the modern consumer, and as the mobility sector strives to maintain its essential appeal, evolving from dependence on finite resources has developed into a necessity. The Riversimple Movement incorporates an awareness of these consumer concerns into every level of its process – curating its leadership structure, design approach and technology with collaboration and environmental responsibility in mind.

Leadership

Riversimple CEO, Hugo Spowers, has nurtured the company from its ambition to pursue the elimination of the environmental impact of personal transport to basing its innovations on feedback from environmental advocates and product users. How has Spowers transformed this sustainability-centred dream into a tenable reality? By supporting the company's transition from a linear design approach to a circular economy model that incentivises each actor in the company's value chain to innovate consciously.

Company Steward, Estelle Clark, describes how a circular economy model has been integrated into Riversimple's infrastructure through the application of a 'Stakeholder Guardian' governance model. Clark identifies the company's stakeholders as the environment, service users, neighbours, staff, investors and commercial partners and perceives each stakeholder as a 'custodian' of their respective sector. What does it mean to be involved in Riversimple's mission as a custodian? The role requires acting individuals to offer insight into how the company can protect the interests of its stakeholders, whom Clark describes as covering spheres across commerce, society and the planet. The question for potential customers of Riversimple centres around execution. In a market pressured by decreasing resource availability, increasing population and increasing demand for mobility – how can a mode of transport meet these standards?

Approach

Riversimple's response arrives in the form of the RASA car (figure 21.1), based on their revolutionary Whole System Design framework. The conception of the RASA

https://doi.org/10.1515/9783110723373-025

car was inspired by the incompatibility of the industry's current production model with the demands of the contemporary consumer market: Riversimple wanted to generate an alternative to the industry's dependence on finite resources, resolve the industry's declining ability to cater to consumers and develop innovations that were not environmentally problematic. How was this vision to be fulfilled? Through the development of a targeted travel solution that reduces environmental pollution, satisfies market demand and increases material and energy efficiency.

The RASA car, fitted with fuel cell technology and a lightweight engine, has been developed with performance and longevity in mind. The vehicle's integrated fuel cell equips it to power four internal motors on demand, eliminating the need for heavyweight performance enhancers, whilst the emissions of the supporting engine are limited to water vapour. In conjunction with this, the RASA car is equipped with an electronic braking system that recovers up to 50% of power expended during brake activation. The combined efficiency and environmental consciousness of this design has made the RASA car a pioneer of Riversimple's performance model, an approach to innovation that maintains the value of products, materials and components over time whilst continuing to satisfy the demands of consumers.

Despite the leap in sustainable design made by the RASA car, the issue of catering to an oversaturated consumer pool persists. Nevertheless, Riversimple has a strategy to combat this issue. Riversimple's service model is based on a rejection of the 'make, use, dispose' approach common across the automobile industry. Rather than selling the RASA car on the understanding that consumers will retain ownership for the duration of its life cycle, Riversimple has opted to lease the RASA car to consumers on a service contract. Customers will have the option of renting the RASA car at a fixed price and variable usage rate, which will cover all operational and fuel-related costs during their contract. Once transport is no longer required by the customer, the RASA car is returned to Riversimple for future rentals, providing a personal transport option that serves the needs of consumers on an immediate, environmentally considerate basis.

The final element that Riversimple must incorporate into their production process is the protection of stakeholder interests. With the RASA car's design being heavily consumer influenced, how does the company plan to approach the needs and concerns of their remaining custodians? Riversimple's answer can be found in the relationships maintained with their suppliers, regulators and environmental ambassadors on a company-wide level. As Riversimple prioritises visibility and traceability across its value network, stakeholders are empowered to hold the company to account for its decision making, design plans and the potential consequences of these for invested parties, causing all of Riversimple's commercial activity to be regulated by those at the frontlines of its impact.

Looking to the future, Riversimple are building towards their goal of visibility and transparency across the value chain and hope to expand their leasing model. This would mean that not only would a customer lease the RASA from Riversimple, but Riversimple would then lease the fuel cell from the fuel cell manufacturer, who in turn would lease the Membrane Electrode Assembly (MEA) from the MEA Supplier and so on until even the platinum was leased from the point of mining. A key challenge to achieving this vision is the development of Blockchain Technology, currently being explored via a partnership with the Universities of Swansea and Exeter via the European funded project Circular Revolution. Riversimple are also pioneering in the way that they are sharing the learning they have gained from circular product and business model development with other start-ups and SMEs transitioning towards circularity. They are doing this via open-source design and through direct support and mentoring.

On reflection

Reviewing Riversimple's journey under a critical lens, several questions remain: Can the RASA car's service contract be considered compatible with the automobile consumer market? Will consumers be willing to exchange the convenience of vehicle ownership for a reduced environmental impact? Are the materials required to build the RASA car economically suited to mass production? Will the RASA car be marketed at a price range accessible to the average earner – allowing its consumption to create the kind of environmental impact envisioned by Riversimple? Each of these factors must be considered to evaluate the potential success of a company like Riversimple and an innovation like the RASA car.

Figure 21.1: Riversimple RASA.
Source: Riversimple.

Sources

http://www.riversimple.com
https://circularrevolution.wales

Georgie Hopkins
22 Rype Office
Sector: Office furniture

Rationale

The 'take–make–waste' linear approach is evident in many sectors of the economy and recycling and is well-documented as being the focus area for action on sustainability. When considering the furniture market, office furniture represents 11% of total world furniture consumption with an annual value of $47 billion. In the United States alone, estimates suggest that 8.5 million tonnes of office furniture ends up in landfill every year. The majority of this waste comprises metal, wood and plastics; while the metal is readily recycled, the wood can be hard to recycle due to varnishes and adhesives. Focusing on strategies such as the reuse, refurbishment and remanufacture of such products is an effective way to retain value and extend material life, as is being implemented by Rype Office.

Leadership

A structural engineer by training, Dr Greg Lavery founded Rype Office in 2015 after recognising the power of the circular economy as a potential game changer. Following extensive research into the application of a circular economy to different industries, Greg identified office furniture as a product sector fit for circular innovation. Focusing on the inner loops of the circular economy model, Rype remanufactures high-quality, used office furniture to 'as new' condition, creating a new market category, sitting between new product (utilising virgin materials and resources) and second-hand product (often compromised in quality and aesthetics). In addition, the company provides a full design service and creates its own furniture from waste, as shown in figures 22.1 and 22.2 providing training and at least a Real Living Wage to the long-term unemployed with disabilities, creating significant social value in the process.

The business focuses firmly on quality, understanding that for commercial success, clients must not be expected to compromise in this area. Greg speaks of their process and technologies enabling them to "supply 'as new' products every single time" with the motivation to lead this new office furniture category. "We aim to surprise and delight customers so they become advocates for remanufacturing." From the beginning, the team has had a culture of questioning what the business can do to be more sustainable, considering best practice and then pushing beyond that. And, to date, that's proving to be a successful formula, with Rype Office having doubled both revenue and employee count year-on-year since launch, now having

https://doi.org/10.1515/9783110723373-026

served over 250 customers. By preventing over 400 tonnes of waste, Rype has saved around 1,000 tonnes of CO_2e of greenhouse gas emissions and provided more than 7,000 h of employment for those furthest from the workforce.

Approach

Rype Office identified the significant value leakage of the linear office furniture market, where current operators pass product responsibility to customers at point of sale, with limited capability to reprocess products. This is largely due to the one-way supply chains and centralised production traditionally used by furniture manufacturers. By establishing regional remanufacturing capabilities, Rype Office is enabling remanufacturing to occur more economically, close to used furniture sources, whilst upskilling local disadvantaged groups and helping them into the workforce.

The Rype Office model captures value across three levels – economically, by saving customers 20–30% compared to wholesale prices for new office furniture; environmentally, with 80% reduction in GHG emissions and waste; and socially, creating employment opportunities for long-term unemployed with disabilities.

A key enabler of this model is Rype Office's commitment to taking full product responsibility, guaranteeing customers to take back any furniture supplied at end-of-use, at no additional cost. In addition to demonstrating a commitment to circular material flows, this also provides the important commercial benefit of securing future material stocks for the company and the potential of repeat business.

Entering an established market with accepted behavioural norms has been a challenge for Rype Office to overcome, needing to build consumer awareness and market adoption. Given that the majority of office spaces require unique furniture configurations and with the perception of lower-quality second-hand product, buying new has traditionally held two advantages, both meeting needs exactly and avoiding compromise. From launch, Rype Office championed its customised design service and financial savings to win customers. However, since 2020, the business has seen a shift in attention, with the social and environmental benefits being recognised and valued more widely, as corporate attention turns to the future low-carbon economy and enhancing social licence to operate. As Greg explained, "Word of mouth, particularly from those who are passionate about addressing change, like we are, is a really great way to move up the adoption curve and win customers in a cost-effective way."

An additional challenge has been the management of unwanted resources from their production facility, such as broken furniture and product that cannot be used. The business has categorised these into seven different waste streams, directing them to relevant channels and keeping a focus on the target of zero waste to landfill.

Looking to the future, Rype Office is currently seeking investment to enable growth into Europe, capitalising on legislative changes and building on the business's first mover advantage. Rype Office has recognised the need to establish robust systems to allow the remanufacturing process to be replicated reliably at each local level.

On reflection

Applying Schumpeterian principles of creative destruction, the Rype business model applies circular economy principles to combine existing resources in a novel way, demonstrating an alternative method of fulfilling an existing market need. As a result, the organisation is creating value across the three pillars of sustainability: economically, environmentally and socially. As such, Rype Office has created a new category of office furniture that better meets all of the triple bottom lines. From Greg's own PhD research, this is consistent with a new paradigm or furniture provision – a circular one that supersedes the worse-performing linear approach and therefore has a promising future.

With ambitious plans to scale the business by moving into new territories, will differing policy and legislative requirements, particularly around waste and recycling, impact these systems or require process adaptation? At a more fundamental level, considering the business commitment to full product-life responsibility, could Rype explore office furniture as a service, enabling a rental model for customers?

Figure 22.1: Rype Office Furniture.
Source: Rype Office.

Figure 22.2: Rype Office Table.
Source: Rype Office.

Source

https://www.rypeoffice.com

Georgie Hopkins

23 Elvis & Kresse

Sector focus: Consumer goods

Rationale

In the circular economy, focus is often centred on the start of the process, designing out waste. However, given the extensive history of the linear economy, there is already considerable material stock within society, representing an opportunity to capture significant value. One well-documented example is building material stocks such as concrete, steel and timber; however, many more niche waste streams also exist. Working with such waste is often not financially viable due to the complexities of decoupling the product itself from other materials, restrictive legislation and the scale of the opportunity. In many instances, virgin material is deemed to be the more financially viable and therefore preferred option.

Leadership

Elvis & Kresse is a social enterprise founded in 2005 with the primary focus on saving London's decommissioned fire hose. Established before widespread recognition of the circular economy as a movement, the aim was to break the linear model and challenge the idea of waste, creating a zero-waste business designed to do more good than harm. The operation is based around three key pillars (as shown in figure 23.1) – to rescue waste material, transform it into something beautiful and donate a share of the profits to charitable causes. Under the entrepreneurial drive of co-founder Kresse Wesling, by 2010 the business had met its initial target, being most famous for rescuing and repurposing all the decommissioned London fire hose available each year. From this success, Elvis & Kresse now handles 15 different waste streams, building new partnerships and circular business models in the process.

Approach

Founded on reverse engineering, Elvis & Kresse completed extensive research to identify the luxury goods market to target with the repurposed fire hose. In the opinion of the founders, it is a sector that represents the structural failings of capitalism with success measured in financial returns, often with limited regard to

https://doi.org/10.1515/9783110723373-027

wider impacts. While this may be similar to many other sectors, this market also offered appropriate margins for financial viability of the fire hose product, given the cost of repurposing the waste material.

Contrary to traditional circular economy thinking, the business demonstrated value creation through capturing value leakage from a very niche existing waste. Due to its composition, decommissioned fire hose is not suitable for shredding, melting or making new again, and hence traditionally goes straight to landfill. In the rescue process, the founders were challenged to work creatively with the material, developing innovative methods for its physical transformation, along with building the business operation alongside. The first bag took over five years to complete and in recognition, now sits within a permanent exhibit at the Victoria and Albert Museum in London.

From this success, Elvis & Kresse has built collaborations with many new partners, handling materials such a printing blankets, parachute silk, coffee sacks and auction banners. Of particular note is their partnership with the Burberry Foundation, challenged to solve the problem of the 35,000 tonnes of waste produced each year by the European luxury market. Elvis & Kresse has designed an innovative system to transform the new leather offcuts into uniform components that can be hand-woven into new products without size or shape restrictions. Incorporated into a number of their luxury bags (as seen in figure 23.2) and wallets, the company has creatively demonstrated how to reengineer what was deemed to be a waste material.

From small beginnings, the business has expanded to currently employ nine people in the UK plus 14 at their production facility in Istanbul, whilst continuing to demonstrate their values, being recognised as a B-Corporation and Living Wage employer. Looking to the future, Elvis & Kresse recently partnered with St Mary's University in designing an open-source, solar-powered micro-forge to address the challenge of 16 million aluminium cans that are littered across the UK each year. Soon to launch their own deposit-return scheme to engage consumers to collect and return cans, the recycled aluminium will then to be crafted into new luxury product, powered by renewable energy and sold with the condition of return to Elvis & Kresse at end-of-use for further repurposing. This form of continuous innovation will become critical as circular design takes hold and flows of waste materials change.

On reflection

Elvis & Kresse fundamentally challenges the perception of success, which is primarily measured in financial terms in our linear society. In contrast, success for this business is through solving social and environmental problems, viewing financial return simply as a necessity to facilitate wider impact.

The unwavering dedication to these values is highly commendable, and while valuable partnerships already exist, the question remains about how the business could maximise the leverage of its creative approach to waste and regenerative systems for the greatest effect? The founders recognise the short time window in which to facilitate a fundamental shift to address climate change. On this basis, could the sharing of their innovative approach be accelerated through greater scaling of the business?

RESCUE   TRANSFORM   DONATE

Figure 23.1: Elvis & Kresse Business Principles.
Source: Elvis & Kresse.

Figure 23.2: Elvis & Kresse Bag.
Source: Elvis & Kresse.

Source

https://www.elvisandkresse.com

Jamie Wheaton
24 Circularity Capital
Sector focus: Finance

Rationale

Acknowledging that traditional linear growth is now being constrained by a series of environmental 'mega-trends,' Circularity Capital is a circularity-focused, private equity firm that highlights how businesses are now confronted with the need to decouple economic prosperity from resource use. Founding partner Ian Nolan argues that: "We are all increasingly realising that much of today's economic activity produces negative externalities, contributing to climate change, loss of biodiversity and general diminution of the natural capital endowment that our generation collectively inherited from our parents, and which we in turn will pass on to our children's generation." On the other hand, businesses may be constrained by barriers such as a lack of expertise or the funds required to transition towards the circular economy.

Circular Capital's aim is "to deliver value for investors by supporting SME growth and innovation in the circular economy" and in doing so enable a transition to the circular economy for SMEs through the provision of two key resources. Namely, through the provision of capital investment as well as access to expertise within the field of circular economy through connections to specialist partners within the field.

Leadership

The previous background of its founding members means that the leadership of Circularity Capital combines experience from both the finance industry and the circular economy. Three of the four founding partners – Ian Nolan, David Mowat and Andrew Shannon – have amassed a combined total of over 60 years' experience in private equity and investment management. The fourth founding member – Jamie Butterworth – is the former CEO of the Ellen MacArthur Foundation, as well as previously developing the Circular Economy 100, an innovation programme established to enable organisations to move towards the circular economy.

Circularity Capital highlights the potential afforded by this combined expertise to SMEs that may be considering a transition to the circular economy, with the combination of finance and circularity acting 'at the heart' of its operations. The leadership team uses the circular economy as a framework to support transition to circularity,

https://doi.org/10.1515/9783110723373-028

whilst adopting an outward-facing approach in identifying investment opportunities within business models that demonstrate an aim to contribute quantifiable, positive societal and environmental impacts.

Approach

Circularity Capital aims to invest in innovative SMEs that specialise within a range of five clearly defined circular models, as outlined in figure 24.1. Firstly, Circularity Capital seeks to invest in businesses that operate 'product-to-product' models, extending asset life or ensuring multiple-use cycles through maintenance and refurbishment. Secondly, Circularity Capital also invests in businesses that operate a 'Product-as-a-Service' model that may benefit access to goods or services as opposed to ownership. Thirdly, Circularity Capital seeks to promote 'product from waste models' in order to maintain material flows from waste into high-value products, whilst 'circular design' models ideally see businesses optimising the design of products or materials towards the possibility of recycling, reuse or biological restoration. Finally, the 'enabling solutions' model envisions businesses using available resources such as data to extend asset life, increase usage cycles and reduce waste and emissions within their output. Businesses that have secured Circularity Capital's support by following such business models have prospered. For example, *Grover* (also included within the Handbook) is a company within Circularity Capital's portfolio which recently secured a further $1 billion in equity and asset-backed financing in order to grow its 'Product-as-a-Service' offering. Another business within the portfolio that provides a 'Software as a Service' offers to maintain the circularity of returned clothing. This business was sold in March 2021 for $70 million to another established business within the retail industry. The large sums secured within such investments or sales demonstrate how Circularity Capital's support can not only facilitate the growth of circular businesses themselves but also allow them to drive industry-wide shifts towards the circular economy.

The approach adopted by Circularity Capital rests upon four 'pillars of support,' which are reflective of the leadership team's expertise. Firstly, Circularity Capital seeks to deploy strategic support and guidance to help businesses deliver their future plans. Secondly, Circularity Capital provides specialist operational support to allow its portfolio members to access suitable operational partners and to allow collaboration in order to fill in capability gaps. This then feeds into the third principle of capability building, allowing access to the world's leading expertise to drive innovation and value. This principle is perhaps the most important of all, allowing investee businesses to access expertise to facilitate training and circular jobs. Finally, Circularity Capital acts as an intermediary, allowing businesses to connect with future customers and partners as well as the potential to bid for further investments.

Uncertainty does however exist within the circular financial market and as such, Circularity Capital requires businesses to offer a number 'key fundamentals' before providing investment and support. Circularity Capital requires prospective businesses to demonstrate a clear circular economic value-creation potential, coupled with strong committed management, hence demonstrating a clear vision for future, circular growth. Prospective businesses should also be operating in sectors with growth as well as demonstrating a proven revenue model. These should then be evidenced within well-defined, executable future strategies. Investments into prospective businesses are also measured against specifically defined environmental, social and governance (ESG) factors, which guides Circularity Capital through investment and post-investment activities. The potential and ongoing progress of businesses' movement towards the circular economy is measured against a set of externally advised, ESG-based metrics, whilst post-investment activities include an annual review of an ESG scorecard system as well as the provision of continuous support.

On reflection

Circularity Capital is at the forefront of investment within the field of circular economy, providing businesses with the opportunity to grow, provided that they can prove their potential to demonstrate real societal and environmental benefits. However, challenges yet may exist when analysing how Circularity Capital's model can be replicated. Banks and other lenders would have to change business models when adapting to models such as 'Product-as-a-Service,' with customers making a series of payments over the lifetime of a product as opposed to making an upfront purchase. Indeed, some business models may appear to demonstrate more potential than others. Whilst models that facilitate the benefits of data in designing out waste would interest the wider retail and hospitality sectors, those based on access over ownership, thus moving away from a traditional sales model, require a more transformational change and are therefore more challenging.

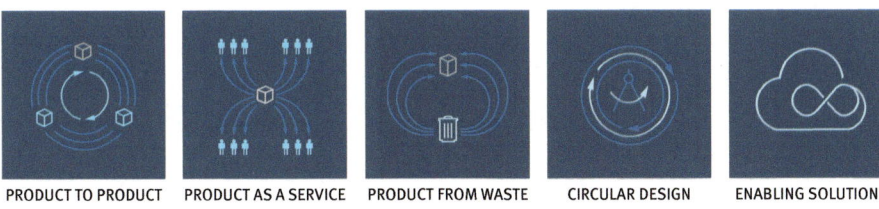

PRODUCT TO PRODUCT   PRODUCT AS A SERVICE   PRODUCT FROM WASTE   CIRCULAR DESIGN   ENABLING SOLUTIONS

Figure 24.1: Circularity Capital Business Model Categorisation.
Source: Circularity Capital.

Sources

https://circularitycapital.com/our-approach
https://www.bloomberg.com/press-releases/2021-03-08/global-blue-acquires-zigzag-global-a-
    leading-e-commerce-returns-platform

Ruth Cherrington
25 Teemill

Sector focus: Fashion and textiles

Rationale

The fashion, textile and apparel industry operates a largely linear system, where resources are extracted and made into products that are used for only a short period of time. It is believed to be the third largest manufacturing industry in the world, after automotive and technology, in addition to being considered the most polluting. There has been increased focus over recent years on the impacts of 'fast fashion.' One UK-focused survey found that consumers kept clothes for an average of 3.3 years before they were discarded or passed on.

The industry uses huge volumes of energy and water, in addition to harmful chemical substances adding to the environmental burden. One argument is that we simply need to buy less. However, even if we halved our current rate, that would still mean significant unnecessary waste and negative environmental, social and economic impacts. With the forecasted rise in global population, it is expected that "the overall apparel consumption will rise by 63%, from 62 million tons [in 2017] to 102 million tons in 2030."

Adopting circular economy principles in the textiles industry could lower greenhouse gas emissions, minimise costs and consumption of virgin material, decrease the risk of price volatility and generate new economic prospects such as 'fashion-as-a-service.' This is the approach taken by Teemill, who have used technology to 'lead the way' on sustainability, making new clothing from recovered material and closing the loop in textiles.

Leadership

Teemill began modestly, but with a goal of reshaping the fashion industry. They wanted to buy things produced from natural materials and powered by renewable energy, but could not source any, so they decided to manufacture their own.

> By rewarding people for keeping the material flowing, we're changing the way people think about their wardrobe. Rather than waste, they see assets and then some really interesting stuff starts to happen. Because our customer is also our supplier, everybody is rewarded for keeping the material flowing (Teemill).

The major focus these days is on sharing the technology they have developed to make circular fashion viable with other start-ups, charities and enterprises so that the

https://doi.org/10.1515/9783110723373-029

industry can change at scale. They release all of their technology, which has been built over the course of ten years, in order to serve as a springboard for future companies.

Approach

An online shopping platform headquartered on the Isles of Wight in the United Kingdom allows people to create and sell their own clothing designs through their marketplace, distributing made-to-order products directly to buyers and sharing earnings with the designers. Products are manufactured in real time once they have been purchased, avoiding overproduction and waste. The company embeds each garment with a Quick Response code within the wash-care label, which can be used to send the item back to get money off their next purchase. Labels are printed with an ink that is slightly more expensive, but can be removed more easily when the product is returned. By ensuring that all products and packaging are made from pure natural materials, they can make new products from the material they recover. They use renewable energy from wind and solar to power their factories in India and they reduce energy further within the factory by using technology that turns off equipment when not in use. They work closely with the organic cotton farmers, as shown in figure 25.1, to ensure the use of toxic pesticides and fertilisers are avoided; co-planting and insect traps are used instead to encourage biodiversity. Water is also carefully managed in local reservoirs and waste water is cleaned and recirculated, waste seeds are used for animal feed and vegetable oil is extracted for food products; every step in the process is carefully considered to close the loop as much as possible. Every product is designed to be sent back when it is worn out (as shown in figure 25.2), and the company vision is to share what they have learnt to change the industry as a whole.

On reflection

This example may be used to demonstrate how fashion companies can incorporate circularity into their operations. It highlights how renewable energy can help fashion last longer. It also demonstrates how to utilise current technologies to improve supply-chain visibility and function in shared-scale economies. The use of data and technology is key to this business model, enabling waste to be designed out, minimising overstocking and maximising material recovery.

Reforming the way garments are created, sold, used, collected and reprocessed and enabling the development of reverse-logistics chains are all required to establish further circular models in fashion. We also need to challenge established consumption habits in western culture, encouraging people to wear and keep clothes longer.

Figure 25.1: Natural Materials and Partnerships.
Source: Teemill, Persmission granted.

Figure 25.2: Teemill Clothing Label.
Source: Teemill, permission granted.

Sources

https://www.teemill.com
https://ellenmacarthurfoundation.org/circular-examples/an-open-access-circular-supply-chain-
for-t-shirts-teemill

Jamie Wheaton
26 Forest Green Rovers
Sector focus: Sport

Rationale

Professional football can inflict a detrimental impact on the environment. Football pitches require water, heating and energy to encourage growth and good conditions, whilst stadium operations such as the use of floodlights, screens and scoreboards can also produce significant CO_2 emissions. The travel of thousands of supporters to and from the matches themselves generates carbon dioxide emissions (supporter travel in the fourth tier of English professional football generated 15,090 tonnes of CO_2e in 2012/2013), whilst the consumption of food or drink may produce significant waste. Professional football, and sport as a whole, is therefore not immune from concerns around sustainability. On the other hand, as Forest Green Rovers (FGR) chairman Dale Vince argues, "sport has a bigger platform than business does." Sport therefore has an opportunity to promote how a movement towards the circular economy can be achieved.

Leadership

Chairman and owner Dale Vince arrived at FGR from a background in sustainability, having founded electricity company Ecotricity during the mid-1990s to provide an alternative energy source to fossil fuel combustion. After buying the club in 2010, Vince immediately started to make top-down changes within its operations, ranging from changes in players' diets to the deployment of solar powers at the club's stadium. The main challenge for Vince was engaging supporters in messages around sustainability "to create a green football club and use it to take the message of sustainability to a new audience of people."

The business model adopted by Vince focuses on the 'mission' of sustainability, with business acting as a means to an end. Rather than generating profit to shareholders, revenues generated from ticket or shirt sales are reinvested back into the club's sustainability activities. The benefits to the business model based on sustainability are two-fold. Firstly, the club attracts more fans on an international scale, with fanbases based as far afield as South Korea and the United States, thus helping to encourage revenue growth. Secondly, Vince believes that strategies such as a vegan player diet help to improve performance on the pitch through player conditioning and the reduction of soft tissue injuries, subsequently providing further increases to revenue.

https://doi.org/10.1515/9783110723373-030

Approach

FGR has adopted a holistic and regenerative approach, helping to adopt circularity within all areas of the football club. Various aspects of FGR's approach have relied on collaboration with other partners. For example, the club has collaborated with its kit provider, PlayerLayer, to produce more sustainable football kits. FGR's players have previously worn shirts made from 50% bamboo charcoal and 50% polyester, whilst they now wear shirts made from 35% recycled coffee grounds and 65% recycled polyester, thus helping to close material loops within the production process. Meanwhile, FGR has also collaborated with Vince's other company, Ecotricity, to provide solar-powered electricity – either through solar panels or a solar tracker – and carbon neutral gas to the entire club.

Meanwhile, electric vehicle charging points have been installed at the New Lawn stadium, encouraging supporters to reduce their own carbon emissions when travelling to and from games. The football pitch itself also demonstrates examples of circular innovation as well as being treated and maintained organically, free from pesticides and herbicides. The grass on the pitch is cut by a GPS-directed 'electric mow-bot' powered directly from sunlight and is also irrigated solely with rainwater gathered beneath the pitch itself as opposed to the use of mains water, thus saving 15,000 L of water which may be required to water a dry pitch.

There have been challenges to FGR's approach to sustainability that have been met with Vince's vision of the possibilities of sustainable futures. Vince's initial intention to remove red meat from players' meals – both for athlete performance as well as Vince's desire to avoid the meat trade – was met with controversy and labelled within British media as the "Red Meat Ban." The challenge was overcome, however, thanks to the principle of stakeholder engagement, with Vince claiming that the club enjoyed the controversy associated with banning food from players' diets, using the opportunity to explain to fans the sustainability-based reasons for doing so. Following seasons then saw a gradual removal of white meat, fish and dairy products from menus completely for both players and fans.

Other barriers that have occurred as part of FGR's sustainability drive have also been regulatory in nature, particularly in relation to the club's proposed Eco-Park stadium to be situated in Gloucestershire. Also incorporating the innovations built at their current stadium, it will be the world's first football stadium to be constructed from wood, thus reducing the carbon footprint associated with building materials, and the site will also benefit from 500 trees and just under 2 km of hedgerows to encourage local biodiversity. However, the stadium was initially refused planning permission in June 2019 owing to concerns around the stadium's location, noise pollution, its impact on the landscape and concerns that it would not provide enough for the local community. Yet, an appeal later that year saw approval granted with the old stadium site also set for regeneration for the local community.

Vince sees FGR at the forefront of professional football's drive towards sustainability. FGR has partnered the English Football League (EFL) to launch the 'EFL Green Clubs' scheme, designed to help the League's 72 clubs transition towards a more sustainable future. FGR saw a tangible improvement in their emissions during last season, seeing their CO_2 emissions from gas and electricity usage fall by 70% and 18%, respectively. In addition, football's global governing body FIFA has declared FGR as the 'greenest football club in the world,' therefore confirming its role in any future attempts at growing circularity and sustainability in sport.

On reflection

FGR plays an important role in the growing awareness of sustainability within sport. Sport is not immune to the challenges of the linear economy, and this has been recognised by the club's chairman. The main challenge is not posed by fan perception – as Vince argues, football fans are "receptive" to the message around sustainability – but rather the challenge of upscaling FGR's business model. FGR is run according to a business model that prioritises the reinvestment of revenue into the club and its message of sustainability as well as engaging with stakeholders such as the club's supporters. How can professional sport – and professional football, in particular – convince club owners to emphasise these values above profit? This could be a barrier, particularly when ownership within the top divisions of English football can be critiqued as favouring personal profit as opposed to values related to circularity.

Sources

www.fgr.co.uk

https://www.unfccc.int/climate-action/momentum-for-change/climate-neutral-now/creating-the-greenest-football-club-in-the-world-forest-green-rovers

https://www.sustainabilityreport.com/2019/11/12/forest-green-rovers-chairman-dale-vince-on-creating-the-most-sustainable-sports-club-in-the-world/

Jamie Wheaton

27 Grover

Sector focus: Technology

Rationale

Speaking in 2020, Grover's founder and CEO Michael Cassau highlighted how the name 'Grover' was a shortened version of 'grow over.' Specifically, Grover is 'growing over' a retail business model that is unsustainable particularly in the world of technology, where "buy now, pay later" deals last longer than the life cycle of the purchased technology itself (e.g. mobile phones and laptops).

The result of such onward growth is a platform that, Cassau argues, merges the best of retail and financial services, where subscribers can rent technology at affordable prices and over flexible periods. By providing technology as a subscription, Grover drives sustainability and movement towards a circular economy through its model (see figure 27.1) based on the return, refurbishment and re-rental of consumer technology devices.

Leadership

Grover was founded by Cassau in 2015 after his own personal experience in the search for flexible, short-term usage of furniture. As an economist, Cassau realised that there was little marginal utility to be gained from the short-term usage of assets and furthermore highlighted the inconvenience in having to buy or use finance options to fund such short-term utility. Cassau subsequently developed a vision for the consumer technology sector, allowing subscribers to flexibly rent technology without the commitment normally associated with a purchase or finance option. The adoption of a 'Product-as-a-Service' (PaaS) model not only facilitates the flexible rental of technology but also the refurbishment and reuse of technology.

The upward trend of Grover's business performance demonstrates the popularity of its service, which emphasises access over ownership. As of December 2021, Grover has a subscriber base of over 1 million customers in Europe, and their US market had already seen rapid growth since Grover's US launch in September 2021. Customers across Germany, Austria, the Netherlands, Spain and the United States have 300,000 devices to choose from (as of December 2021), all of which were new when purchased and are refurbished between each rental until they break. The funding round announced in December 2021 means Grover could bolster its stockroom to 1.5 million devices, particularly in its growing US market.

https://doi.org/10.1515/9783110723373-031

Approach

Grover's model is best described as a 'pay-as-you-use' service, as opposed to the linear model of production and consumption. Grover currently advertises over 3,000 products within nine different categories (mobile phones and tablets, computers, drones, audio technology, wearables, cameras, gaming, smart home and home entertainment), available to subscribers for a rental period that is controlled by the customer. Grover's services are predominantly aimed towards subscribers who want to access the latest technology without the commitment associated with ownership. This, Cassau argues, allows subscribers to avoid the financial burden of purchasing technology that would not be worth buying or indeed financing with a loan over the long term. The service has proven to be popular with subscribers, with one in three choosing to rent their device longer than their initially selected rental period.

However, it is not only B2C subscribers who are supplied by Grover's services. Grover also supplies B2B clients, with companies seeking to source technology flexibly in order to quickly perform business functions and onboard new employees. The ability to supply B2B clients with rented technology allows other companies to move towards a circular economy, with businesses also benefitting from Grover's stock of modern but refurbished technology. Indeed, all technologies available through Grover are advertised "as good as new." This advertisement does not impact on customer service, however. As Cassau himself argues, "customers really don't care," likening the rental of technologies to a subscriber to the use of a "rental car or hotel room."

In summary, once a subscriber has returned their technology to Grover, it is then refurbished to an 'as new' condition before being rented to another subscriber. In contrast to the traditional linear model of production and consumption, the refurbishment and subsequent rental of technology significantly increases the product value and the length of its life cycle. According to Cassau, the average asset owned by Grover will last for an average life cycle of two to three years, which may typically see it used by three to four different subscribers. At the end of its life cycle, the asset is then sold to the open market for a reduced price or indeed recycled, thus increasing the value of the asset's materials. Subscribers also benefit from 'Grover Care' in which Grover covers 90% of the repair costs in the event of any accidental damage. Through its regenerative model, Grover aims to prevent 24,000 tonnes of electronic waste by 2025.

Grover's business model has paved the way for its growth into new territories as well as initialising partnerships with global tech giants. Based in Germany, Grover launched its services in the Netherlands in 2020, before launching its services in the United States in September 2021. Grover also provides rental services in Austria and Spain. Samsung has launched a subscription service with Grover, with subscribers able to rent technology from Grover through each brand's respective online shops, whilst Microsoft and Grover also collaborated to support education by facilitating easy access to technology for school students in Germany during the COVID-19 pandemic.

Also in 2021, Grover announced the launch of its first financial services product, the Grover Card, to bring embedded finance into their offering. The Grover Card is a Visa Debit Card, issued by Solarisbank, and allows users to transform daily spending into Grover Cash rewards, offering a 3% return on everyday purchases that can be redeemed to lower or completely cover the cost of tech rental.

On reflection

The traditional production and consumption model associated with the linear economy of technology can be keenly felt in our environment. Indeed, the failure to maintain digital technologies within resource loops results in negative externalities such as e-waste dumps, which can be found in locations around the world. As of 2021, the population globally produces 50 million tonnes of e-waste every year which, at the same rate of growth, could increase to an annual figure of 120 million tonnes by 2050. This will increase the release of harmful chemicals that damage public health and impact the environment. Grover's mission to provide regenerated technology on flexible terms to subscribers who wish to avoid commitment to a large purchase or finance scheme helps to reduce the need for such e-waste dumps in addition to the use of resources required to produce new digital products more generally. Moreover, the PaaS model is already popular particularly within the digital industry, with directly positive results for sustainability. Spotify and Netflix both allow access to digital content for a small, flexible monthly fee, thus reducing the need to produce discs. Grover represents another example of the PaaS model, and Cassau highlights the similarities between Grover and Netflix through the diverse nature of choices available on each platform. The challenge that is presented towards the wider potential is the need to convince consumers that access is indeed a more valuable option over ownership. Nonetheless, with Grover's recent growth into new markets as well as the popularity of other platforms such as Spotify and Netflix, this is a barrier that may be easily felled over the coming years.

CHOOSE　　　　　　USE　　　　　　RECIRCULATE

Figure 27.1: Grover's process.
Source: Grover.

Sources

www.grover.com
www.youtube.com/watch?v=YiOJTaC7ruw
www.circularitycapital.com/podcast-1/2021/1/25/podcast-1-37c9g-ynnnn

Isabelle Housni
28 ReStore project
Sector focus: Furniture

Rationale

ReStore is an organisation in Trondheim, Norway that aims to reduce waste from household goods and furniture, generated by university students in the local area and more widely. Trondheim is a city with a big university presence where students moving to the area to study need to furnish their student accommodation. At the end of the academic period, however, this furniture is discarded due to the lack of market for the large volume of items no longer needed by the departing students.

ReStore has worked to close the gap between supply and demand by storing the unwanted furniture over the holiday periods and then offering it to new students moving to the area for free. This prevents the large quantities of furniture already bought and still with a long lifespan remaining from ending up in landfill, while also promoting reuse, which is another tenet of the Circular Economy. This project has since gone further, expanding to include bicycles and their repair as well as the repair of donated furniture, in related projects known as ReSykkel and ReBuild.

Leadership

The organisation ReStore was founded in 2019 by three students at the Norwegian University of Science and Technology (NTNU) who experienced first-hand the struggle to find second-hand furniture upon their arrival, yet then saw the discarded student furniture going to landfill at the end of the semester. This inspired the organisation ReStore, which has since had five new leaders at the helm supporting the project's Circular Economy aims.

The project partnered with Trondheim's student welfare organisation, known as SiT, who sponsored the project in its initial stages, providing seed-funding and storage premises after the idea was pitched to them with a long-term plan to show its viability. The plan needed to show that a system with longevity and that could be taken over by others could be implemented, which has proven to be a strength of the project. Successive leaders have introduced new initiatives such as the creation of ReSykkel, a project that retrieves and repairs bicycles; as well as the most recent project addition, ReBuild, which incorporates another fundamental part of the Circular Economy – the repair of donated items.

https://doi.org/10.1515/9783110723373-032

The founders of the ReStore project felt that finding a good first partner who believed in the project, in this case SiT, was instrumental to its success. This support from SiT also encouraged other organisations to support the project. Support schemes such as the Climate-KIC Incubator program and the Greenhouse Incubator program that both target Circular Economy initiatives were able to help the ReStore project in its early stages. The project also benefitted from Norway's government-mandated support schemes where banks must donate a proportion of funds to not-for-profit organisations.

Approach

The ReStore project works by providing a space to store student's unwanted furniture at the end of their academic stay and donating it to students moving to the university. The ReStore project also offers a pickup service for larger items such as sofas, as many students do not have a car to transport these themselves. It started receiving donations of furniture just 3 months after the 10-page proposal was seen by SiT in May 2019.

While the ReStore project was widely recognised as necessary by students and alumni of the university, the first challenge was to find suitable space to store the furniture and access capital to fund it. In addition to providing seed funding, SiT provided the organisation with its first premises when other landlords were uninterested, providing an unused kindergarten to store furniture as well as lending them a van in order to collect furniture from students.

The Student Welfare Organisation was able to contribute to the ReStore project due to a shared aim to reduce student waste, which ultimately led to reduced fees for SiT, as they are responsible for the proper disposal of furniture from student villages and thus they also saved money by sending less furniture to landfill. There continues to be a collaboration between these two organisations, ReStore and SiT, due to their shared aims of reusing and reducing unnecessary waste.

In its first summer, the project was able to save approximately 12 tonnes of carbon from the furniture that was diverted from landfill and instead reused, excluding the emissions that would have been generated in transporting the furniture to landfill and buying new furniture.

The ReSykkel project sits in its own premises in a shopfront in Oslo and specialises in picking up abandoned and unwanted bicycles, including broken ones. When possible, these are serviced or repaired, using parts from unrepairable bicycles when needed. These are then sold for affordable prices. By extending the life of the bicycles, reusing parts and preventing unnecessary waste, they aim to help the environment and change consumer behaviour.

Since its inception, ReStore has expanded into other Norwegian cities such as Ålesund, with plans to also move into Gjøvik, with the help of the Student Welfare organisation, SiT. The project has also recently started ReBuild, which aims to repair broken furniture that has been donated, building collective knowledge as the project progresses with the help of their members. They have received support from Trondheim Kommune, the local authority, by participating in Re-Use Week, known as Ombruksuka, where they repaired furniture throughout the week.

On reflection

An interesting point discovered in the creation of the ReStore project was that many students attending university in Trondheim had previously identified the same problem with furniture and had intended to do something similar, however had been unable to act upon it. This was largely due to the need for a long-term approach, in order to embed a functioning system into the student ecosystem, as opposed to a quick fix that fell apart once the original creators had left the university.

This problem is very reflective of wider society, where environmental problems and their solutions can be identified; however, the steps to act on them can appear prohibitive when there are no structures in place to support them over the long term. In the case of the ReStore project, having organisations and government initiatives with similar goals helped put a foundation in place that could be succeeded by others, supported by new students on the SiT.

Sources

https://www.resykkel.com/
https://universitas.no/sak/67327/free-your-stuff-and-restore-it/
http://rethinkrestore.no/

Isabelle Housni
29 Packshare
Sector focus: Packaging/digital

Rationale

Packshare has created a digital platform to facilitate the reuse of packaging within local communities. Businesses, with packaging they no longer need, are connected through the Packshare website with other local businesses and individuals who require packaging, creating a circular model where packaging is not thrown away or wasted.

Co-owners of the company, Roo Pescod and Louisa Street, founded the tech company Packshare in 2018 after Louisa was inspired to create more sustainable supply chains for packaging when working for a business packing products for delivery. She found that one business would be throwing away packaging that the business next door was buying and wanted to provide a solution.

This is not a problem unique to Cornwall, England, where the business was founded, as the most recent data from the UK government, in UK Statistics on Waste, shows a total recycling rate for packaging at 62.2% and dropping to 47.4% for the recycling of plastic packaging.

Also contributing to waste in packaging supply chains are changing consumer behaviours. When packaging is sent to larger retail stores with stock, there are systems in place to ensure it is collected at the site to be reused or recycled, but due to shopping habits shifting online and smaller e-commerce businesses becoming more prolific, much more of this packaging is instead going to individual homes and is increasingly more likely fall out of the supply chain and be thrown away. Packshare thus offers a useful way to redirect packaging back into active use and supply.

Leadership

Since Packshare was started by its two co-founders in 2018, they have added another two members to the team in new programming and marketing roles. As Packshare is a start-up with a clear purpose, activity can be led by all members of the team. The business and its impact are predominantly driven by the number of members who sign up to the scheme. In order to be successful in its aim, Packshare requires large numbers of businesses or individuals registered to both give and receive packaging, which can be encouraged through marketing and word of mouth.

The digital infrastructure provided by Packshare is also an enabler for the circular economy, creating a place where interested parties can register and view others

https://doi.org/10.1515/9783110723373-033

who are already engaged to receive packaging. Thus, all members of the small team comprising Packshare are well-suited to encourage and facilitate the onboarding of clients, with skills suiting the project.

Approach

There are two types of users of Packshare: one with packaging to give and the other as a recipient of packaging. All interaction using the Packshare site is free, although users can choose to donate if they would like to.

If an individual or a business has packaging to give to others, they can type in their postcode and detail the type of packaging they have from a drop-down menu, which will then reveal a list of businesses nearby with details such as the distance from the postcode entered and the types of packaging they require, as well as contact information and opening hours. Users with packaging to give can also find businesses near them using the site's interactive map.

Once an appropriate business has been identified, the giver of packaging can then drop the packaging to their location, using the information provided to them. While there is currently no option to ask the business to collect it instead, they can be contacted using the information provided to see if this can be arranged.

On the other side, users who would like to receive packaging must create an account on Packshare. Once they have provided an email address and a password, they can proceed to fill out details of their business, opening hours, delivery instructions and the type of packaging they would like to receive. This then adds them to the interactive map and to the directory.

One of the limitations of the Packshare system is that currently the onus is on the user with packaging to take it to the user who would like it, and this can be a potential barrier to uptake. For example, a business might not have the time needed to drop off packaging to other locations, particularly if there are none in the immediate area. This is currently an area under development, and Packshare is creating a directory of business profiles for those with packaging to give, rather than solely those who would like to receive it.

By adding this new functionality, along with the ability for businesses to outline the type of packaging they have so that they might be sought out by those who need packaging, the flexibility of the system can be increased. Future plans also include expansion beyond the UK, where Packshare is currently operating, in order to facilitate small businesses from around the world to be able to easily share and reuse packaging.

On reflection

While there is a focus on businesses registering on Packshare, both to receive and to give packaging, if messaging were altered to encourage more individuals to sign up, there could be even more impact and packaging saved from landfill. E-commerce spending has been increasing annually, which means more packaging sent to consumer's homes, therefore it seems increasingly viable to focus on individuals as well. While individuals are allowed to use Packshare where appropriate, the lack of targeted messaging to this audience might be a barrier.

Additionally, introducing additional points for packaging to be dropped off and collected from, instead of just business addresses, might increase the number of users and thus packaging that is reused. While the default is for those who have packaging to drop it off to those who need it, having a collection/drop-off point might be a more accessible way for some clients to drop off packaging. For example, using a public location like a village hall or a large supermarket might encourage or remind individuals and businesses that there is another way for them to deal with unwanted packaging in a more sustainable way.

Sources

https://www.packshare.org
https://assetspublishing.service.gov.uk/government/uploads/system/uploads/attachment_data/
    file/1002246/UK_stats_on_waste_statistical_notice_July2021_accessible_FINAL.pdf

Isabelle Housni
30 Lendwithcare
Sector focus: Finance

Rationale

Lendwithcare is a circular model for offering microfinance, such as small loans and banking services, to low-income clients. For low-income individuals, it can be very hard to access capital to fund business ventures, particularly in countries that struggle with poverty and do not have significant banking infrastructure in place. CARE, the impartial charity delivering humanitarian aid and development, has noted that often other microfinancing institutions (MFIs) do not cover the most rural communities who are usually in the most need, due to high costs of reaching them and the comparatively small demand when compared with larger populations in towns and cities.

Lendwithcare aims to rectify this by creating a community of those who can donate small amounts of money and connecting them with entrepreneurs requiring microfinancing. This money helps small business owners across the world finance their ventures. When the loan is repaid, it is then available to fund another loan request on the platform; thus, the circularity adds more benefit to a regular loan model.

Leadership

Lendwithcare is run by CARE international, a charity founded in 1945 that provides humanitarian support globally. The driving force behind Lendwithcare is comprised of a small core team of five members, with a wider network supporting them, including other CARE International UK departments.

As Lendwithcare operates within the financial sector, arranging approximately 900,000 loans from 60,000 lenders to 129,000 entrepreneurs, they work in partnership with 15 MFI partners. These MFI partners facilitate loans and are managed by the Lendwithcare team. The team of five also handle fundraising at different levels, marketing and communications, as well as arranging visits to the different areas being supported.

In addition to the wider CARE charity, there is also the support care team who provide help to lenders navigating the site as well as its operations. There are also five honorary advisors who use their experience across different sectors and roles to offer specialist advice, including technology, entrepreneurship, accounting and not-for-profit expertise to help with strategic planning, fundraising and governance. Additionally, volunteers help run many areas of operation, which helps embed the altruistic culture Lendwithcare fosters with its aims.

https://doi.org/10.1515/9783110723373-034

With expertise held across the entire network of Lendwithcare, in addition to the philanthropic aims of the charity, all areas are motivated by a similar goal of providing support to those in need. As the organisations behind Lendwithcare, such as its partner MFIs and CARE, are vast with many employees as well as different business aims and operations, it is benefitted by having a smaller central team solely responsible for Lendwithcare that can leverage its large network while ensuring that its own aims are fulfilled. By having a separate support care team that assists individuals with lending on the platform, they are able to ensure that the team is well placed to help support as many loans as possible.

Approach

Lendwithcare offers microfinancing to low-income clients. While the organisation focuses a lot of its marketing and communication on small loans, this also includes other financial services such as savings accounts, money transfers and even insurance.

Lendwithcare has partnered with MFIs local to the areas being reached, such as Pakistan, Cambodia, Georgia, Ecuador, Peru, Philippines, Zambia, Malawi, The Occupied Territories of the West Bank and Gaza, Vietnam, Zimbabwe and Rwanda. These partners must have a focus on improving the lives of people in poverty, as well as satisfying necessary criteria set out by the not-for-profit.

Entrepreneurs get in contact with one of the local MFIs that work with Lendwithcare and once their idea has been assessed for viability, their profile will be uploaded to the Lendwithcare website. Prospective lenders can then access the site and choose an entrepreneur to support, where the loan request is fully detailed. Information provided includes details regarding the entrepreneur requesting financing, as well as the intended business venture, the amount required, repayment terms and which MFI partner will be underwriting the loan. The average repayment term is 12 months.

Lenders can choose to lend a minimum of £15 up to the whole amount. This money goes from the Lendwithcare platform to the MFI partner, who then sends it to the entrepreneur being supported, at each repayment. The entrepreneur who received the loan then follows a regular repayment schedule in order to return the original investment. The structure is similar to a crowdfunding service; however, the money is in a controlled environment where it is expected to be repaid so that it can be reused.

One challenge that Lendwithcare has as an organisation is loans being defaulted on. While the rates of this are low, cited at under 1%, it is something that can occasionally happen for any number of reasons. In this case, the lenders are informed that the repayments will no longer happen, as well as being given the reason for this. A mitigating factor is that, due to the small financial amounts given individually that collectively reach the target amount, an individual usually does not have a large loss. To avoid as many instances of loan defaults as possible, MFIs give each entrepreneur training and support to ensure they are in the strongest position possible.

Finally, once repayments are made, funds can be lent again to a new entrepreneur using the platform, withdrawn or donated to the CARE charity underpinning Lendwithcare. There is even a mechanism in place to ensure that money left in dormant accounts (12 months of inactivity) are donated to the charity, under their terms and conditions. This again maximises the benefit of all the money donated as opposed to a traditional loan.

By bringing together different skilled organisations from the charity and the financial sector with a common aim of supporting areas that are underserved, and combining this with a digital platform that is easy to access and use, Lendwithcare is able to transform the market for microloans, making them more accessible to the public and leveraging a community of people far away from the target audience. By being able to partner with even more MFIs, the areas supported can be increased and thus even more people can be helped.

On reflection

Lendwithcare has offered a recognisable but new model for microfinancing, combining elements of crowdfunding with circular economy principles. While it is important that there are well-structured and experienced institutions overlooking the financial aspect, many were unable to operate to the same level during the COVID-19 pandemic that affected the world. It may be useful to explore additional ways for entrepreneurs to be able to reach the Lendwithcare platform.

The structure that Lendwithcare uses to provide microfinancing abroad could also be used locally to provide smaller financing options for SMEs or sole traders. For this to be adopted as a model domestically, it would need to have backing from a local financial institution; it could provide the same benefits as crowdfunding platforms, but instead be set up as a loan that is intended to circulate between new entrepreneurs once repaid. This could become a springboard for innovation and productivity, by providing the funds to realise a venture that might otherwise not be funded (e.g. offering opportunity to individuals with poor credit). While there would need to be structure such as provided by MFIs, it could allow people to contribute to projects in their local communities and further afield on a continuous basis.

Source

https://lendwithcare.org

Jamie Wheaton

31 Páramo

Sector focus: Fashion and textiles

Rationale

Páramo was founded in 1992 by Nick Brown who, when seeking to develop a brand of more comfortable outdoor clothing, discovered 'Creaciones Miquelina,' a small workshop in Colombia which provided practical help to women trapped within exploitative settings. It is this partnership that has seen Miquelina's work certified as Fair Trade by the World Fair Trade Organization, whilst Nick also underlines how the jackets are produced with pride within the Miquelina workshop. Indeed, Nick underlines how outdoor explorers who wear Páramo may have no idea as to how the clothing itself has transformed the lives of those who have sought refuge within the workshop.

This ideal informs the scope of Páramo's mission as well as its partnership with Miquelina, the creation of 'high-competition clothing that changes lives both in the outdoors and in a poor community in Bogotá.' Indeed, over 80% of Páramo's annual production occurs at Miquelina, with both partners benefitting from aligned long-term social and commercial objectives.

Leadership

Páramo benefits from its founder's experience in the creation of waterproofing manufacturer Nikwax, the first company in the world to produce water-based waterproofing products to replace solvent-based aerosols during the 1980s. These waterproofing products are incorporated within Páramo's clothing today.

Nick Brown's concern for socio-environmental development is implicit within Páramo's approach, proving that successful growth does not presuppose the disposal of ethical production. Indeed, "Páramo is a company of people who want to make a difference as well as make a living, and our policies and programmes reflect this." Páramo's ethical strategy is executed through five key areas, ensuring that the company can make a positive impact on both society and the environment itself. In doing so, Páramo promotes ethical manufacturing through the creation of circular jobs and the provision of training, whilst also promoting the circularity of its materials in addition to the removal of harmful substances from its manufacturing process.

https://doi.org/10.1515/9783110723373-035

Approach

Páramo's approach underlines how the manufacturing and supply of clothing can be conducted in an ethical and regenerative fashion. Páramo's first key approach is based on 'fair trade manufacturing' – namely, the provision of employment and training for vulnerable women sheltered in Miquelina. Creaciones Miquelina was established by Sister Esther Castaño in 1977. However, with only 2 sewing machines available for 20 women, Miquelina was soon in need of support. As Neil recalls, "as soon as we got involved with Miquelina, we started to see what they were doing. And more importantly, we began to understand the significance of what we were doing for them." This has continued into the present day, with almost 50,000 high-quality garments created in Miquelina during 2015. This continued partnership allows Páramo to continue its role within the development of skills of those who have been rescued by exploitation. The circularity of finance is demonstrated by the growth of Miquelina, which now boasts 44 years of experience, a workforce of 200 women within its workshop, as well as a total of 500 people to have benefited from the shelter during its existence.

Circularity is also ensured by Páramo's recycling scheme, which prolongs the value and usage of its clothing. The scheme encourages customers to return any Páramo product in exchange for a discount on new Páramo products. Garments that are returned are either repaired and resold at reduced prices with a 12 month warranty, or are alternatively recycled if beyond repair. Any recycling that does occur is done so through a chemical recycling process, thus preserving the quality of the material. This ensures that materials are maintained and regenerated within resource loops, as well as reduces the environmental impact associated with sending clothing to landfill. Páramo also makes the transition to the circular economy easy for its customers who can choose to return garments by post or in-person through participating retailers. Páramo ensures sustainability through its third principle – 'gear that lasts.' The incorporation of Nikwax analogy fabrics, which provides protection from water without the need for taped seams or membranes, makes Páramo's clothing more durable through the reduced risk of damage. The need to recycle is therefore reduced by the durability of garments that Páramo argues can last for at least 15 years.

Páramo makes further contributions to sustainability through its fourth principle – 'support for conservation.' Páramo and Nikwax have both been calculating and offsetting primary carbon dioxide emissions since 2007 as well as donating to the World Land Trust's (WLT) Carbon Balanced Programme to offset pre-2007 emissions. This donation is used to regenerate damaged areas of habitat and prevent further deforestation by extending nature reserves in Ecuador. Páramo also makes additional donations to the WLT Action Fund in order to protect ecosystems that are under imminent threat. Páramo's final principle is the manufacture of clothing that is 'guaranteed PFC free.' In doing so, Páramo ensures that its garments are water-repellent without the need for treatment from polyfluorinated compounds (PFCs). Indeed, the use of PFCs can incur a detrimental environmental impact during the manufacturing process. This has presented a challenge for Páramo, which admits that it cannot know which chemicals are applied to materials it may source from across the world. In response, Páramo now applies a strict methodology in testing clothing. Suppliers must meet a strict manufacturing specification, whilst Páramo tests its products for fluorine as well as applies its own Nikwax fabric finish.

On reflection

Páramo's journey is inspirational, encouraging circularity not only in terms of the products that are produced but also for the capital, revenue and skills that are recirculated through its partnership with Miquelina. The challenge here would be how to replicate this circularity across the wider retail industry. Páramo's founder credits ethical manufacturing as benefitting the high-standard, durable goods it sells to its customers. Therefore, the business model is based above all on quality rather than quantity. Could this business model be adapted by textile-based retailers whose profit margins are based upon the mass selling of goods at a faster rate? Páramo's recycling scheme could also be difficult to replicate across the industry, with other retailers possibly lacking the profit margins required to reward customers for trading in their old garments. Nonetheless, these are unique selling points that help Páramo to stand out from the crowd in their contribution towards a circular economy.

Source

https://www.paramo-clothing.com

Ruth Cherrington

32 Circular & Co

Sector focus: Product design

Rationale

Billions of cups of take-away hot beverages are sold each year, most of which are sold in single-use cups. Because the cardboard cup is coated with a plastic layer, disposable paper coffee cups are generally non-recyclable, making it difficult to separate these components throughout the recycling process. Even if they are free from plastic, most refuse collection businesses regard them as contaminated waste. Coffee is a food stain, not a contaminant, and should not cause any technical issues in mainstream cardboard recycling. However, the perceived quality is not worth the risk for businesses, especially when compared to the large amounts of high-quality cardboard they receive from outer cardboard packaging.

The 'Circular Travel Cup' is produced from discarded paper coffee cups collected from coffee shops, supermarkets and other locations. The cup has a one-handed push-and-click top that can be opened. Additionally, the 360-degree lid allows users to sip or pour coffee from any side, making drinking simpler and safer when driving or travelling. Single-use cups may be requirements in certain environments due to hygiene or convenience. As a result, single-use cup recycling initiatives are critical for both consumers and companies to lessen their environmental effect. These recycling initiatives should be considered alongside reusable coffee cups and biodegradable alternatives are waste-reduction possibilities.

Leadership

Former Dyson designer Dan Dicker and a team of recycling professionals created a new coffee cup with the goal of developing a superior and sustainable reusable cup. In 2003, he founded his firm Circular & Co., manufacturing circular products and collaborating with a network of industry professionals to achieve true circular solutions. He wanted to encourage individuals, communities and businesses to shift their consumption habits and prioritise circular design. With a much bigger voice and industry presence, the firm continues its objective to provide value to waste by designing products led by a more determined brand. The company's founder is convinced that circular design will eliminate waste and pollution, and that it offers a genuine opportunity for practical change at a time when the world is in dire need of answers.

https://doi.org/10.1515/9783110723373-036

Approach

Development of the 'Circular Travel Cup' came after the launch of a paper cup recycling initiative that had the goal of reducing the amount of waste generated by the millions of paper cups used in the UK each year. The initiative recognised that the problem of cup recycling could not be solved by a single organisation and that, in order to achieve its goals, coordination across the supply chain was necessary. Understanding how to capture the material was a significant component of the company since segregating material at source is a critical feature of any good recycling strategy. Significant progress has been made highlighting the UK's fastest growing waste stream to the attention of companies, the public and the media. It has also attracted a rising number of well-known brands and businesses and is currently the leading voice in cup recycling.

The product design and business model incorporates several components of the circular economy concept. Firstly, the product incorporates materials recovered from discarded products. Secondly, the product is designed to be durable and last more than 10 years. Products are designed for longevity and they provide support to repair them where possible. For example, the cups have a seal that is smooth on one side and stepped on the other and is designed to fit exactly on the inside corners of the lid. The company provides easy, free seal replacements if the seals are lost or broken. The material is free of harmful industrial chemicals and is 100% recyclable. Thirdly, discarded products are collected through take-back systems and reverse logistics. To ensure that they capture as many resources as possible, they have created a 'take-back scheme' to offer these materials a second use. A discount is provided to anyone who returns their product, and some items are eligible for a free component switch.

On reflection

The circular economy concept starts from the outset to avoid waste and pollution from initiation. Recycling will not be enough to offset the massive volume of waste we create in the face of our present environmental issues. While recycling is obviously important, we must guarantee that goods and materials are intended to be reused, refurbished and remanufactured from the start.

The partnership is this case study between waste management companies, product designers and coffee suppliers is one of the UK's first dedicated cup collection and recycling operations. The cups are sent to a specialised reprocessing facility, where the plastic film and paper are separated, allowing the components to be recovered and repurposed into new goods. The discarded materials are used to create new items. The 'circular cup' is made with a virgin polypropylene inner and recycled polypropylene outer, combined with single-use paper cups. One of the barriers to a truly closed-loop system is the lack of food-safe recycled material. Contamination is a technical challenge that currently limits how circular any economy can become. We need to ensure that hazardous or toxic materials are not recirculated from legacy products.

This case study highlights the challenges around infrastructure and knowledge needed to allow materials to be recovered and recycled into new products. It illustrates the need for partnerships and collaboration to change opinion and challenge long-held cultures and habits. Further research is still needed to understand how to encourage consumers to adopt reusable products and services. For example, despite the fact that 69% of individuals own a reusable cup, just one out of every six people uses it every time they buy a hot beverage. Several coffee shops have started trial programmes where customers may pay a modest deposit for a reusable cup they can use for both hot and cold drinks (see figures 32.1 and 32.2). Customers will be able to use their cup and then return it to a kiosk or store, where their deposit will be refunded. This comes after the UK announced plans to decrease single-use waste, including introducing a small 'tax' on disposable cups to discourage their use. The money raised from this 'tax' has gone towards initiatives encouraging reuse, new recycling infrastructure and litter prevention around the UK. The use of reusable cups has more than doubled in certain establishments since the cup 'tax' was implemented.

Figure 32.1: Circular & Co Reusable Thermal Cup.
Source: Circular & Co.

Figure 32.2: Circular & Co Reusable Thermal Flask.
Source: Circular & Co.

Sources

https://circularandco.com
https://circularandco.com/reusable-coffee-cups

Isabelle Housni

33 Terragr'eau

Sector focus: Agriculture and food systems

Rationale

Terragr'eau is a French project created to protect the Evian watershed, in the town of Évian-les-Bain, from runoff caused by industrial farming in the local area. Evian water is natural spring water drawn from the heart of the French Alps; however, 60% of the surrounding land around it is used for agriculture whose practices threaten the quality of the water.

One of the environmental impacts to the Evian watershed area, or impluvium, is nitrate pollution caused by both artificial fertilisers and animal waste that are extensively used in agriculture and that leach into water sources, polluting surface and groundwater. This can affect the quality of the water, which is subject to strict rules determining the levels of various chemicals allowed within it, amongst other standards such as microbial and radioactive limits.

The sustainable solution provided by Terragr'eau was to install an anaerobic digestion unit and composting plant to take in organic waste from the local area and convert it into biogas, a renewable energy source. Using the agricultural effluent, as well as waste from the cheese industry and green waste collected from local sites, the Terragr'eau project has turned organic waste that posed an environmental threat into a renewable energy source, demonstrating the benefits of a circular economy.

Leadership

The project was a result of a long consultation period between interested parties, such as farmers, Evian the mineral water company and local combined public authorities, known as Communauté de Communes du Pays d'Evian (CCPE). This Evian inter-communal authority coordinated and led the project after the feasibility study, which commenced in 2006. The project was also supported by the Danone Ecosystem Fund, whose aim is to provide inclusive and innovative solutions to ecosystems in locations in which they operate, with Danone owning the Evian water brand.

In addition to organisations local to the area, the operator and designer of the digester, SAS Terra'greu, is made up of three companies – Serpol (an environmental management company), Biovalis (a sewage treatment company) and Methanergy (biogas recovery plants design and installation). The group SAS Terra'greu operates

https://doi.org/10.1515/9783110723373-037

the now independent site. The project was built collectively with input from all parties, including local farmers, and has offered collective benefits to all involved.

The project received €10.8 million in funding, supplied by various organisations including public aid, such as ADEME, the French Environment and Energy Management Agency and the European Regional Development Fund. It also received funding from water company Evian, the CCPE (the inter-communal authority for Evian), the operator SAS Terragr'eau and the Danone Ecosystem fund, all of which were crucial to the project's success. This investment was also vital for the farmers in the area who would be unable to finance such a large project themselves, and this collaboration has enabled the continuation of agriculture in the area, but more efficiently and sustainably with harmful waste captured and reused.

Approach

Despite being in the works a decade before, the Terragr'eau digester was commissioned in 2016 and became an independent site by 2018. The Terragr'eau site is comprised of two anaerobic digesters that can treat 40,000 tonnes of organic waste per year, of which 36,000 tonnes are agricultural effluent.

The liquid biomass is heated to 38 °C and mixed continuously for 29 days, heated using the biogas produced in the digester. The site has digestate tanks for sorting and storing waste material with a total volume of 6,750 m^3, which can be generated in 24 h of production. There is also gas storage at the top of the digester and a 400 nm^3/h flare.

The purification unit on-site transforms the biogas produced by the fermentation process, 65% methane, into biomethane after activated carbon is used for filtration as part of the pre-treatment. To be used in the wider natural gas network, the biomethane must be in line with the gas network operator's specifications. Once it has been purified, it is checked for compliance and odorised so it can be detected in leaks, before being injected to be used as natural gas that can be used as an energy source.

In addition to the operator group, SAS Terragr'eau, the farmers who consulted on the project also created a collective using the same name, Terragr'eau, to control the digestates produced as a by-product of the gas. This is spread over their farms, covering a total area of 3,100 ha. Seventy-seven farmers in the area were benefitted by gaining organic digestate used as a nutrient-rich fertiliser.

Overall, the Terragr'eau project brought a 10% reduction of CO_2 emissions in the Évian-les-Bains area, as well as providing 2.2 million euros worth of biogas per year, of which 825,000 m is biomethane. This provides 9.2 MkWh/year of energy and shows that implementing a system to capture biowaste can provide a renewable source of energy, reduce emissions in the area while providing an organic product that can be used as fertiliser. Furthermore, nearly 12,000 individuals in the area supplied by the Evian spring have had their drinking source protected.

On reflection

The Terragr'eau project was a large-scale operation that needed significant amounts of investment, time and the collective efforts of many organisations to be achieved. While this is worthy of merit, this also presents a barrier to smaller scale operations wishing to emulate Terragr'eau's achievement. About 27% of the funding for the project came from public aid organisations, whereas the rest was provided by local authorities, businesses and funds earmarked for innovative projects. This suggests that projects requiring huge infrastructural investments need the input of multiple stakeholders to be enacted. While the contributions from local farmers were also integral to the Terragr'eau project, it was only possible within a wider network of support. Thus coordination between public bodies, businesses and charities as well as regional authorities should be encouraged, while also carving a space for individual contributions in order to maximise impact.

Sources

https://www.eclaira.org/initiative/h/terragreau–anaerobic-digestion.html
https://www.grdf.fr/entreprises/carte-de-france-des-references/biomethane/terragr-eau
http://ecosysteme.danone.com/projectslists/terragreau/
https://www.reussir.fr/lait/methanisation-un-epandage-collectif-des-digestats-avec-la-sica-
    terragreau

Ruth Cherrington
34 LUSH cosmetics
Sector focus: Fast-moving consumer goods (FMCG)

Rationale

Fast-moving consumer goods (FMCGs) are products that are sold in a short period and for a low price. FMCGs account for 60% of total consumer spending, 35% of material inputs into the economy and 75% of municipal waste. Packaged food, beverages, toiletries, cosmetics and medicines are all examples.

The challenge is in understanding how to deliver these goods with the desired quality, whilst minimising the use of finite resources and the generation of waste. Rethinking the product might include redesigning the basic concept, looking at how it is created, or determining whether the same value can be offered in a new size or shape. Due to the fast-moving nature of FMCGs, packaging has an important role to protect the product quality. Product design innovation can alter packing requirements while preserving or even increasing the user experience. By applying circular design principles, it may be possible to modify the system to make packaging reusable, recyclable or compostable or to eliminate the need for packaging altogether.

LUSH is a British cosmetics company that use upstream innovation to rethink its products and eliminate packaging waste. They avoid the need for bottles, containers and tubes for many of their goods by developing cosmetics and personal care items (like shampoo and soap) in solid form, rather than liquid. Since 2007, the company has sold more than 38 million package-free 'naked' shampoo bars throughout the world, avoiding the use of more than 90 million plastic shampoo bottles.

Leadership

The company was founded in 1995 by six partners emerging from the demise of a previous mail-order business. It was formed on the idea that fresh, innovative, radical cosmetic goods should be created to fill a market need. The company's core values are based on these ideas. They have a rigorous supplier policy that prohibits them from using items that have been tested on animals. Another example of their environmental commitment is the campaign against over-packaging, which highlights an understanding of how much waste the average person generates in their lifetime. As a result, *around 66% of their items sold each year are 'naked.'* The founder has received royal acknowledgement for their contributions to the beauty industry and has been recognised as one of the 'most influential' individuals on multiple occasions.

https://doi.org/10.1515/9783110723373-038

Approach

Package-free or 'naked' cosmetics are a declaration of love to minimalism and environmental protection. Solid personal care products eliminate the need for bottles, containers and tubes. By reducing product weight, carbon emissions connected with transportation can be reduced as well. A shampoo bar, for example, can use up to 15 times less space than a liquid shampoo (based on the same number of uses). In comparison to liquid shampoos, LUSH's annual sales of shampoo bars save 450,000 L of valuable water.

The company wants to produce items for every need, allowing people to buy just what they need and ultimately reduce waste. The business goes beyond packaging by stripping back products to combine the best quality, fresh and essential ingredients. They minimise the use of synthetic materials and have invented self-preserving formulas to reduce their use of chemical preservatives. Over 82% of the 'all-year-round range' is self-preserved. For example, salt is sourced from suppliers who protect and sustain the salt marshes and migratory birds that live in these environments. The majority of the supply chain consists of direct relationships with manufacturers and growers to control the quality of the supply chain and pass along core business values.

They have been working hard to create a revolution in the cosmetics industry, but this has not come without challenges. In 2012, the company produced a graphic video showing a performance artist undergoing animal laboratory tests in the window of Lush Regent Street London, to raise awareness of their worldwide Fighting Animal Testing campaign and EU petition. The video was watched by three million viewers; however, the initiative resulted in the closing of the store after the landlords objected to the content of the video.

The business wants to be an example to other cosmetic companies on how to make great products, with ethical ingredients, whilst considering the environment and fulfilling their customer needs. They are also continually innovating their product range. They are working hard to achieve the goal of providing customers with the most waste-free shopping experience possible. In addition to their naked products, they are producing novel packaging materials and reusable plastic packaging items made from 100% post-consumer recycled feedstock for refilling. In 2021, they launched a new deposit return scheme for their plastic packaging. The idea is that customers buy their items with the understanding that they are renting the packaging and can easily return it when they are ready, therefore, the company is responsible for waste reduction and resource recycling.

On reflection

The company has been driven by the desire to change the cosmetics industry by providing innovative package-free products. This unique approach makes customers feel like they are getting the best products that are fair to animals, the environment and people in the supply chain. They have continually looked at their products and processes to improve and incorporate further elements of a circular economy.

To address the problems of specifying components and providing customers with guidance on how to apply the product, the company has used the power of creativity and technology. They created their own software, a product identification tool that allows customers to scan a product with their phone to acquire the same information as a physical label. Customers are additionally engaged with the items through interactive material about the ingredients and the stories behind them, which is provided through the application software. Data may play a key role in future developments, to further understand from the customer how a cosmetics routine without any packaging can be embraced more widely. Further adoption of the business model will require a significant customer behaviour shift, to accept new methods of consumption and reuse.

Figure 34.1: Lush hand soap.
Source: LUSH.

Sources

https://www.lush.com
https://ellenmacarthurfoundation.org/articles/circular-economy-products

Jamie Wheaton
35 Shark Solutions
Sector focus: Construction

Rationale

Shark Solutions, based in Denmark, is the market leader in the production of recycled polyvinyl butyral (rPVB). Regenerated from glass, rPVB is developed from waste streams for companies so that they may produce more environmentally friendly products and move towards a more sustainable business model. At least 5% of the world's 1.5 billion cars requires a new windshield every year, with the old glass, alongside architectural glass, being sent to landfill. According to Shark Solutions, this process makes 'no sense' given the negative environmental impact incurred by the disposal of large quantities of materials that could hold significant economic value. Shark Solutions reduces the waste stream by recycling the old glass as well as the adhesive that was previously used to guarantee its strength.

Leadership

Shark Solutions was founded from an idea that first arose in 2005, when a team of entrepreneurs sought to discover how glass could be separated from the polyvinyl butyral adhesive that ensures that car windshields remain intact in the event of an accident. The breakthrough came in 2013, enabling Shark Solutions to grow its operations in Europe as well as the United States. In addition to the sale of recycled glass, the CEO of Shark Solutions, Jens Holmegaard, has also identified usage possibilities for rPVB in other industries such as the paint and carpet industry, where it could replace current less sustainable adhesives. In doing so, Shark Solutions is driving a wider transition to circularity.

Approach

Shark Solutions' approach is based on the separation of damaged car windshields, as shown in figure 35.2, and laminated building glass from polyvinyl butyral. Once separated, the recycled adhesive, rPVB, can be used as an adhesive in other industries such as flooring, paints, textiles, artificial grass and building materials.

The rPVB generated from the separation of recycled glass (figure 35.1) ensures circularity according to four key criteria when compared to alternatives such as styrene acrylics, polyurethane or VAE latex. As opposed to the market alternatives, rPVB is

https://doi.org/10.1515/9783110723373-039

sustainably sourced through its generation from waste streams, whilst it is also non-toxic, avoiding the use of chlorine or phthalates. Additionally, it meets future standards, is fully recyclable to use within subsequent life cycles in the circular economy and is also liable to meet future industry standards on sustainability. Indeed, rPVB beats all of its competitive products in relation to these aspects and is also favoured by customers who lean towards more sustainable products.

Shark Solutions is also able to demonstrate its active role in reducing emissions by saving glass from entering waste streams; Shark Solutions recycled a windshield every 3.8 s in 2020, saving up to 35,000 tonnes of CO_2 emissions. The company also contributes to achieving a number of Sustainable Development Goals, including SDG9 (industry, innovation and infrastructure) through its provision of sustainable alternatives to toxic plastics, SDG12 (responsible consumption and production) through its circular business model, SDG13 (climate action) through its saving of CO_2 emissions, SDG15 (life on land) through preventing waste in landfill and SDG17 (partnerships for the goals) through forming industry partners to encourage circular transition.

Shark Solutions also requires its suppliers to embrace a 'code of conduct,' requiring that business partners adhere to the same environmental, social and corporate governance standards. In particular, as well as complying with local environment regulations, partners should promote safe and environmentally sound development, manufacturing, transport, use and disposal of products, as well as reduce emissions and impact on biodiversity. Partners should support the protection of human and workers' rights, as well as treat employees with respect in a safe workplace. They should also abide by all international trade laws and regulations, as well as consider business integrity as the basis of business relationships.

Eight years had passed since the start of experimentation when the group of entrepreneurs had finally succeeded in separating glass from polyvinyl butyl. The main barrier to access, according to Holmegaard, is the lack of cost-effective processes in separating the materials. Now, however, Shark Solutions enables a wider shift to circularity within numerous sectors through use of its SharkDispersions, designed as a single-use binder for indoor paint solutions or textile coatings, and SharkPellets, which can be used in an extrusion application for textiles, coatings and other composites. Meanwhile, in the United States, Shark Solutions also provide a mobile Shark Glass Separator, which can process 20–30 tonnes of laminated glass per hour. Indeed, from what turned out to be one lengthy wait to find the ideal result from entrepreneurial experimentation has translated into a successful circular business.

On reflection

The success story embodied within Shark Solutions replicates the struggle that can be encountered with entrepreneurialism as a whole. Indeed, the group of entrepreneurs who first started out with their idea had to persevere and invest time and money into exploring whether laminate could be separated from recycled glass. The journey eventually took them 8 years. This should therefore provide hope to entrepreneurs who aim to start a circular business but may be deterred by the time it may take to move from discovery to take-off. The next barrier of entry to circularity is that of the funding required to investigate and launch products. However, numerous funding options are available to innovators who can secure them, such as crowdfunding or in exchange for shares.

Figure 35.1: rPVB generated from the separation of recycled glass.
Source: Shark Solutions.

Figure 35.2: Broken windscreens.
Source: Shark Solutions.

Sources

https://www.shark-solutions.com
https://circularitycapital.com/shark-solutions-portfolio

Georgie Hopkins
36 gDiapers
Sector focus: Fast-moving consumer goods (FMCG)

Rationale

Plastic pollution is an exceptional problem for which the true scale and scope of impact are still not fully known. Research regularly exposes new environmental and social challenges caused by plastic in our oceans and waterways, especially harming marine life, damaging habitats and impacting food chains. Most notably, an Ellen Macarthur Foundation report of 2016 determined that without action by 2050, there could be more plastic in the ocean than fish. The PEW report of 2020 determined that around 11 million metric tonnes of plastic waste enters our oceans every year, with that volume predicted to escalate to around 29 million metric tonnes by 2040 unless action is taken. Needing a raft of approaches to address the crisis, the circular economy is seen to be an essential element.

Leadership

The specific focus of gDiapers is tackling the challenge of disposable nappies that contribute to the unsustainable growth of plastic pollution. The business was founded in 2005 by Kim and Jason Graham-Nye as the couple became parents and began to understand first-hand the long-lasting impact of these products. Globally, more than 380,000 nappies enter landfills or global waterways **every minute of every day**, remaining on babies for as little as 3 h before being discarded, taking over 500 years to break down within the environment. The founders are tenacious and unrelenting in working towards their goal to entirely eliminate plastic disposable nappies from the planet.

Approach

As a start-up, the journey has not been easy, with a particular challenge of competing in a market with large-scale commercial operators in the field, some of whom produce over 200 million disposable nappies a day. Starting out, the company first developed a hybrid nappy, which was part reusable and part compostable, providing more flexibility than traditional cloth nappies. While this has been successful, there was always a sense of limitation being seen as a premium product, which is

https://doi.org/10.1515/9783110723373-040

both more expensive than disposable nappies and also not quite as convenient. The company worked to envisage ways to lower costs, enabling customers not to be penalised for making the sustainable choices, removing the need to 'afford to be green.'

The Holy Grail is to provide a product that is better on all levels than the status quo. However, measurement and scale challenge this. While it is possible to understand the volume of the nappy waste problem, there is no direct cost associated with this waste stream from which to offset the environmental benefits of gDiapers. For many cities and regions, the tangible cost of disposal charged to a company or an individual was zero, often because that measurement does not yet exist. With the source material, plastic, being low cost and in some countries there being no waste disposal costs, gDiapers was competing on an uneven playing field, with producers pushing responsibility entirely onto consumers through the existing linear model.

Considering the business through the lens of a circular economy enabled Kim and Jason to reassess the fundamental business model and, working in collaboration with the Ellen Macarthur Foundation CE100, they developed a blueprint for a circular nappy solution 'nappies to nature.' Most notably, they posed a catalytic question of 'What do we need to do to make our nappies free to consumers?' and challenging themselves with how to make the product higher value at the end of life than at the start of life. The process of developing a new product that was not only more convenient for parents but that also accounted for the full life cycle of the material flow took a number of years to design, test and patent but finally resulted in the creation of gCycle.

gCycle is a circular solution grounded in cradle-to-cradle design offering consumers a service of delivery, collection and composting of nappies, practicing extended producer responsibility, closing the loop and ensuring the circularity of the product (as depicted in Figure 36.1). This scheme is successfully running across childcare centres in Hobart, Tasmania, where a partnership with Veolia enables nappies to be collected in garden waste and fed into the city waste streams, with the compost selling at AUS $75 per cubic metre. However, building networks to support such an innovative business concept has been a challenge and gCycle struggles to get buy-in from suited partners such as wastewater processing companies in the United States and Europe. When asked to collaborate on a relatively small-scale project which seen to be outside of their remit, the companies can be reluctant to recognise the value they could unlock. For gCycle, this is an area where policymaking and regulation change are essential to accelerate value capture.

Extending from this, the company originally assumed they would deliver and scale the model in more developed economies such as the UK and the United States, then translate the model to reach further into developing nations, but the opportunity appears to have reversed with scope for leapfrogging transformation stages. For example, in India, less than 5% of parental population use disposable nappies and yet this equates to more than 50 million nappies discarded each day. Disposable nappies as a product did not exist in India more than a decade ago, yet with the growing middle class, the scope for acceleration of use is significant. Through work in Indonesia, the company has identified an advantage of being both a significantly better product and much more environmentally sound, than currently available low-cost, poor-quality plastic nappies (which cause high rates of nappy rash). Contrary to the product's challenge faced in more developed markets, in these regions, consumer adoption has been much easier to gain, accelerating the reach of the project. Testament to this, gCycle has set up and manages an entire circular service in Indonesia, with the long-term aim to build the provider partnerships needed to enable this to become self-sustaining. The company has identified an opportunity to position their service in these early-stage markets, setting expectation around use, functionality and compostability from the outset, particularly relevant given the widespread increased awareness of plastic pollution.

On reflection

gDiapers has used reframing to devise a very innovative theory of change, developing a circular economy business model to work towards its vision to entirely eliminate plastic disposable nappies from the planet through their new gCycle nappy and service. The challenges remain common also with success dependent upon collaboration with a wider business ecosystem that needs to be motivated and empowered to change, along with consumer behaviour. Measurement and financial cost–benefit analysis have been highlighted as a key challenge, particularly where the current cost of disposal is not considered in financial terms, which extends the scope of measurement and cost analysis needed in order to demonstrate not just environmental but also economic sense. Focusing on less developed countries with lower adoption rates for the use of plastic disposable nappies is a great insight, although presumably such untapped opportunities have not escaped the attention of the traditional market providers. Speed is likely to be of the essence, grabbing first mover advantage in these spaces, along with potential for legislative and regulatory assistance, as seen in the South Pacific island of Vanuatu where the government was planning to ban the sale and use of disposable nappies since December 2020.

deliver
new diapers are
delivered to your door

collect
used diapers are
collected each week

regenerate
waste composts and
regenerates the earth

Figure 36.1: gDiapers principles.
Source: gDiapers.

Sources

https://www.gdiapers.com
https://www.pewtrusts.org/en/research-and-analysis/articles/2020/07/23/breaking-the-plastic-
    wave-top-findings
https://economictimes.indiatimes.com/industry/cons-products/fmcg/mamy-poko-catching-up-
    with-leader-pg-in-diaper-biz-in-india/articleshow/72467723.cms?from=mdr

Georgie Hopkins

37 Ricoh

Sector focus: Technological solutions

Rationale

The shift from linear 'take–make–waste' to more circular business models can be a challenge for incumbent operators, but is equally becoming increasingly critical on commercial as well as environmental grounds due to two key factors. Most notably, as the cost of raw materials increase, securing resources for manufacturing and production will become more challenging. Strong global demand (especially from China), the rise in oil prices and a shortage of global shipping capacity are noted to have put additional pressure on supply chains since the global pandemic. Secondly, both with concerns of wider brand reputation and changing legislation around extended producer responsibility, companies are needing to be more attentive to the full life cycle of products they produce.

Leadership

Established in 1934, the global company Ricoh is the world's leading manufacturer of office automation equipment, such as photocopiers and printers. As a thought-provoking case study demonstrating the realities of making such a shift, Ricoh has developed two business models, building a remanufacturing operation alongside the traditional sales route. This process has required a significant investment in systems and processes to ensure both operational efficiency and financial viability. Worth noting is also that while the remanufacturing business model demonstrates an effective circular business model, it is assessed by Ricoh primarily on an economical basis.

Approach

Having grown over the last decade, the model of 'selling a printed page' now represents around 50% of the UK office automation revenue. The two key attributes of this model are the long-term funding requirements of the model and the resource-heavy nature of maintaining and recovering assets at the end of use. However, these are balanced by the benefit of maintaining control of the hardware, consumables and parts throughout their entire lifecycle. This final point is key given the

https://doi.org/10.1515/9783110723373-041

company has a policy of zero waste to landfill, which can be challenged by the linear sales model given that control of the assets is transferred to the customer, with no specified end-of-life plan.

To focus attention on the cycles of products, Ricoh developed the Comet Circle almost 30 years ago, detailing company activity in product life cycles and that of third parties. Despite its age, this model bears many similarities to the material loops seen in the more recently depicted framework of a circular economy. While this provides a broad theory, as a global company with geographically dispersed assets, there is a resource implication of moving product around the globe, which requires careful management. As such, sitting behind the Comet Circle is the Ricoh Asset Cascade, supported itself by a proficient enterprise resource planning tool. The Asset Cascade aids the commercial judgement of teams on the ground to determine the most viable route for appropriate 'feedstock,' in the form of hardware, consumables and parts, returning to the Ricoh system. Essential to viability of the lease model proposition and 'selling a printed page,' every component is monitored with tracking and use-meters providing appropriate granularity of detail to accurately assess the retained value in individual parts.

A key tension in this system is the company-enforced parameter of economic viability; unless the finances add up, the activity is unlikely to be continued. Taking one example of a consumable toner bottle, a reused bottle can ordinarily offer an 8% cost saving over a virgin bottle, which is significant when considering that Ricoh fills many hundreds of thousands of toner bottles every month. However, the recovery cost of the reusable bottle constitutes over 60% of the cost so a change in transport costs, as is a reality post-pandemic, can quickly cause this to become an economically unviable operation. Following company protocol, this activity would then be challenged from a viability perspective, which in itself brings challenges over logistics and operations, in addition to conflicting with social and environmental messages. Recovery channels are hard to turn off once set up, demonstrating the finely balanced economic model sitting against a business trying to be environmentally sound.

Looking to the future, legislation may steer a change to Ricoh's primary logic and force an update to their business model, away from a requirement of economic viability. The EU has recently published guidance for Public Procurement for a Circular Economy, encouraging the application of circularity within tenders for government procurement. While this is optional at the moment, within the sphere of office automation, Italy has already applied a requirement of 30% circularity to all consumables, whilst other EU nations are looking to implement the code in 2022, with an as-yet-unknown percentage potentially against hardware as well as consumables. With such moves, the continuity of circular practices may become a requirement in order to tender for lucrative government contracts, even at a financial cost, which gives rise to the question of who will bear that cost.

On reflection

Fundamentally, Ricoh is operating in a market that is both saturated and shrinking. Printing volumes, particularly post-pandemic, continue to reduce. One argument would suggest that their remanufacturing operation does not signify true circular business transformation. However, as the core of the Ricoh operation, it is fundamental in order to maintain commercial viability in the short term and provide opportunity for a broader company transition.

Cannibalisation is also a genuine issue with this model, which Ricoh monitor carefully. Given the high fixed-asset costs associated with production of new products, the volume of remanufacturing is actively adjusted so as not to impact essential production volumes. In this sense, remanufacturing remains a secondary operation of the business, which when combined with the economic parameter, perhaps leaves this Ricoh operation more in the sphere of resource efficiency than true circularity.

Source

https://ukproducts.ricoh.com

Isabelle Housni
38 Riverford Organics
Sector focus: Agriculture and food systems

Rationale

The current food system has supported a fast-growing population through the linear economy model, extracting nutrients from soils for mass production, using large amounts of energy and resulting in significant waste in the process. This is resulting in devastating effects on the environment, including negative health impacts from air pollution, water contamination, loss of biodiversity from pesticide use and antimicrobial resistance in animals.

An organic farm based in Devon, UK, has embraced a system that manages the land and livestock that works with nature. Riverford Organic Farmers has strived to grow its vegetables in a more sustainable way, where it protects the soil quality and choose varieties to grow that are suited to the soil, climate and time of year rather than using high-yield varieties that necessitate the use of fertilisers. Regenerative food production has been outlined as an approach to enhance the environment through systemic benefits, including increasing the amount of wildlife that can live in the area, supporting overall planetary health.

Leadership

Organic farming at Riverford was started by Guy Singh-Watson in 1986, who had seen first-hand the risk to human health of spraying pesticides. This experience led him to shirk current practice in order to adopt organic methods of growing. The company was owned and run by Singh-Watson, but with none of his children wanting to take over and unwilling to sell out to distant venture capitalists, he sold the business to his staff and transferred it to employee ownership in 2018. The model of employee ownership protects the core values, avoiding the need for outside investment managed by external stakeholders who could push the business away from the founding ethics.

Riverford Organic Farmers promotes increased communication between producers and consumers to share an understanding of seasonal variability and changes to supply. This relationship enables adoption of circular economy principles by educating consumers on the realities of growing and the environmental harm that can come with mass production. Inspired by the principles of organic farmers, the impact of everything done by the business is carefully considered, including fair treatment of farmers, materials used in packaging and how the food is transported. This holistic

https://doi.org/10.1515/9783110723373-042

view also compliments the circular economy by encouraging more sustainable practice throughout that mimic and support the existing biological systems found in nature.

Approach

Riverford Organic Farmers offers a variety of organic produce to customers in England and Wales, serving fruit, vegetables and meat from their organic butchery in Devon. These are sold in various pre-set box options, or a customer can choose to build their own. These can be delivered regularly, with the customer able to tailor the frequency to their preference.

Non-organic farming allows routine use of hundreds of synthetic pesticides, contributing to a global human and ecological health burden. In contrast, organic farmers are only permitted to use pesticides delivered from natural ingredients such as garlic and clove oil, in addition to working with nature by encouraging predatory insects to manage pests and spread of disease.

The food grown and provided by Riverford Organic Farmers has a focus on slow growth to develop flavour. This applies to the livestock as well, which are also organic and free-roaming, with breeds picked that develop slowly as opposed to mass farming practices where animals with quick growth are prioritised, or are given hormones as allowed in some countries.

One of the challenges faced by Riverford Organic Farmers is how to cater to its growing consumer base. Debate exists on whether organic farming systems can provide enough food for a growing population; however, Riverford Organic Farmers has so far overcome this concern by collaborating with other like-minded farmers.

The one farm, started in 1987, has since grown to include three regional sister farms, and a farm in France to deliver a national veg box scheme to around 50,000 customers a week. It also works with other organic farms further afield in Italy, Spain and even the Dominican Republic, although Riverford Organic Farmers ensures that produce imported is not delivered by air freight, which it found has 40–50 times the CO_2 emissions of sea freight it uses, as well as road transport where appropriate, such as from France.

As a certified B-Corp, the company has pledged to have all its delivery vans to be electric by 2025, to further decrease emissions by 70% compared to the previous diesel ones. Additionally, as part of its Climate Action project and journey towards becoming a net-zero business, Riverford Organic Farmers has changed the fuel its heavy goods vehicle lorries run on, which means a 70–90% reduction in emissions. It is now switching to HVO (hydrotreated vegetable oils) made from waste oils, not virgin oil crops, which means no land used for growing food is impacted. It is also guaranteed to be free of palm oil.

Additionally, the company is conscientious about the packaging it uses, with all packaging either recyclable paper or home compostable. It also has systems in place to collect and reuse packaging, with the reusable veg boxes made out of 98% recycled materials.

Finally, Riverford Organics is also committed to a tree planting initiative, having planted 800 native trees to promote biodiversity and capture carbon. It is also trialling agroforestry, where trees are used to provide sustainable food and increase soil health.

On reflection

While the mission behind Riverford Organics is a positive one, in its aims to grow organically as well as educate consumers on seasonality and sustainability, 20% of produce from Riverford Organics is grown abroad and imported. This is because the UK consumers have grown accustomed to the different fruits and vegetables they have been exposed to, which cannot be grown in the UK without using methods that have significant environmental impact. The company has looked at different impact assessments and found that importing these fruits and vegetables by sea and road, never air freight, has less environmental impact than growing this produce in the UK using methods such as artificial heat and fertilisers. While it still offers fruit and vegetable boxes that are completely locally grown, there is a limit to how much consumer taste can be encouraged to change when the preferred, less environmentally friendly options continue to be accommodated.

Sources

https://www.riverford.co.uk
https://www.ft.com/content/b5481020-a4bf-11e9-a282-2df48f366f7d
https://www.riverford.co.uk/ethics-and-ethos/bcorp

Georgie Hopkins
39 Oxwash
Sector focus: Technological solutions

Rationale

Microplastic pollution is a specific area of environmental challenge, dominating the floating plastics found in seas and now found in 90% of all surface waters worldwide. Since 2011, it has been known that most microfibre pollution comes from washing clothes, with an International Union for Conservation of Nature report determining that 35% of all microplastic pollution came from laundry effluent, driven by both domestic and commercial laundering and drying processes. Beyond this, washing and drying of laundry is responsible for a significant volume of carbon emissions and water usage, with estimates suggesting domestic laundry equates to 440 kg of CO2e per person, per year. In addition, there is significant ecological impact of releasing enzymes, bleaches, perfumes and colourants from detergents into our environment, particularly harmful to waterway ecosystems.

Leadership

While perhaps not the most glamorous market segment, Oxwash is working to develop clinical-grade laundry and wet cleaning services by applying ground-breaking technological developments using chemistry that is free of the harsh chemicals found in traditional dry cleaning solutions. Founded in 2018 by ex-NASA scientist, Dr Kyle Grant, along with Oxford engineer, Tom de Wilton, the business aims to put a sustainable spin on laundry using water reclamation, renewable energy, microfibre filtration and ozone disinfection.

Approach

The company was created to respond to the evident challenge, having been designed with circularity in mind and a focus on net-zero carbon emissions. Oxwash's pilot model in Oxford focused on local logistics, which was then rolled out to Cambridge and London during the pandemic. Oxwash offers domestic and commercial laundry services, with zero-emissions collection and delivery to customer premises, as shown in figures 39.1–39.3.

https://doi.org/10.1515/9783110723373-043

Referred to as lagoons, each operation uses water reclamation or load-weighting technology to reduce water usage by up to 70% compared to domestic washing machines. With added microfibre filtration, over 95% of fibres shed during washing is captured, preventing leakage into waterways. Cleaning comes in the form of biodegradable detergents that are dispended using dynamic dosing technology, ensuring optimal performance. And possibly the most cutting-edge element, adapted from use in spacecraft, is the application of ozone as a disinfecting agent. Generated using renewable energy and oxygen from the air around us, ozone provides a powerful disinfecting agent against microorganisms, working even better at 20 °C than 40 °C or higher. The compound advantage of these combined technologies, set alongside rigorous performance standards, provides potential to transform many commercial settings, expanding scope and potential overall impact of the Oxwash business model.

A key challenge of such transformational business models is gaining traction within the market, which for domestic users will entail a much larger shift in behaviour than commercial partners, for whom the concept of laundry-as-a-service is more accepted. To give some context, the company promotes that one 8 kg load of domestic washing through Oxwash saves 32 L of water, 1.4 kg of carbon emissions and 1.5 h of customer time. At this point, the economics of using the service are higher than traditional launder-at-home methods; however, this draws back to the challenge of putting a value on the externalities of clothes washing, currently a cost that is not borne by the consumer directly.

Equally, access to funds for the intensive start-up and expansion phases can be challenging, particularly in this sector, which requires significant investment into both new technology and capital expenditure. This is becoming less of a hurdle, however, with record levels of investment being attracted to sustainable investments in particular since the global pandemic. Having successfully completed four funding rounds in the last 2 years, raising over £5.7 million, Oxwash is poised to expand both across the UK and globally, bringing on-demand, sustainable laundry and dry cleaning to even more domestic and commercial customers. With a particular focus on Circular Fashion Rental, Oxwash is positioning themselves as a key solution provider for powering sustainable garment aftercare.

On reflection

Having successfully demonstrated proof of concept and established the business, Oxwash is in the growth phase, rolling out across new cities and regions. When considering the large-scale opportunity of commercial contracts, regulation could prove to be a barrier in certain settings. While the ground-breaking technology enables effective disinfecting at cooler temperature washing, technical regulations will need to be aligned to facilitate the use of Oxwash services. By way of example, Infection Prevention Control within NHS healthcare settings includes established protocol for specific high-temperature washing requirements that potentially sit at odds with Oxwash's advanced technology.

Figure 39.1: Oxwash cycles.
Source: Oxwash.

Figure 39.2: Oxwash laundry.
Source: Oxwash.

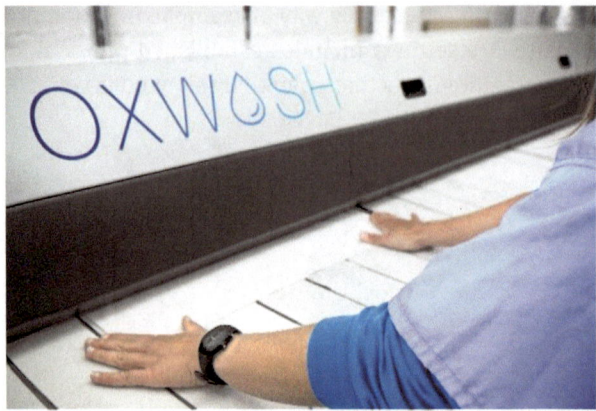

Figure 39.3: Oxwash laundering process.
Source: Oxwash.

Sources

https://www.oxwash.com/
https://startups.co.uk/news/oxwash-secures-funding/

Ruth Cherrington
40 Triodos Bank
Sector focus: Finance

Rationale

The focus on growth and excessive consumption is unsustainable in our current economic model. Natural resources are finite, and the waste generated by the linear economic model has a significant negative impact on our environment. As a result, Triodos Bank, a Dutch-registered European investment firm, has embraced the circular economy concept, promoting the need to change our present economic system into a regenerative system by design. It exclusively loans to socially and environmentally conscious organisations and enterprises. It claims to only finance initiatives that benefit the community, improve the environment and protect human liberty. As a result, it actively looks for 'ethical' borrowers from individuals, charities and companies. It is a modern example of a European social bank that has evolved to address the unique demands of the circular economy, which frequently experience challenges acquiring financing from traditional sources.

Leadership

Triodos is an independent bank founded in 1980 and owned by public shareholders, with shares kept in a trust that safeguards the bank's social and environmental goals. When it joined with the first British ethical bank in 1995, it entered the UK market. This first ethical bank in the United Kingdom emerged in 1974 from a spiritual science that was established to demonstrate that it may be beneficial and restorative to both individuals and mankind as a whole. This ethos is core to their mission today 'to help create a society that protects and promotes quality of life and human dignity for all.'

Approach

The bank has established a standard set of banking services, including individual savings accounts, as well as current and investment accounts for social enterprises, charities and organisations. The bank's policy is to equal or beat the rates given by mainstream banks and building societies, and to place itself in the bottom quartile of rates. It promotes the business by emphasising that account holders will benefit

https://doi.org/10.1515/9783110723373-044

from both attractive interest rates and social benefits. The funds support social and environmental projects, and the terminology used in the titles of its savings and investment products aims to reflect this.

Because the bank evaluates credit applications using social and environmental criteria, it is likely to attract depositors who are wary of mainstream commercial banks. They particularly promote their ethical lending portfolio, other business relationships and institutional arrangements, which are aspects of conventional banks that may be at odds with the depositors' values.

Depositors have the option of specifying exactly how their money should be used by the bank. One of the social saving accounts, for example, has an interesting feature that allows depositors to set their own interest rate up to the maximum stated rate. Interest waived by depositors is passed on to specific projects or prioritised areas by charging a reduced interest rate on loans, essentially subsidising the borrowers' activity. The 'organic saver' account provides donations to organic-related businesses that promote organic farming, growing and processing.

The business portrays itself as a trustworthy financial institution. It distinguishes itself from other banks in two major ways. For starters, it only lends to initiatives that are not only financially feasible but also judged socially beneficial. Second, the openness of its lending portfolio is exceptional. It gives a list of current initiatives in which funds are being invested, demonstrating a commitment to the concept of depositor responsibility. This gives a valuable paradigm for financing future circular economy initiatives.

On reflection

Complete circular business models are very rare to discover in publicly traded firms. As a result, the bank evaluates a company's business model, strategy and resource management to determine if it is a leader in the transition to a circular economy. This evaluation process is conducted for every application and illustrates the core ethos of the business. As an investor, they see their job as supporting the supply side of this change. Circular concepts, which are frequently stated in terms of business models, may be developed for practically any industry, with resource-dependent businesses having the most substantial advantages.

Source

https://www.triodos.co.uk

Ruth Cherrington
41 Co Cars
Sector focus: Mobility

Rationale

Cars, bikes and other forms of transport are designed to get you from one place to another. To have this service without all the costs of ownership and maintenance is a more efficient use of resources, energy and information and is one of the core concepts of a circular economy. Mobility as a service is a concept that describes a shift away from ownership and towards access to transport as a service. It harnesses a range of tools, including accessible mobile applications, cashless payments, data analytics, and on-demand and pre-scheduled transit.

Co Cars, a social enterprise located in Exeter in the South West of England, operates a low-carbon, on-demand shared mobility network throughout the city and across the wider South West region. In 2020, it reported that its active car club members had resulted in a 30% reduction in car usage, a 15% reduction in transport expenses and a 20% rise in walking and cycling as a result of using their services. Furthermore, between 2015 and 2020, it was responsible for removing 150 private cars from the road, saving 149 tonnes of carbon dioxide emissions every year.

Leadership

Mark Hodgson, the Founder and Managing Director, is in charge of business growth with important stakeholders such as municipal governments, housing developers, multinational corporations and other enterprises. The businesses (both car and e-bike sharing) are informed by a deep understanding of the shared mobility sector.

Only a few years ago, people could not imagine sharing beds, yet now Airbnb is a proven business model with a global network of 4 million hosts in practically every country. A similar outcome is expected for transport, as many people increasingly realise owning a car is not cost-effective given the number of miles they drive. Mark Hodgson comments: "With over 90% of car journeys being under a 70-mile range, the rise of electric and autonomous cars is only going to gather pace in the coming years. We simply don't need cars that can do 400 miles on a tank of fuel anymore (or at least not many of us)."

https://doi.org/10.1515/9783110723373-045

Approach

The business revolves around a network of shared cars and e-bikes (see figure 41.1) across the South West of England, principally in Exeter. Through agreements with municipal governments, housing developers and private enterprises, the business will have over 50 electric cars in Exeter alone by Q1 of 2022. Cars can be booked via mobile application, allowing you to choose your preferred car, date and times to organise a journey. By removing six to ten private vehicles from the road, each low-emission shared vehicle helps to alleviate congestion and pollution. Prices are affordable and include fuel, insurance, tax, servicing, repairs and roadside breakdown.

Co Bikes, Co Cars shared e-bike network, is expanding across Exeter to over 150 bikes, to be found at such locations as train stations, University of Exeter, the city centre and other key locations. Bikes can be rented and returned by app, with users able to buy bundles of time to make cycling an even more cost-effective choice.

Mobility as a service has the potential to reduce vehicle ownership. It has the potential to improve the efficiency and use of transport providers that contribute to a region's overall transit network. The cost of public transportation might be greatly reduced if an efficient network is combined with new technologies such as autonomous cars.

On reflection

The idea for the company was formed in 2005, when the founder received a £3500 grant, which he used to lease one car. It was parked behind the library, with a key safe and a basic internet booking system. It was one of the country's earliest vehicle-sharing clubs at the time, and it was (and still is) a social company.

The company has scaled out since then to include more electric cars, and also runs a successful and rapidly expanding e-bike network. Consequently, it was awarded the first ever double accreditation by CoMo, a national shared transport charity and representative body. Mobility is a critical component of a functioning economy, but current mobility solutions ignore the economic, environmental and societal consequences that can be harmful to health and well-being. Wider adoption and diffusion of this service to more locations could provide accessible, affordable travel for a larger community.

Figure 41.1: Co Cars and Co Bikes.
Source: Co Cars.

Sources

https://www.co-cars.co.uk
https://grow-media.co.uk/top-stories/mark-hodgson-a-sustainability-road-map/

Georgie Hopkins

42 Oddbox

Sector focus: Agriculture and food systems

Rationale

Across the globe, over one-third of food is wasted, equating to 74 kg of food per capita each year. The significance of this challenge cannot be overstated, with food waste being the third biggest contributor to greenhouse gas emissions. Such wastage increases food insecurity, burdens systems and contributes to climate change, biodiversity loss and pollution. In response, Sustainable Development Goal 12.3 aims to reduce food loss and halve food waste by 2030, providing significant environmental, social and economic benefits.

Leadership

Cofounded by Emily Vanpoperinghe and Deepak Ravindran in 2016, the inspiration for Oddbox came from visiting fresh fruit markets in Portugal, where the often wonky-looking fruit and vegetables on offer caught their eye, appearing very different to what is commonly presented in the UK supermarkets. On researching the significant volume of produce rejected before leaving farms in the UK, the duo recognised an opportunity to make a difference, and from this Oddbox was born: a social impact business fighting food waste in UK farms. This passion to make a difference is engrained within the business, with Oddbox now a recognised B-Corp, actively demonstrating that business can be a force for good.

Approach

Upwards of 30% of fresh fruit and vegetables produced in the UK is rejected by supermarkets for not meeting their specification; be that 'too bendy,' 'too ugly,' 'the wrong colour,' 'too big' or 'too small.' The company works directly with farmers, taking excess seasonal stock and cutting out the supermarkets by offering direct home delivery to the end consumer.

As a CE enabler, the founders of Oddbox identified value leakage in the supply chain and have provided a missing capability to capture it. From the farmers' perspective, working with Oddbox provides a safe destination for food at risk of being left in the field or need to be discarded. For customers, they receive the convenience of

https://doi.org/10.1515/9783110723373-046

nutritious fruit and vegetables delivered to their home, with the added benefit of knowing that they are buying ethically and helping to prevent food wastage. In a further move to zero waste, Oddbox partners with two charities, donating any leftover produce at the end of each week, further alleviating food poverty.

The success of the model rests to a large extent on a fair and equitable distribution of value. This joint approach, which compensates farmers whilst also generating additional financial, social and environmental benefits, helps to build long-term partnerships between all actors in the value chain. However, it was not plain sailing in the early years, and the company was challenged to build partnerships with growers when volumes were low and there was scepticism as to whether consumers would accept wonky produce.

Combining social action, the business acts as an educator, using engagement as an opportunity to demonstrate how every customer is making a difference every week and build momentum with the early adopting customers, growing their market reach. This has been possible through working alongside experts to define metrics and continuously refine the measurement of impact, both in terms of carbon emissions and water usage, producing personalised impact reports that are shared with customers throughout the year.

Since launching in 2016, the business has delivered over 4 million boxes (figure 42.1), saving over 23.5 tonnes of food. Having started in London, the service has already expanded to include the Southeast and areas of Northern UK. Through building a dedicated community, Oddbox has shown sixfold year-on-year growth since inception. Having raised £3 million in March 2020, the company recently secured a further £16 million in Series B funding to further facilitate expansion plans across the UK.

On reflection

Oddbox is a great example of an enabling organisation having identified value leakage in the system and provided the infrastructure to capture value.

The company has a vision to become the hub for wonky and surplus produce in Europe, and there is no doubt the fruit and vegetables are available to be redistributed. However, a question may exist around seasonality and there being a suitable mix of produce to be able to provide customers with expected variety in each box. Additionally, with extended business reach, there will be a greater need for shipping relatively low-value, bulky produce. Could there be a situation where distribution costs outweigh the wider benefits of the business model? That is a possibility that may be of particular focus since the Covid pandemic and rising transportation costs.

Figure 42.1: Oddbox.
Source: Oddbox.

Sources

https://www.oddbox.co.uk/
https://www.uktech.news/news/london-based-sustainable-food-delivery-firm-oddbox-raises-16m-
    to-fight-food-waste-20210825

Ruth Cherrington
43 Fairphone
Sector focus: Electronics

Rationale

Waste from Electrical and Electronic Equipment is one of the quickest growing waste streams in the world. Globally, 44.7 million tonnes of electronic waste were created in 2016, with mobile phones accounting for 435,000 tonnes. Only 20% of electronic waste is reported as being collected and recycled under suitable conditions, while the remaining 80% is either placed into the residual waste stream, inappropriately discarded (resulting in leakage to the environment), traded or treated in unsuitable conditions. At both the manufacturing and disposal ends, the repercussions of this linear electronics system have an influence on the environment and human health. In the mining and production of items, large amounts of energy and hazardous chemicals are required, and the demand for resources has been connected to hazardous working conditions.

Mobile phone and smartphone technologies are progressing rapidly; the average mobile phone is replaced every 18 months, with consumers likely to keep old models in a drawer at home rather than dispose of them. They contain valuable materials, such as tin, tungsten, tantalum and gold, often referred to as 'conflict minerals' as they are mainly sourced from countries where civil war has been ongoing for decades and groups fight to control mines and trade routes. Valuable materials are contained in such small amounts that they are not yet recycled at industrial levels and only generate interest for their reuse or reclamation of battery materials.

Leadership

In September 2009, a public relations professional and an industrial designer in the Netherlands launched a campaign to raise awareness about the link between smartphones and conflict minerals. They wanted to highlight the issues around mineral exploitation, working closely with the supply chain to ensure fair labour conditions for the workforce. They asked the public to participate in the development of a fair phone free of conflict minerals. They intended that any device that resulted would be a hypothetical concept that would be shown at a local museum. They never intended to bring a phone to market.

The campaign had transformed into a social enterprise by 2013, with the new goal to create a phone that prioritises human values. A crowdfunding appeal was launched to generate income for production of their first model, and surprisingly

https://doi.org/10.1515/9783110723373-047

customers were willing to pay with little expectation in return, investing in the social aspects and the mission. By the end of the year, they had sold 25,000 non-existent smartphones. The first Fairphone model was released in 2013, and in 2021 they had more than 100,000 Fairphone owners (figure 43.1). The founders and business leaders aim to build a deeper understanding between people and their products, driving conversations about environmental impact and responsibly sourced materials.

Approach

A circular economy for consumer electronics would keep devices in use for as long as possible, either by the original owner or someone who finds new value and use in them. Devices will eventually be passed along to professionals who will restore them properly, reuse or remanufacture the important components within, and separate and recycle the materials.

Modular design is a good way to make it easier to repair, remanufacture and upgrade products. Making it simple to remove only a portion of a product simplifies disassembly, minimising the cost and difficulty of replacing broken components. The modular construction in this case enables the screen to be replaced easily and inexpensively if it is broken. Furthermore, modular systems are easier to customise and adapt to the ever-changing demands of users, avoiding items from becoming outdated and guaranteeing that they are used for a long time. Fairphone also sells spare parts and offers online tutorials to keep phones in use for as long as possible.

Fairphone is transparent about the cost breakdown of the phone, focussing ambitions to create a more responsible economy and declaring only a small profit (if any) from the sales. By supporting the reuse and repair of mobile phones, exploring new electronics recycling possibilities and decreasing electronic waste globally, the business is taking a step closer to a circular economy. The goal is to maximise the usage of resources in consumer electronics, while also supporting recycling programmes to ensure that precious materials are responsibly sourced and their value retained within the system by encouraging reuse.

On reflection

The social enterprise emerged within a challenging industry to provide innovative new products (and later service) to address the social problem arising from the supply chain. As a result, they have used the principles of circular design within the environmental dimension to reduce the impact from technical resource extraction. They have created an entire new business paradigm shift, focussing on a product-to-service shift and therefore retaining the quality of the supply chain throughout.

Dismantling a phone before recycling increases the recovery of metals, particularly valuable metals, as well as the utilisation of polymers for both material recovery and energy generation. Aside from the potential benefits, there are significant product and industry-related constraints in design for recycling. For example, the combinations of metals, compounds, fillers, polymers and functional materials need careful consideration to reduce loss of elements. The complex material combinations in current phones make deconstruction very challenging. Although it is unclear if material recovery and circularity of a complicated product like a smartphone can ever be realised, it is encouraging that these concepts are gaining traction and shows the potential for wider adoption and diffusion. Further study is needed to properly appreciate the advantages of modularity. We know, for example, that the modular design makes these devices easier to deconstruct, but the measurements to evaluate the cost, environmental and social benefit of these phones have yet to be established.

Figure 43.1: Fairphone.
Source: Fairphone.

Sources

https://www.fairphone.com/en/2015/09/09/cost-breakdown-of-the-fairphone-2/
https://www.ellenmacarthurfoundation.org/explore/circular-design

Ruth Cherrington
44 Ooho from Notpla
Sector focus: Plastics and technological solutions

Rationale

Every year, countless single-use plastic containers are discarded due to the rapidly expanding takeout food industry. Due to contamination from food leftovers, recycling takeout packaging is often impracticable; therefore, it ends up in the municipal solid waste stream. Upstream innovation in a circular economy is tracing a problem back to its source and addressing it there. It implies that instead of figuring out how to deal with an increasing mountain of waste, we try to avoid it in the first place. Wherever feasible, reusable containers should be used; however, this is not always possible, leaving a fraction of containers that are only used once. A London-based start-up, Notpla, has developed an edible and biodegradable material for drinks and sauces as a solution for these challenging takeaway items. The company designs natural packaging made from seaweed, which is one of nature's most renewable resources, with a daily growth rate of up to 1 m. It does not compete with food crops, does not require freshwater or fertiliser and helps to de-acidify our waters.

Leadership

Notpla was started after the two co-founders met studying on an innovation design engineering, master's degree at university. After their initial video showcasing the product went public in 2013, they were invited to join Europe's largest climate-focused accelerator. They began testing the material at events, festivals and takeout restaurants with the help of scientists and chemical engineers. They quickly began developing a scalable manufacturing process.

They received money through a crowd financing site in April 2017, which allowed them to expand the business, manufacture the first machine and establish a manufacturing base in London. They received additional funding in the summer of 2018 to speed up the development of the second product and expand the chemical, commercial, engineering and design teams. They raised an additional £4 million in December 2019 to extend their product line, focusing on creating advanced packaging solutions that disappear, naturally. They still describe themselves as a sustainable packaging start-up and one of their core values is replicating what nature does best. They are driven by the interest in finding innovative alternatives to single-use plastic, which will enable the transition towards a circular economy by redesigning material loops.

https://doi.org/10.1515/9783110723373-048

Approach

Nature inspired the concept, since fruits like oranges have their own biodegradable peel that serves as the ideal natural container. They mimicked nature by making their own biodegradable packaging out of edible components. The membrane can be thrown to the ground if needed and degrades in less than 6 weeks just like fruit peel. It provides a viable alternative to the flexible packaging commonly used in fast food and beverages, which has a low recycling value and can leak into the environment. In 2017, they perfected the strength, structure and shelf-life of the membrane technology.

They have formed various partnerships to test the potential application in various markets. For example, in 2019, they teamed up with the manufacturer of a well-known isotonic carbohydrate-electrolyte sports drink to conduct a large experiment. They handed out over 30,000 sports drinks to marathon runners. This proved an innovative solution to decrease long-term plastic consumption at mass-participation athletic events and it is now being promoted as the preferred hydration option for similar events.

During an 8-week trial with 10 London restaurants, they also replaced 46,000 sauce sachets. However, they needed to develop advanced production equipment and machinery to produce and fill single-use sachets on an industrial scale to fulfil demand. The seaweed material has the potential to be as cost-effective as plastic and aluminium as an input material at scale, but the production process proved problematic. Significant human interaction limited maximum manufacturing capacity to 300 units/day, indicating that, despite great commercial interest, they were unable to match the volume demands of restaurants, takeout, and retailers at a cost-competitive level. They applied for funding from the UK government to de-risk, enable and support innovation.

In 2021, they received further funding from the UK Government to support research and development to refine its composition, upscale and automate the technology to demonstrate viability as a biodegradable plastic-wrap replacement product within the cosmetics industry. Their long-term goal is to produce packaging for a variety of single-use liquid markets on an industrial scale, in a consistent, high-quality, low-cost way to satisfy both suppliers and customers.

On reflection

Biodegradable plastics have been around for over a century. Although the majority of them are still in their infancy, the sector has seen a revival in recent decades. Only a few groups of biopolymer materials are currently regarded as competitive. When compared to their traditional equivalents, the primary disadvantages of biodegradable polymers in terms of market diffusion are performance and cost.

In the past, few businesses were worried about the impact of packaging on the environment. Now that packaging has become an item that everyone's fears may tarnish a brand's image or values, the market potential has shifted. The founders of this company have discovered a niche application in which to launch their product. However, they are still in the early stages of the diffusion process, when new products are costly and performance attributes are prioritised over economic considerations. Only after periods of commercial experimentation, and formation of an industrial basis, standardisation and mass production, manufacturing costs are expected to drop significantly. They have successfully utilised crowdfunding and government grants to support product development and are now looking at other markets for further applications.

There are wider societal questions around the requirements for 'on-the-go' products and whether we can reduce our current levels of consumption to minimise the impacts associated with these goods and services. There is also apprehension about a potential rebound effect, in which improvements in particular items are negated by an increase in material consumption and usage. Despite the ongoing discussion about the social dimensions of consumption and the potential need for adequacy-oriented lifestyles, this product hopes to eliminate the customer trade-off between convenience and sustainability.

Source

https://www.notpla.com

Jamie Wheaton

45 Repair Café

Sector focus: Plastics and technological solutions/fashion
and textiles

Rationale

The concept of the Repair Café emerged from Amsterdam in 2009 when journalist
Martine Postma sought to change perceptions linked to the disposal of goods associated with a linear economy. Having already guided her local community on how
to produce less waste, Postma realised that to repair assets or goods was an important way to reduce waste. "If something is broken, the first reaction should be: this
should be mended." Postma therefore sets out to change perceptions towards the
repair and revalorisation of damaged belongings in order to reduce waste by prolonging life cycles.

Leadership

Repair Café is a community-based approach that quickly started to spread after its
origination in the Netherlands. As cafés started to open across Europe in 2010,
Postma founded the Repair Café International Foundation as a non-profit organisation that provides support to local groups within the Netherlands and further afield
who wish to start their own Cafés. Repair Cafés have subsequently been founded
around the world, in countries such as India, Japan, Belgium, Germany, France, the
UK and the United States. Individual Cafés are run by volunteers and are funded by
participants, with organisers such as Upcycle Kernow suggesting a small donation
per customer to meet costs.

Approach

The Repair Café International Foundation operates according to three main goals:
to bring repairing back into local society, to maintain and educate others in repair
expertise and to promote social cohesion in the local community by connecting its
members from all backgrounds. The concept of a Repair Café therefore relies on the
support and participation of the local community and, in doing so, contributes to
the circular economy through two key aspects: the maintenance of broken goods

https://doi.org/10.1515/9783110723373-049

and the reduction of waste streams, and the education given to participants on regenerating assets.

The maintenance and regeneration of damaged goods is an obvious yet simple contribution that may enable communities to shift towards circularity. Customers who wish to have assets repaired are asked to bring necessary parts, or if said parts cannot be sourced, then they can be provided by the skilled volunteers. As Upcycle Kernow also highlights, a wide range of goods can be restored within Repair Cafés, including but not limited to 'textile and sewing, mechanical and electrical, mobile phones, jewellery, bikes and ceramics.' Customers therefore benefit from a wide range of skillsets provided to them for a small donation, whilst their assets are saved from waste streams. The education given by skilled volunteers, however, is an equally important aspect in the Repair Cafés' promotion of circular habits. This education is a cyclical, hedonic process that mutually benefits both volunteers and customers. The linear economy, the Repair Café International Foundation argues, results in the collective loss of repair skills. Individuals would rather dispose of their products when broken than repair them. Repair Cafés are intended to restore those lost skills. Volunteers who bring their repair skills 'get the appreciation they deserve,' whilst they also pass their skills on, thus enabling individuals to see their possessions in a new light. The education given also promisingly starts at a young age, with some Repair Café volunteers visiting schools to give repair lessons. Education is used as a promising tool to promote shifts towards a circular economy and the reduced need for raw materials.

The Repair Café movement has gathered worldwide momentum. As of 2021, according to the Repair Café International Foundation, 2,214 Repair Cafés were in operation around the world, operated by an estimated number of 33,000 volunteers. The running of said Cafés is also ensuring that an estimated total of 40,000 items are being repaired every month. Previous research from the Repair Café International Foundation demonstrates how items that are repaired can vary from coffee machines and laptops, to clothing, to clocks and to lamps.[i] Repairing items also play a role in reducing CO_2 emissions. One study, based on combined data from 13 different UK-based Repair Cafés, found that an average of −10 kg CO_2e is saved per 1 kg of product that is successfully repaired.[i]

Importantly, however, the Repair Café provides a safe space for members of the community to socialise, providing a vital community service. Upcycle Kernow advertises its Repair Café as an opportunity to "Bring your fix-it skills, bring your broken stuff, get things fixed, learn new skills, share your knowledge, eat cake, drink tea, [and] meet new people." Furthermore, for those who cannot attend in person, Repair Cafés also use social media to facilitate connections between individuals

---

i https://circularandco.com

and skilled repairers. Indeed, Repair Cafés are encouraging social cohesion in both the community and also online.

On reflection

Repair Cafés are a strong example of the growth of grassroots movements, encouraging a wider paradigm shift. What was initially help freely given by Martine Postma to her local community has spread globally, helping communities extract the most value from goods that would otherwise have been disposed of within a linear economy. Repair Cafés also prove that education is the key driver towards a circular economy. If communities are educated to promote circularity above all else, then the economic value of greater quantities of goods would be maintained for longer periods of time whilst waste streams would be significantly reduced. Repair Cafés also theoretically may contain a blueprint where skilled volunteers may in the future become skilled workers, where circular jobs are created to service the local community and shift general trends away from the disposal of goods. The skills being taught to young people in schools provide hope that the ability to repair broken or damaged goods will form future trends.

Sources

https://sugru.com/projects-inspiration/fix-repair/meet-martine-postma-founder-of-the-repair-cafe
https://www.repaircafe.org
https://www.upcyclekernow.org/community-events

Jamie Wheaton
46 Gerrard Street

Sector focus: Digital

Rationale

Gerrard Street, based in the Netherlands, operates a subscription service for its high-quality modular headphones, a strategy that was inspired by focus groups held by its founders. The focus groups discovered that most customers share the same frustrations when using headphones that even the smallest cable breakage can render the product worthless. Not only does this leave customers with a dilemma (should they invest large sums of money into premium yet fragile headphones?) but this also contributes to an electrical waste problem that Gerrard Street itself acknowledges is a significant burden on the environment. Gerrard Street's headphones-as-a-service model seeks to reduce the waste stream caused by even the smallest damage to wiring.

Leadership

The business model is centred around the flexible renting of modular headphones that are easy to disassemble and repair with no glue used within their production. The innovative model therefore helps to shift the paradigm from ownership to accessibility. As Gerrard Street's management argues, "Circular economy design is fully focused on how to design for reuse and/or recycling in combination with a product that consumers will love. In contrast, linear design thinking is only focused on designing a product that will sell." Combining durable design, high-quality sound and a flexible model allows Gerrard Street to maximise use cycles for each component part.

Approach

Gerrard Street's transformation towards the circular economy is based upon its alternative and more regenerative approach towards headphones. Indeed, "How many times have you thrown away a device because something small could not be repaired?" is a question asked by Gerrard Street towards prospective customers or subscribers. The disposal of technologies that may only need a small repair, it argues, contributes to a global disposal of 53.6 million metric tonnes of electronic

https://doi.org/10.1515/9783110723373-050

waste per year, a figure that could grow to 74 million metric tonnes by 2030, as well as including US $57 billion of gold, silver, copper, platinum and other high-value metals that were sent to waste during 2019. Gerrard Street therefore recognised the financial and environmental value that could be saved by reducing such disposal.

As an alternative to headphones that fall prey to breakages in wiring, Gerrard Street operates a modular design that allows 85% of the design's components to be reused. Modular components also form a durable, standardised design, meaning that fewer virgin materials are extracted in the production of new headphones, and the subscription model allows Gerrard Street to recover and refurbish headphones or their components at the end of each usage cycle. Gerrard Street therefore succeeds in its goal of reducing electrical waste and maintaining the value of components through the generation of numerous usage cycles. The process for the return of broken parts is simple. Subscribers can order a part for free from their online account with said part arriving via post the next day. After replacing the faulty part with its replacement, the former is then sent back in a return envelope to be refurbished or recycled as appropriate.

Gerrard Street also encourages an alternative approach to the traditional ownership model through a flexible subscription. Removing barriers to ownership which normally exist within a price charged for the ownership of premium headphones, a small monthly fee (around €10 per month) allows it to reach a wider customer base. Subscribers also have the freedom to amend their monthly subscription, whether cancelling or changing headphone models, as well as purchasing their own set of headphones for a larger, one-off sum.

The ability to purchase headphones may seem counter-intuitive to Gerrard Street's mission of ensuring numerous usage cycles. However, all customers, whether they have subscribed or purchased their headphones, are entitled to free repairs. Headphone owners, like subscribers, can order replacement parts by following a similar process. The offering of a 'free repairs forever' service thus incorporates owners into the same biosphere as subscribers, allowing Gerrard Street to maintain a closer connection to subscribers and owners alike. The inclusion of all customers within the biosphere ensures that they are all valued within the decisions made as part of the business model.

On reflection

Gerrard Street embodies a transition to circular economy for two key reasons. Firstly, for its adoption of technology based on modular assembly as opposed to the use of glue. Not only does this address the concerns raised by the focus groups held by Gerrard Street's owners, but their easy assembly and disassembly facilitates the easy regeneration of parts. Secondly, Gerrard Street's headphones-as-a-service business model would represent a significant paradigm shift if upscaled, the reasons for which are twofold. Indeed, the encouragement of customers to pursue access over ownership on more affordable terms is a model that would reduce production as well as ensure the circularity of goods. However, the transformation of a business model that values owners as well as subscribers ensures that Gerrard Street maintains its service to customers during the post-purchase phase, ensuring the circularity of headphones for an infinite number of cycles. Other businesses who may wish to adopt this model would need to account for the life cycles of goods, and the subsequent impact on margins, that would be associated with servicing customers 'forever.'

Sources

https://ellenmacarthurfoundation.org/circular-examples/gerrard-street
https://gerrardstreet.nl

Jamie Wheaton

47 Patagonia

Sector focus: Fashion and textiles

Rationale

Patagonia was founded in 1973, but has embraced a CE-driven approach since 2005 when – inspired by the concept of cradle-to-cradle – it decided to create a line that would never find itself in landfill.

The rationale for embracing CE for Patagonia is twofold. Not only does Patagonia see itself at the forefront of driving circularity within the textiles industry, but it also seeks circularity in order to maintain the value of textiles and thus increase profit margins. Indeed, "Capping products may sound like an ax to profit, but in a circular economy, it's about generating more money from the same products and leaning into materials like recycled wool that are actually cheaper than their virgin counterparts." Patagonia therefore seeks to maintain both environmental and economic prosperity through its shift to circularity.

Leadership

Patagonia's approach to sustainability reflects the philosophy held by its founder, Yvon Chouinard, who argues in his book *Let My People Go Surfing:* "Patagonia exists to challenge conventional wisdom and present a new style of responsible business. We believe the accepted model of capitalism that necessitates endless growth and deserves the blame for the destruction of nature must be displaced." The drive towards circularity is therefore an approach inspired by the top of Patagonia's management, reflected within policies that aim to challenge the *modus operandi* of capitalism that Chouinard argues can be destructive to global populations, both financially and environmentally.

Approach

Patagonia's strategy towards CE drivers is multi-faceted. Firstly, its focus towards sustainable usage of materials has evolved since the inception of its Common Threads Garment Recycling Programme in 2005. The initial programme did not result in a success. The clothing line in question would have been made by customers' old Capilene baselayers, recycled by a third-party recycling company in Japan. Logistically, however,

https://doi.org/10.1515/9783110723373-051

this proved too difficult. Patagonia could not source sufficient quantities of baselayers to make durable products, whilst the third-party recycling company moved to China, where rules on waste were stricter. Other strategies, however, have proven more successful. Patagonia's 'Responsibili-Tee' is a 100% recycled, Fair Trade certified, sewn tee-shirt, made with 4.8 plastic bottles and 0.3 cotton scrap. Patagonia also continues to develop its 'Tee-Cycle' t-shirt, which is its attempt at designing products from old t-shirts, thus allowing Patagonia to own all its waste across the garment's life cycle.

Recycling is also an activity that customers of Patagonia are encouraged to embrace. In 2011, the Common Threads Recycling Programme evolved into the Common Threads Initiative, where customers could buy and sell used Patagonia products on a dedicated eBay store. This initiative then evolved into 'Worn Wear,' allowing customers to shop for used items in stores as well as eventually online. However, whilst Worn Wear has facilitated the repair of 130,000 items, it only accounts for $5 million of Patagonia's business, with new products still the most popular choice.

Chouinard's focus on a more sustainable and ethical form of capitalism is also reflected in Patagonia's treatment of workers within its supply chains. Patagonia was a founding member of the Fair Labor Association in 2001, whilst it also joined Fair Trade USA's programme in 2014, with the latter driving Patagonia to embed Fair Trade principles within its own organisation. As one member of Patagonia's senior leadership team argues: "although we purchase clothes based on fabric and fit in the store, the people who make our clothes are a huge part of the equation and are often forgotten." Patagonia thus seeks to ensure a circularity of capital and resources, with its overseas suppliers benefitting from childcare facilities, improved water supplies and bonus payments. The process, Patagonia argues, is not as simple as simply paying the factory more money but ensuring – with the help and scrutiny of external bodies – that workers receive fair treatment.

Patagonia is also actively pursuing strategies to minimise its carbon footprint, and in doing so "not only fighting for the places . . . we love but also for everyone." The overall strategy to this end is sixfold. Firstly, Patagonia is moving towards using solely renewable electricity for its premises (including stores, distribution centres and offices), whilst it also intends to reduce energy use throughout its supply chain. This could be a complex goal, with such changes requiring local infrastructure and policy change depending on factory location. Patagonia also aims to use only renewable or recycled materials in its products by 2025, whilst it will also grow its Worn Wear programme to support further recycling. Patagonia also aims to invest in other global carbon-capture projects in order to reduce emissions. These strategies, Patagonia hopes, will help to see it become carbon-neutral by 2025.

Finally, Patagonia also funds external, grassroots-level environmental groups with its own 'Earth Tax,' or 1% of its sales. As an alternative to funding corporation- or foundation-backed environmental groups, Patagonia aims to rebalance the charitable sector by donating to charities across the world to work in initiatives such as removing dams, restoring forests and rivers, finding solutions to mitigate climate change, protect green land and marine life, protect endangered species and plants, and support local and sustainable agriculture. Through this approach, Chouinard's drive to rethink capitalism is spread through organisations that are seeking to win their own environmental battle.

On reflection

Patagonia's drive to achieve circularity reflects its founder's desire to encourage us to rethink how we can create a more ethical and regenerative capitalist system. Interestingly, Chouinard is an actor who is different from the *homo economicus* that would normally prosper by making rational decisions based on the economic choices presented to them. Instead, Chouinard is an actor who seeks to root his decisions within the practicalities that arise from the overlap between the natural and social order. Economic actors should base their decisions not only on the economic information available to them but also on the environmental scenario that is presented to them. This is a base for which actors can pursue a shift to the circular economy. Economic actors make decisions by simultaneously considering environmental and economic choices. The key challenge would be encouraging this thinking amongst both established corporations who prosper within the linear economy as well as actors seeking to start anew.

Source

https://eu.patagonia.com/

Isabelle Housni

48 Whirli

Sector focus: Fast-moving consumer goods (FMCG)

Rationale

Play is commonly regarded as a crucial aspect of learning and development, and toys are especially useful since they are such an integral component of learning through play. The world toy market has grown by over 13% since 2010. The UK has the largest toy market in Europe, as well as the fourth largest globally. Despite the strong demand for toys, figures from the UK government have shown that an incredible 8.5 million units of brand-new toys are thrown away to landfill each year in the UK (East Sussex County Council, 2020). Whirli aims to solve this problem of discarded toys by having a circulation of toys that users can subscribe to. Instead of owning, users can 'borrow' toys and return them for new items, reducing the amount of toys that go to landfill when done with, while still encouraging toys to be played with and enjoyed.

Leadership

Whirli was founded in 2018 by Nigel Phan, who wanted to try and reduce the environmental impact that toys have on the planet, after witnessing first-hand the quantity of toys designed to entertain a toddler. As a newer business, it is easier to embed core values throughout the team as it expands; however, these values are also actively pushed from the top-down.

Nigel Phan divides his time between the different areas of the business to ensure that processes and principles are understood and any obstacles to these are identified. Customer feedback and business data are used to direct stock and remove items from the library that do not have the necessary durability for the circular model.

Whirli has recently expanded and moved to a new warehouse that focusses operations on sustainability and the circular economy. The Head of Operations, Craig Simpson, is actively looking to reduce the Carbon Footprint of the business through different measures. These measures include: locating the warehouse closer to the couriers thereby reducing the distance to drive; introducing packaging that can be reused, composted or recycled; using motion sensor low-energy lights throughout the premises; and having the warehouse powered by renewable energy. The next intention is to look at how electric vehicles can further reduce emissions for the business.

https://doi.org/10.1515/9783110723373-052

This shows that in addition to strong leadership, input and ideas from the staff and customers also help to drive the business and its goals forward. There is also the opportunity for the entire team to meet informally fortnightly to discuss their experience at the company, again giving an opportunity for the staff to put forward ideas.

Approach

Whirli uses a subscription model with different price points, starting at £9.99 per month up to £29.99 per month. These different price points provide various amounts of Whirli Tokens accordingly, with 80 Whirli Tokens (which are supposed to be worth approximately £80) given at the lowest price point, to 240 Whirli Tokens at the highest price point, with the addition of bundles.

Table 48.1: Whirli price bundles.

Toy sack	Toy box	Toy trunk	Toy chest
£9.99	£14.99	£19.99	£29.99
80 Whirli tokens	Up to 120 Whirli tokens	Up to 160 Whirli tokens	240 Whirli tokens and additional bundles

Additionally, free deliveries can be added onto the total for £6 per month, as opposed to £3.49 per delivery or return, which is helpful for those wanting to swap toys at a higher frequency.

Whirli tokens can be used to choose toys to be delivered and played with, however, the tokens are not replenished monthly but instead when a toy is returned. When a toy is done being played with and enjoyed, it can be sent back with the tokens returned that can then be used to pick a new toy that will be delivered.

Subscribers can pick toys from a library of over 1,000 toys, which have been curated to include currently popular, durable, high-quality toys. There is also no time limit on enjoying and playing with the toys; they can be returned once they have finished being played with – research commissioned by Whirli through Sapio Research found that parents believe 23% of the toys bought for their children were neglected after a single week.

Once a toy is returned, it is sterilised in Whirli facilities before being sent to another family.

If a child wishes to keep a toy, it can be purchased for below retail price and any toy that is kept for over 8 months becomes the subscriber's permanently.

While reasonable wear and tear is included as part of the subscription plans, if a toy is damaged or broken, it may need to be purchased or replaced, depending on the extent of the damage. However, the company ensures this will always be at a price below retail value.

While currently Whirli caters to newborns and up to age 8, the subscription model has room to expand to older ages and further expand their library. Future expansion from Whirli can include larger geographical locations with it currently serving the UK, although more warehouses in different countries might be needed for this, to ensure it will continue to have a low carbon footprint from deliveries.

Unfortunately, there is sometimes a stigma surrounding pre-loved items, which can hinder adoption of Circular Economy principles that push reuse as a priority. The structure offered by Whirli should help to change negative impressions and attitudes – such as assuming 'pre-owned' means unclean or inferior – by allowing subscribers to choose from current and popular toys as well as ensuring they are subject to quality control. By offering a selection that is appealing to its audience, minds can be changed young to embrace a model of renting rather than owning.

On reflection

The model from Whirli is great at preventing toys from becoming waste by creating a model where instead of being thrown away they can be delivered to a new family to enjoy. One way to expand on this concept is to do the same for damaged and broken toys, whether this be through returning them back to the original manufacturer or by creating an in-house repair team.

Additionally, toys that are found to be too flimsy or not durable are removed from the toy library, which is good for Whirli customers – but not for the planet. If Whirli were to share this data with the original manufacturer, having built up a good relationship with them, they may be able to get these design flaws rectified and thus prevent future toys from going to landfill from other consumers as well.

Source

https://whirli.com/

List of figures

https://doi.org/10.1515/9783110723373-053

List of tables

https://doi.org/10.1515/9783110723373-054

Index

https://doi.org/10.1515/9783110723373-055